STERLING
Test Prep

AP Chemistry

Practice Questions

8th edition

Customer Satisfaction Guarantee

Your feedback is important because we strive to provide the highest quality prep materials. Email us comments or suggestions.

info@sterling–prep.com

We reply to emails – check your spam folder

8 7 6 5 4 3 2 1

ISBN-13: 978-1-9547253-1-7

Sterling Test Prep materials are available at quantity discounts.

Contact info@sterling–prep.com

Sterling Test Prep
6 Liberty Square #11
Boston, MA 02109

Published by Sterling Test Prep

Printed in the U.S.A.

Thousands of students use our study aids to achieve high AP test scores!

Scoring well on the AP exams is essential to earn placement credits and admission into competitive colleges, which will position you for a successful future. This book prepares you to achieve a high score on the AP Chemistry exam by developing the ability to apply your knowledge and quickly choose the correct answer. Solving targeted practice questions builds your understanding of fundamental chemistry concepts and is a more effective strategy than merely memorizing terms.

This book has 900 high-yield practice questions covering AP Chemistry topics. Chemistry instructors with years of teaching experience prepared this material by analyzing the exam content and developing practice material that builds your knowledge and skills crucial for success on the exam. Our editorial team reviewed and systematized the content to match the current College Board AP Chemistry curriculum. We are experts on preparing students for standardized tests and college admissions.

The detailed explanations describe why an answer is correct and – more important for your learning – why another attractive choice is wrong. They provide step-by-step solutions for quantitative questions and teach the scientific foundations and details of essential chemistry topics needed to answer conceptual exam questions. Read the explanations carefully to understand how they apply to the question and learn important chemistry principles and the relationships between them. With the practice material contained in this book, you will significantly improve your AP score.

We wish you great success in your academics and look forward to being an important part of your test preparation!

220315gdx

AP Chemistry Review provides a comprehensive review of topics tested on the AP Chemistry exam. The content covers foundational principles and concepts necessary to answer exam questions.

This review book will increase your score.

Visit our Amazon store

AP Chemistry, Biology and Physics online practice tests

Our advanced online testing platform allows you to take AP practice questions on your computer to generate a Diagnostic Report for each test.

By using our online AP tests and Diagnostic Reports, you will:

Assess your knowledge of subjects and topics to identify your areas of strength and weakness

Learn important scientific topics and concepts for comprehensive test preparation

Improve your test-taking skills

To access these and other AP questions online
at a special pricing for book owners, see page 495

Advanced Placement (AP) prep books by Sterling Test Prep

Biology Practice Questions

Biology Review

Physics 1 Practice Questions

Physics 1 Review

Physics 2 Practice Questions

Physics 2 Review

Environmental Science

Psychology

U.S. History

World History

European History

U.S. Government and Politics

Comparative Government and Politics

Human Geography

Visit our Amazon store

College Level Examination Program (CLEP)

Biology Review

Biology Practice Questions

Chemistry Review

Chemistry Practice Questions

Introductory Business Law Review

College Algebra Practice Questions

College Mathematics Practice Questions

History of the United States I Review

History of the United States II Review

Western Civilization I Review

Western Civilization II Review

Social Sciences and History Review

American Government Review

Introductory Psychology Review

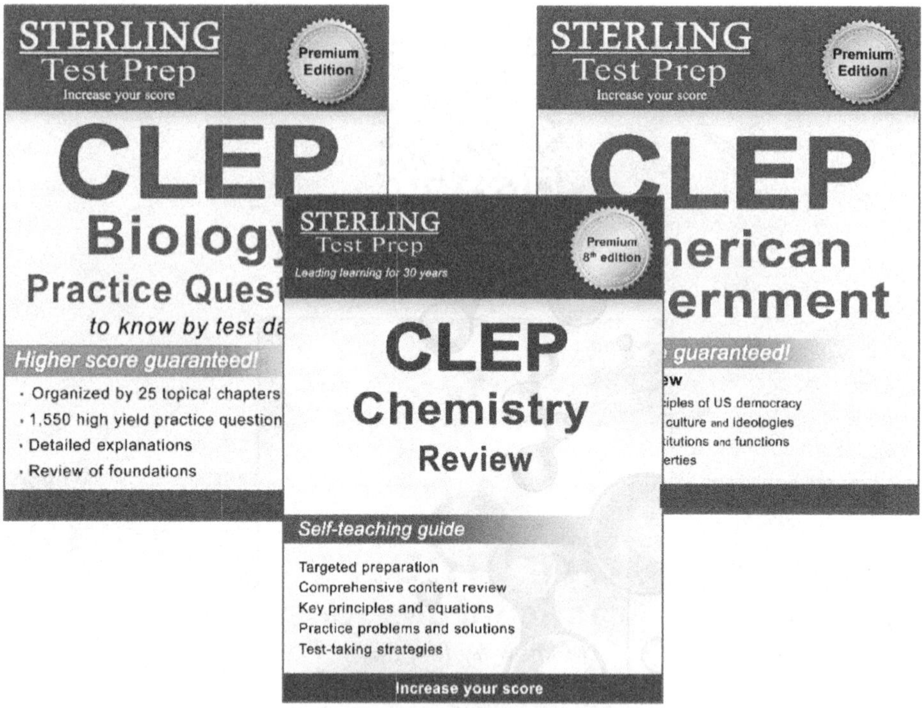

Visit our Amazon store

Table of Contents

Table of Contents (*continued*)

AP Chemistry Preparation and Test-Taking Strategies

Test preparation strategies

The best way to do well in AP Chemistry is to be good at chemistry. There is no way around knowing the subject; proper preparation is the key to success. Prepare for the test to answer with confidence as many questions as possible.

Study in advance

Devote 3 to 6 months studying. The information is manageable by studying regularly during the weeks before the test.

Cramming is not a successful tactic. However, do not study too far in advance. Studying more than six months ahead is not advised and may result in fatigue and poor knowledge retention.

Develop a realistic study and practice schedule

Cramming eight hours a day is unfeasible and leads to burnout, detrimental to performance.

Commit to a realistic study and practice schedule.

Remove distractions

During this preparation period, temporarily eliminate distractions.

However, balance is key, and it is crucial not to neglect physical well-being and social or family life. Prepare with full intensity but do not jeopardize your health or emotional well-being.

Develop an understanding over memorization

When studying, devote time to each topic.

After a study session, write a short outline of the concept to clarify relationships and increase knowledge retention.

Make flashcards

Avoid commercial flashcards because making cards helps build and retain knowledge.

Consider using self-made flashcards to develop knowledge retention and quiz what you know.

Find a study partner

Occasionally studying with a friend preparing for the test can motivate and provide accountability.

Explaining concepts to another improves and fine-tunes your understanding, integrates knowledge, bolsters competence, and identifies deficiencies in comprehension.

Take practice tests

Do not take practice tests too early.

First, develop a broad and detailed understanding of concepts. In the last weeks, use practice tests to fine-tune your final preparation. If you are not scoring well on practice tests, you want time to improve without undue stress.

Alternate studying and practicing to increase knowledge retention and identify areas for further study. Taking practice tests accustoms you to the challenges of test-taking.

Test day strategies

Be well-rested and eat the right foods

Get a full night's sleep before the test for proper mental and physical capacity. If you are up late the night before, you will have difficulty concentrating and focusing on the test.

Avoid foods and drinks that lead to drowsiness (carbohydrates and protein). Avoid drinks high in sugar, causing glucose to spike and crash.

Pack in advance

Check what you are allowed to bring to the test. Pay attention to the required check-in items (e.g., printed confirmation, identification).

Pack the day before to avoid the stress of not frantically looking for things on test day.

Arrive at the testing center early

Starting off right is an advantage. Allow time to check in and remain calm before the test begins.

Map and test your route to the center in advance and determine parking locations, if applicable. If you are not familiar with the test location, visit before the test day to practice and avoid travel errors. Plan your route correctly to arrive at the center without additional challenges and unnecessary stress.

Maintain a positive attitude

Avoid falling into a mental spiral of negative emotions. Too much worry leads to underperformance.

If you become anxious, chances are higher for lower performance in preparation and during the test. Inner peace helps during preparation and the high-stakes test.

To do well on the test requires logical, systematic, and analytical thinking, so relax and remain calm.

Focus on your progress

Do not be concerned with other test-takers. Someone proceeding rapidly through the exam may be rushing or guessing on questions.

Take breaks and breathe deeply

Do not skip the available timed breaks. Refreshing breaks help you finish strong.

If time allows, eat a light snack to replenish your energy. Your mind and body will appreciate the available breaks.

The best approach to any test is *not* to keep your head down the whole session. While there is no time to waste, take a few seconds between questions to breathe deeply.

Momentarily clear your thoughts to relax your mind and muscles.

Time management strategies

Timing

Besides good preparation, time management is the critical strategy for the exam.

Timed practice is not the objective at this stage. While practicing, note how many questions you would have completed in the allotted time.

Average time per question

In advance, determine the average time allotted for each question.

Use two different approaches depending on which preparation phase you are working on.

During the first preparation phase, acquire, fortify, and refine your knowledge.

During your final practice stage, use the time needed to develop analytical and thought processes related to specific questions.

Work systematically

Note your comprehension compared to the correct answers to learn the material and identify conceptual weaknesses. Do not overlook the value of explanations to questions; these can be a great source of content, analysis, and interdependent relationships.

During the second preparation phase, do not spend more than the average allotted time on each question when taking practice tests.

Pace your response time to develop a consistent pace and complete the test within the allotted time. If you are time-constrained during the final practice phase, work more efficiently, or your score will suffer.

Focus on the easy questions and skip the unfamiliar

Easy or difficult questions are worth the same points. Score more points for three quickly answered questions than one hard-earned victory.

Answer all familiar and easy questions to maximize points if time runs out.

Identify strengths and weaknesses

Skip unfamiliar questions on the first round as challenging questions require more than the average allotted time.

On the second review, questions that you cannot approach systematically or lack fundamental knowledge will likely not be answered through analysis.

Use the elimination and educated guessing strategy to select an answer and move on to another question.

Do not overinvest in any question

Some questions consume more time than the average and make you think of investing more time. Stop thinking that investing more time in challenging questions is productive.

Do not get entangled with questions while losing track of time. The test is timed, so you cannot spend too much time on any question.

Look at every question on the exam

It is unfortunate not to earn points for a question you could have quickly answered because you did not see it.

If you are in the first half of the test and spending more than the average on a question, select the best option, note the question number, and move on. You do not want to rush through the remaining questions, causing you to miss more answers.

If time allows, return to marked questions and take a fresh look. However, do not change the original answer unless you have a reason to change it. University studies show that hastily changing answers often replace correct and incorrect answers.

Multiple-choice questions

Answer all questions

How many questions are correct, not how much work went into selecting the answers matters. An educated guess earns the same points as an answer known with confidence.

On the test, you need to think and analyze information quickly. The skill of analyzing information quickly cannot be gained from a college course, prep course, or textbook. Working efficiently and effectively is a skill developed through focused effort and applied practice.

Strategic approach

There are strategies, approaches, and perspectives to answer multiple-choice questions and help maximize points.

Many strategies are known by you and seem like common sense. However, these helpful approaches might be overlooked under the pressure of a timed test.

While no strategy replaces comprehensive preparation, apply probability for success on unfamiliar questions.

Understand the question

Know what the question is asking before selecting an answer. It is surprising how many students do not read (and reread) carefully and rush selecting the wrong answer.

The test-makers anticipate these hasty mistakes, and many enticing answers include the specious choices. A successful student reads the question and understands it precisely before looking at the answers.

Focus on the answer

Separate the vital information from distracters and understand the design and thrust of the question.

Answer the question posed and not merely pick a factually accurate statement or answer a misconstrued question.

Rephrasing the question helps articulate what precisely the correct response requires. When rephrasing, do not change the meaning of the question; assume it is direct and to the point as written.

After selecting the answer, review the question and verify that the choice selected answers the question.

Answer the question before looking at choices

This valuable strategy is applicable if the question asks for generalized factual details. Form a thought response first, then look for the choice that matches your preordained answer.

Select the predetermined statement as it is likely correct.

Factually correct, but wrong

Questions often have incorrect choices that are factually correct but do not answer the question.

Predetermine the answer and do not choose merely a factually correct statement. Verify that the choice answers the question.

Do not fall for the familiar

When in doubt, it is comforting to choose what is familiar. If you recognize a term or concept, you may be tempted to pick that choice impetuously.

However, do not go with familiar answers merely because they are familiar. Think through the answer and how it relates to the question before selecting it.

Words of caution

The words *"all," "none,"* and *"except"* require attention. Be alert with questions containing these words, as the answer may not be apparent on the first read of the question.

Double-check the question

After selecting an answer, return to the question to ensure the selected choice answers the question as asked.

Fill the answers carefully

Many mistakes happen when filling in answers. Filling the answers correctly is simple but crucial. Be attentive to the question number and enter the answer accordingly. If you skip a question, skip it on the answer sheet.

Quantitative multiple-choice questions

Know the equations

Many exam questions require scientific equations, so understand when to use each. As you work with this book, learn to apply formulas and equations and use them in many questions.

Manipulate the formulas

Know how to rearrange the formulas. Many questions require manipulating equations to calculate the correct answer.

Familiarity includes manipulating the terms, understanding relationships, and isolating variables.

Estimating

Estimating helps choose the answer quickly for quantitative questions with a sense of the order of magnitude. Often, estimation enables the correct answer to be identified quickly compared to the time needed for calculations.

Estimating is especially applicable to questions where the answer choices have different orders of magnitude. It saves time to estimate instead of computing the solution.

Evaluate the units

For quantitative problems, analyze the units to build relationships between the question and the correct answer. Understand what value is sought and eliminate wrong choices with improper units.

Make visual notes

Write, draw, or graph the question to determine the information provided, the objective, and the concept tested. Even if a question does not require a graphic answer, a chart or table often allows a solution to become apparent.

Experiments questions

Determine the purpose, methods, variables, and controls of the experiment. Understanding the presented information helps answer the question.

With multiple experiments, understand variations of the same experiment by focusing on the differences. For example, focus on the changes between the first and second experiments, second and third, and first and third. Direct comparison between experiments helps organize the information and apply it to the answer.

Elimination strategies

If the correct answer is not immediately apparent, use the process of elimination.

Use the strategy of educated guessing by eliminating one or two answers. Usually, at least one answer choice is easily identified as wrong. Eliminating even one choice increases the odds of selecting the correct one.

Process of elimination

Eliminate choices:

- Use proportional estimations for quantitative questions to eliminate choices that are too high or low.
- Eliminate answers that are *almost right* or *half right.* Consider *half right* as *wrong* since these distractor choices are purposely included.
- If two answers are direct opposites, the correct answer is likely one of them. However, note if they are direct opposites or another reason to consider them correct. Therefore, eliminate the other choices and narrow the search for the correct one.
- For numerical questions, eliminate the smallest and largest numbers (unless for a reason).

Roman numeral questions

Roman numeral questions present several statements and ask which is/are correct. These questions are tricky for most test-takers because they present more than one potentially correct statement.

Roman numeral questions are often included in combinations with more than one answer. Eliminating a wrong Roman numeral statement eliminates all choices that include it.

Educated guessing

Correct ways to guess

Do not assume you must get every question right; this will add unnecessary stress during the exam. You will (most likely) need to guess some questions.

Answer as many questions correctly as possible without wasting time that should be used to maximize your score.

For challenging questions, random guessing does not help. Use educated guessing after eliminating one or two choices.

Playing the odds

Guessing is a form of "partial credit" because while you might not be sure of the correct answer, you have the relevant knowledge to identify some wrong choices.

There is a 25% chance of correctly guessing for random responses since questions have four choices. Therefore, the odds are guessing 1 question correctly to 3 incorrectly.

Guessing after elimination of answers

After eliminating one answer as wrong, you have a 33% chance for a lucky guess. Therefore, your odds move to 1 question right to 2 questions wrong.

While this may not seem like a dramatic increase, it can make an appreciable difference in your score.

Confidently eliminating two wrong choices increases the chances of guessing correctly to 50%!

When using elimination:

- Do not rely on gut feelings alone to answer questions quickly.

  Understand and recognize the difference between *knowing* and *gut feeling* about the answer. Gut feelings should sparingly be used after the process of elimination.

- Do not fall for answers that sound "clever," and choose "bizarre" answers.

  Choose them only with a reason to believe they may be correct.

Eliminating Roman numeral choices

A workable strategy for Roman numeral questions is to guess the wrong statement.

For example:

A. I only

B. III only

C. I and II only

D. I and III only

Notice that statement II does not have an answer dedicated to it. This indicates that statement II is likely wrong and eliminates choice C, narrowing your search to three choices. However, if you are confident that statement II is the answer, do not apply this strategy.

Free-response questions

Free-response questions typically require processing information into existing conceptual frameworks.

There are some personal choices for writing the desired response.

Free-response question skills

Be able to present and discuss relevant examples. Practice clarifying or evaluating principles.

Perform detailed analysis of relationships and respond to stimulus materials such as charts or graphs.

Understand the question

As with the multiple-choice questions, understand what the question asks. Assume that the question, as written, is direct and to the point.

Mental rephrasing should not add or alter the meaning or essence of the question.

Answer the questions in order of competence

You are not bound to answer the questions according to their sequence. Survey questions and decide which require minimal effort or time.

Avoid getting entangled and frustrated to use time efficiently and maximize points.

Do not write more than needed

Additional work beyond the question's stated directives does not earn a higher score or result in extra credit.

For time management and keeping responses relevant, answer the question but avoid superfluous responses.

Organize thoughts

Before writing, brainstorm the questions' topics. Outline your thoughts on scratch paper during the composition process. Essential definitions, ideas, examples, or names are valid details when relevant.

Organized thoughts produce a coherent response. With practice, balance time between brainstorming and writing.

Follow the structure

Structure responses to match the order specified in the question for a grader-friendly answer.

The reader should not need to search for topics used in the grading criteria. A well-structured essay has complete sentences and paragraphs.

A formal introduction and conclusion are unnecessary; go directly into answering the question.

Answer questions in their entirety

It is essential to answer questions thoroughly, not just partially. For example, some questions ask to identify and explain. Performing only one step is inadequate and will be graded accordingly.

Task verbs

Below are *task verbs* common for free-response questions – underline these directives as you see them in practice questions and on the exam.

Refer to these tasks and verify when completed. Do not overlook them when writing a comprehensive response.

Compare – provide a description or explanation of similarities or differences.

Define – provide a specific meaning for a word or concept.

Identify – provide information about a specified topic without elaboration or explanation.

Describe – provide the relevant characteristics of a specified concept.

Develop an argument – articulate a claim and support it with evidence.

Explain – provide information about how or why a relationship, process, pattern, or outcome occurs, using evidence and reasoning.

Conclude – use available information to formulate an accurate statement demonstrating understanding based on evidence.

Explaining *how* typically requires analyzing the relationship, process, pattern, or outcome.

Explaining *why* typically requires analyzing motivations or reasons for the relationship, process, or outcome.

Write legibly

The reader must decipher your writing so the response can be scored appropriately, so write legibly. Practice writing under a time limit to produce a readable response. If this issue may exist, ask a friend to read a sample response to understand the words expressed.

Consider printing key phrases or highlighting them cleanly (e.g., asterisk, underline) in your response.

Use plain language

All claims should be directly stated. You do not want the graders to guess how something demonstrates a point.

Regardless of if they correctly guess your intentions, you will be graded critically for ambiguities. Present relevant information clearly and concisely to demonstrate the argument's primary points.

Use facts to bolster arguments

Written responses should include specific facts and avoid unsubstantiated claims. Do not use long, meandering responses filled with loosely-related facts regarding specific concepts. Avoid contradictions, circular definitions, and question restatements.

Review and correct answers

If questions are completed and time remains, review each response. These thoughtful revisions can earn crucial points.

Assess if anything is needed to be added or requires correction. Put a simple strikethrough through the error for a mistake, so the grader disregards that portion. If you add content, insert an asterisk (*) and refer the reader to the end of the essay.

Diagnostic Tests

Diagnostic Test 1

This Diagnostic Test is designed to assess your proficiency on each topic and NOT to mimic the test. Use your test results and identify areas of strength and weakness to adjust your study plan and enhance your fundamental knowledge. The length of the Diagnostic Tests is optimal for a single study session.

#	Answer:				Review	#	Answer:				Review
1:	A	B	C	D	___	**31:**	A	B	C	D	___
2:	A	B	C	D	___	**32:**	A	B	C	D	___
3:	A	B	C	D	___	**33:**	A	B	C	D	___
4:	A	B	C	D	___	**34:**	A	B	C	D	___
5:	A	B	C	D	___	**35:**	A	B	C	D	___
6:	A	B	C	D	___	**36:**	A	B	C	D	___
7:	A	B	C	D	___	**37:**	A	B	C	D	___
8:	A	B	C	D	___	**38:**	A	B	C	D	___
9:	A	B	C	D	___	**39:**	A	B	C	D	___
10:	A	B	C	D	___	**40:**	A	B	C	D	___
11:	A	B	C	D	___	**41:**	A	B	C	D	___
12:	A	B	C	D	___	**42:**	A	B	C	D	___
13:	A	B	C	D	___	**43:**	A	B	C	D	___
14:	A	B	C	D	___	**44:**	A	B	C	D	___
15:	A	B	C	D	___	**45:**	A	B	C	D	___
16:	A	B	C	D	___	**46:**	A	B	C	D	___
17:	A	B	C	D	___	**47:**	A	B	C	D	___
18:	A	B	C	D	___	**48:**	A	B	C	D	___
19:	A	B	C	D	___	**49:**	A	B	C	D	___
20:	A	B	C	D	___	**50:**	A	B	C	D	___
21:	A	B	C	D	___	**51:**	A	B	C	D	___
22:	A	B	C	D	___	**52:**	A	B	C	D	___
23:	A	B	C	D	___	**53:**	A	B	C	D	___
24:	A	B	C	D	___	**54:**	A	B	C	D	___
25:	A	B	C	D	___	**55:**	A	B	C	D	___
26:	A	B	C	D	___	**56:**	A	B	C	D	___
27:	A	B	C	D	___	**57:**	A	B	C	D	___
28:	A	B	C	D	___	**58:**	A	B	C	D	___
29:	A	B	C	D	___	**59:**	A	B	C	D	___
30:	A	B	C	D	___	**60:**	A	B	C	D	___

This page is intentionally left blank

1. Cobalt is element 27 on the periodic table. ^{60}Cobalt is sometimes used as a therapeutic in the treatment of cancer. How many neutrons and protons are contained in the nucleus of the ^{60}Cobalt isotope?

A. 27 neutrons and 27 protons
B. 33 neutrons and 27 protons
C. 27 neutrons and 33 protons
D. 33 neutrons and 33 protons

2. Which of the following is the weakest form of interatomic attraction?

A. induced dipole-induced dipole
B. dipole-induced dipole
C. covalent bond
D. ion-dipole

3. Which property primarily determines the effect of temperature on the solubility of gas molecules?

A. The ionic strength of the gas
B. The molecular weight of the gas
C. The polarity of the gas
D. The kinetic energy of the gas

4. If the molecular weight of a compound is 219 g/mol, what is the molecular formula of the compound with the following percent composition?

C = 49.3%   O = 43.8%   H = 6.9%

A. $C_5O_2H_5$
B. $C_4O_2H_5$
C. $C_2O_3H_3$
D. None of the above

5. How much energy is required to change 17.8 g of solid Cu to molten Cu at its melting point of 1,083 °C? (Use heat of fusion for Cu = 205 J/g)

A. 3,649 J
B. 187.0 J
C. 1,433 J
D. 2,247 J

6. If the temperature of the system is raised for the following reaction with $\Delta H < 0$, the equilibrium is [ ].

$$CO\ (g) + 2\ H_2\ (g) \rightleftarrows CH_3OH\ (g)$$

A. variable
B. unaffected
C. shifted to the left
D. shifted to the right

7. What was the molarity of the original solution, if a 24 mL sample of a $CuSO_4$ solution was dried and 0.88 g of copper (II) sulfate remained? (Use the molecular weight of Cu = 63.55 g/mol, S = 32.06 g/mol and O = 16 g/mol)

A. 0.46 M
B. 0.57 M
C. 0.43 M
D. 0.23 M

8. Which of the following pairs of 0.25 M solutions react to form a precipitate?

A. KOH and $Ba(NO_3)_2$

B. NaI and KBr

C. K_2SO_4 and CsI

D. $NiBr_2$ and $AgNO_3$

9. For the electrochemical cell shown below, which of the following changes causes a *decrease* in the cell voltage?

$$Ni\ (s)\ |\ Ni^{2+}\ (1\ M)\ ||\ H^{+}\ (1\ M)\ |\ H_2\ (2\ atm)\ |\ Pt\ (s)$$

A. Decreasing the concentration of Ni^{2+} ion

B. Lowering the pH of the cell electrolyte

C. Decreasing the mass of the nickel electrode

D. Increasing the pressure of H_2 to 4 atm

10. Almost all of the mass of an atom exists in its:

A. electrons

B. nucleus

C. outermost energy level

D. first energy level

11. What is the molecular geometry of PH_3?

A. tetrahedral

B. octahedral

C. trigonal pyramidal

D. linear

12. Ten liters of O_2 gas is under a pressure of 260 torr. If the volume of this gas is increased to sixteen liters at a constant temperature, what is the new pressure?

A. 133 torr

B. 163 torr

C. 86 torr

D. 199 torr

13. What is the total number of moles of gas in a sample that contains 32.0 g of CH_4, 24.0 g of O_2, 32.0 g of SO_2 and 44.0 g of CO_2?

A. 2.30 moles

B. 4.25 moles

C. 5.65 moles

D. 3.10 moles

14. A sample of aluminum absorbed 9.22 J of heat, and the temperature increased from 21.2 °C to 28.5 °C. What is the mass of the aluminum? (Use the specific heat c of aluminum = 0.90 J/g·°C)

A. 2.6 g

B. 5.6 g

C. 3.8 g

D. 1.4 g

15. Which of the following is the equilibrium expression?

$CS_2\ (g) + 4\ H_2\ (g) \leftrightarrow CH_4\ (g) + 2\ H_2S\ (g)$

A. $K_{eq} = [CS_2]\cdot[H_2]^4 / [CH_4]\cdot[H_2S]^2$

B. $K_{eq} = [CH_4]\cdot[H_2S]^2 / [CS_2]\cdot[H_2]$

C. $K_{eq} = [CH_4]\cdot[H_2S]^2 / [CS_2]\cdot[H_2]^4$

D. $K_{eq} = [CS_2]\cdot[CH_4] / [H_2]^2\cdot[H_2S]^4$

16. Which of the following statements describes a saturated solution?

A. contains dissolved solute in equilibrium with undissolved solid
B. precipitates rapidly if a seed crystal is added
C. contains as much solvent as it can accommodate
D. contains only electrolytes

17. Which of the following statements is correct?

A. In a basic solution, $[H_3O^+] > 10^{-7}$ and $[^-OH] < 10^{-7}$
B. In a basic solution, $[H_3O^+] > 10^{-7}$ and $[^-OH] > 10^{-7}$
C. In a basic solution, $[H_3O^+] < 10^{-7}$ and $[^-OH] < 10^{-7}$
D. In a basic solution, $[H_3O^+] < 10^{-7}$ and $[^-OH] > 10^{-7}$

18. Which group contains only metalloids?

A. Pm, Sm and Nd
B. Si, B and Sb
C. Cu, Ag and Au
D. Po, Cs and Ac

19. Which atom is NOT bound to hydrogen in a hydrogen bond?

I. S II. N III. F

A. I only
B. II only
C. III only
D. I and II only

20. As the volume of a sealed container decreases at a constant temperature, the gas begins to deviate from ideal behavior. Compared to the pressure predicted by the ideal gas law, the actual pressure is:

A. higher, because of intermolecular attractions between gas molecules
B. higher, because of the volume of the gas molecules
C. lower, because of intermolecular attractions among gas molecules
D. lower, because of the volume of the gas molecules

21. Methane (CH_4) can be used as automobile fuel to reduce pollution. What is the coefficient of oxygen in the balanced equation for the reaction?

$$\_CH_4(g) + \_O_2(g) \xrightarrow{\text{Spark}} \_CO_2(g) + \_H_2O(g)$$

A. 1
B. 2
C. 3
D. 4

22. What is the enthalpy for the reaction $N_2 + O_2 \rightarrow 2\ NO$? (Use bond dissociation energy for $N_2 = 226$ kcal/mol, $O_2 = 199$ kcal/mol and NO = 145 kcal/mol)

A. 135 kcal/mol
B. 344 kcal/mol
C. 290 kcal/mol
D. –290 kcal/mol

23. Which of the changes shifts the equilibrium to the right for the following reversible reaction?

$$SO_3(g) + NO(g) + heat \leftrightarrow SO_2(g) + NO_2(g)$$

A. decreasing temperature
B. increasing volume
C. increasing $[SO_3]$
D. increasing $[SO_2]$

24. Which of the following statements is/are true?

I. A gaseous solution consists of a gaseous solute and a gaseous solvent
II. A solid solution consists of a solid solute and a solid solvent
III. A liquid solution results from a gaseous solute and a liquid solvent

A. I only
B. II only
C. I and II only
D. I, II and III

25. Which of the following indicators is/are orange at pH 5.2?

I. phenolphthalein II. methyl red III. bromothymol blue

A. I only
B. II only
C. III only
D. I and II only

26. Nickel-cadmium batteries are used in rechargeable electronic calculators. Which substance is oxidized in the following balanced reaction for recharging a NiCd battery?

$$Cd(OH)_2(s) + Ni(OH)_2(s) \rightarrow Cd(s) + NiO_2(s) + 2\ H_2O(l)$$

A. $Ni(OH)_2$
B. $Cd(OH)_2$
C. Cd
D. NiO_2

27. Which types of subshells are present for the n = 2 energy level?

A. *d*
B. *p*
C. *s*
D. both *s* and *p*

28. Which compound shown below has a covalent bond within the molecule?

A. CsF
B. HF
C. NaF
D. $MgCl_2$

29. Which of the following statements best describes a solid?

A. Indefinite shape, but definite volume
B. Definite shape, but indefinite volume
C. Definite shape and volume
D. Indefinite shape and volume

30. How much phosphorus is required to produce 412 g of phosphorus trichloride when the following unbalanced reaction is run to completion?

$$P_4\,(s) + 6\,Cl_2\,(g) \rightarrow PCl_3\,(l)$$

A. 93 g
B. 148 g
C. 186 g
D. 296 g

31. What are the products for the following unbalanced decomposition reaction?

$$LiHCO_3\,(s) \rightarrow$$

A. Li_2CO_3, H_2O and CO_2
B. Li_2CO_3, H_2 and CO_2
C. Li, H_2O and CO_2
D. Li, H_2 and CO_2

32. Which of the following factors increase(s) the rate of a chemical reaction?

I. Decreasing the concentration of the reactants
II. Adding a catalyst to the reaction vessel
III. Decreasing the temperature of the reaction

A. I only
B. II only
C. III only
D. I and II only

33. Determine the boiling point of a 3 molal aqueous solution of NaCl. (Use the boiling point constant of water = 0.512 °C/m)

A. 96.93 °C
B. 99.66 °C
C. 106.14 °C
D. 103.07 °C

34. What are the products of the following neutralization reaction?

$HC_2H_3O_2\,(aq) + Ca(OH)_2\,(aq) \rightarrow$

A. $Ca(C_2H_3O_2)_2$ and H_2
B. $Ca(HCO_3)_2$ and H_2O
C. $Ca(C_2H_3O_2)_2$ and H_2O
D. $CaCO_3$ and H_2O

35. Which classification of elements do silver, iron, mercury, and rhodium belong to?

A. transduction metals
B. actinides
C. alkaline earth metals
D. transition metals

36. How do the Lewis-dot structures of elements in the same group of the periodic table compare?

A. The number of electrons in the Lewis-dot structures equals the group number for each element of the group
B. Elements of the same group have the same number of valence electrons
C. The number of valence shell electrons increases by one for each element from the top to the bottom of the group
D. The structures differ by exactly two electrons between vertically consecutive elements

37. A sample of an unknown gas was isolated in a gas containment bulb. The volume of the bulb was 1.8 liters, the temperature was 23.5 °C, and the manifold pressure was 578 torr. What is the volume of this gas at STP? (Use standard temperature = 273 K and standard pressure = 760 torr)

A. 1.26 L
B. 1.49 L
C. 1.67 L
D. 2.26 L

38. If 12 moles of $N_2O_4\,(l)$ are added to an unknown amount of $N_2H_3(CH_3)\,(l)$. When the following unbalanced reaction runs to completion, which is the limiting reagent if 26 moles of H_2O are produced?

$$5\,N_2O_4\,(l) + 4\,N_2H_3(CH_3)\,(l) \rightarrow 12\,H_2O\,(g) + N_2\,(g) + CO_2\,(g)$$

A. $N_2O_4\,(l)$
B. $H_2O\,(g)$
C. $4\,N_2H_3(CH_3)\,(l)$
D. $N_2\,(g)$

39. Using the following bond dissociation energies, determine the energy change for the following combustion reaction:

$2\,O_2 + CH_4 \rightarrow CO_2 + 2\,H_2O$

C–C: 350 kJ/mole   O=O: 495 kJ/mole

C–H: 410 kJ/mole   O–H: 460 kJ/mole   C=O: 720 kJ/mole

A. 880 kJ/mole
B. –742 kJ/mole
C. –650 kJ/mole
D. –1,248 kJ/mole

40. In which of the following equilibrium systems does the equilibrium shift to the left when the pressure of the system is increased?

A. $H_2\ (g) + Cl_2\ (g) \leftrightarrow 2\ HCl\ (g)$
B. $2\ SO_2\ (g) + O_2\ (g) \leftrightarrow 2\ SO_3\ (g)$
C. $4\ NH_3\ (g) + 5\ O_2\ (g) \leftrightarrow 4\ NO\ (g) + 6\ H_2O\ (g)$
D. $N_2\ (g) + 3H_2\ (g) \leftrightarrow 2\ NH_3\ (g)$

41. Which of the following is the solvent for a homogenous mixture consisting of 24% ethanol, 32% propanol and 44% water?

A. Ethanol
B. Water
C. Propanol
D. Ethanol and propanol

42. If the concentration of H_3O^+ is 4.5×10^{-3} M, what is the molar concentration of ^-OH?

A. 3.44×10^{-10} M
B. 1.11×10^{-7} M
C. 1.5×10^{-3} M
D. 2.2×10^{-12} M

43. How many e^- are needed to balance the following half reaction $Cl_2O_7 \rightarrow HClO$ in an acidic solution?

A. 12 e^- to the left side
B. 6 e^- to the right side
C. 2 e^- to the left side
D. 3 e^- to the right side

44. Which of the following electron configurations represents an excited state of an atom?

A. $1s^22s^22p^63s^2$
B. $1s^22s^22p^5$
C. $1s^22s^22p^2$
D. $1s^22s^23s^1$

45. Which of the following elements is the most electronegative?

A. S
B. F
C. Li
D. Cs

46. Charles's law involves which of the following?

A. Indirect proportion
B. Varying mass of gas
C. Constant volume
D. Varying temperature

47. Which is NOT a general guideline for balancing an equation?

A. Balance ionic compounds as a single unit
B. Balance polyatomic ions as a single unit
C. Begin balancing with the most complex formula
D. Write correct formulas for reactants and products

48. Use the bond energies provided to determine if the following synthesis reaction is exothermic:

$H_2 + Cl_2 \rightarrow 2\ HCl$

H–H: 436 kJ/mol   Cl–Cl: 243 kJ/mol   H–Cl: 431 kJ/mol

A. Exothermic with less than 75 kJ of energy released
B. Endothermic with less than 75 kJ of energy absorbed
C. Exothermic with more than 75 kJ of energy released
D. Endothermic with more than 75 kJ of energy absorbed

49. Which statement is true when the equilibrium constant $K_{eq} = 3.4 \times 10^{12}$?

A. Mostly products are present
B. Amount of reactants equals products
C. Mostly reactants are present
D. None of the above

50. How much NaCl (*s*) is required to prepare 100 ml of a 4 M solution? (Use the molecular weight of NaCl = 58 g/mol)

A. 23.4 g
B. 14.5 g
C. 11.2 g
D. 19.2 g

51. Which of the following statements is true for determining if an acid or a base is classified as *weak* or *strong*?

A. The concentration of the acid or base
B. The extent of dissociation of the dissolved acid or base
C. The solubility of the acid or base
D. Ability to be neutralized by buffering solution

52. What is the average atomic mass of an element that contains three isotopes of 16.0 amu, 17.0 amu and 18.0 amu with relative abundances of 22%, 44%, and 34%, respectively?

A. 16.6 amu
B. 17.1 amu
C. 17.3 amu
D. 17.6 amu

53. What type of bond joins adjacent water molecules?

A. dipole
B. hydrogen
C. ionic
D. covalent

54. Which of the following alkanes has the highest boiling point?

A. $CH_3C(CH_3)_2CH_2CH_3$
B. $CH_3CH(CH_3)CH_2CH_2CH_3$
C. $CH_3CH_2CH_2CH_2CH_2CH_2CH_3$
D. $CH_3CH(CH_3)CH(CH_3)CH_3$

55. Which of the following reactions is the correct equation for the reaction of magnesium with hydrochloric acid to yield hydrogen and magnesium chloride?

A. $2\ Mg + 6\ HCl \rightarrow 3\ H_2 + 2\ MgCl_2$
B. $Mg + 2\ HCl \rightarrow 2\ H + MgCl_2$
C. $Mg + 3\ HCl \rightarrow 3\ H + MgCl_2$
D. $Mg + 2\ HCl \rightarrow H_2 + MgCl_2$

56. Given the data below, calculate ΔG° for the reaction:

$$H_2\,(g) + I_2\,(s) \rightarrow 2\ HI\,(g)$$

$H_2\,(g)$: $\Delta H°_f = 0$ kJ mol^{-1} $\Delta S° = +130.6$ J mol^{-1}·K^{-1}

$I_2\,(s)$: $\Delta H°_f = 0$ kJ mol^{-1} $\Delta S° = +116.12$ J mol^{-1}·K^{-1}

$HI\,(g)$: $\Delta H°_f = +26$ kJ mol^{-1} $\Delta S° = +206$ J mol^{-1}·K^{-1}

A. +118.4 kJ
B. +2.7 kJ
C. –6.2 kJ
D. +52.2 kJ

57. Which of the following is true after a reaction reaches chemical equilibrium?

A. The amount of reactants and products is constant
B. The amount of reactants and products is equal
C. The amount of reactants is increasing
D. The amount of products is increasing

58. Colloid particles settle out of solution during coagulation because they:

A. bind to the solution and precipitate
B. dissociate and precipitate
C. bind together and precipitate due to gravity
D. dissolve and precipitate due to gravity

59. In the following reaction, which reactant is a Brønsted-Lowry acid?

$$NaHS\,(aq) + HCN\,(aq) \rightarrow NaCN\,(aq) + H_2S\,(aq)$$

A. NaCN
B. H_2S
C. NaHS
D. HCN

60. How many e^- are gained or lost in the half-reaction: $Na \rightarrow Na^+$?

A. 1 e^- is gained
B. ½ e^- is lost
C. ½ e^- is gained
D. 1 e^- is lost

Use the answer key to check your answers. Review the explanations in detail, focusing on questions you didn't answer correctly. Note the topic of those questions. Complete this BEFORE taking the next Diagnostic Test.

Notes for active learning

Diagnostic Test 2

This Diagnostic Test is designed to assess your proficiency on each topic and NOT to mimic the test. Use your test results and identify areas of strength and weakness to adjust your study plan and enhance your fundamental knowledge. The length of the Diagnostic Tests is optimal for a single study session.

#	Answer:				Review	#	Answer:				Review
1:	A	B	C	D	___	**31:**	A	B	C	D	___
2:	A	B	C	D	___	**32:**	A	B	C	D	___
3:	A	B	C	D	___	**33:**	A	B	C	D	___
4:	A	B	C	D	___	**34:**	A	B	C	D	___
5:	A	B	C	D	___	**35:**	A	B	C	D	___
6:	A	B	C	D	___	**36:**	A	B	C	D	___
7:	A	B	C	D	___	**37:**	A	B	C	D	___
8:	A	B	C	D	___	**38:**	A	B	C	D	___
9:	A	B	C	D	___	**39:**	A	B	C	D	___
10:	A	B	C	D	___	**40:**	A	B	C	D	___
11:	A	B	C	D	___	**41:**	A	B	C	D	___
12:	A	B	C	D	___	**42:**	A	B	C	D	___
13:	A	B	C	D	___	**43:**	A	B	C	D	___
14:	A	B	C	D	___	**44:**	A	B	C	D	___
15:	A	B	C	D	___	**45:**	A	B	C	D	___
16:	A	B	C	D	___	**46:**	A	B	C	D	___
17:	A	B	C	D	___	**47:**	A	B	C	D	___
18:	A	B	C	D	___	**48:**	A	B	C	D	___
19:	A	B	C	D	___	**49:**	A	B	C	D	___
20:	A	B	C	D	___	**50:**	A	B	C	D	___
21:	A	B	C	D	___	**51:**	A	B	C	D	___
22:	A	B	C	D	___	**52:**	A	B	C	D	___
23:	A	B	C	D	___	**53:**	A	B	C	D	___
24:	A	B	C	D	___	**54:**	A	B	C	D	___
25:	A	B	C	D	___	**55:**	A	B	C	D	___
26:	A	B	C	D	___	**56:**	A	B	C	D	___
27:	A	B	C	D	___	**57:**	A	B	C	D	___
28:	A	B	C	D	___	**58:**	A	B	C	D	___
29:	A	B	C	D	___	**59:**	A	B	C	D	___
30:	A	B	C	D	___	**60:**	A	B	C	D	___

This page is intentionally left blank

1. Which of the following is/are a characteristic of metals?

I. conduction of heat   II. high density   III. malleable

A. I only
B. II only
C. III only
D. I, II and III

2. What is the total number of valence electrons in a molecule of SOF_2?

A. 18
B. 20
C. 26
D. 19

3. What is the name given to the transition of a compound from the gas phase directly to the solid phase?

A. deposition
B. sublimation
C. freezing
D. condensation

4. Which are the spectator ions in the following balanced reaction?

$$2\ AgNO_3\ (aq) + K_2SO_4\ (aq) \rightarrow 2\ KNO_3\ (aq) + Ag_2SO_4\ (s)$$

A. silver ion and sulfate ion
B. potassium ion and nitrate ion
C. potassium ion and sulfate ion
D. silver ion and nitrate ion

5. Calculate the mass of a sample of gold if it requires 488 J to be heated from 21.8 °C to 34.4 °C? (Use the specific heat of gold $c = 0.130$ J/g·°C)

A. 597 g
B. 383 g
C. 472 g
D. 298 g

6. The reaction shown is an example of which step in a free radical reaction?

$$\cdot OH + H_2 \rightarrow H_2O + H\cdot$$

I. initiation   II. propagation   III. termination

A. I only
B. II only
C. III only
D. II and III only

7. Which substance shown below does NOT produce ions when dissolved in water?

A. manganese (II) nitrate
B. CsCN
C. CH_2O
D. KClO

8. What are the products from the complete neutralization of sulfuric acid with aqueous sodium hydroxide?

A. $Na_2SO_4\ (aq)$ and $H_2O\ (l)$
B. $NaHSO_4\ (aq)$ and $H_2O\ (l)$
C. $NaHSO_3\ (aq)$ and $H_2O\ (l)$
D. $Na_2S\ (aq)$ and $H_2O\ (l)$

9. A galvanic cell consists of an Ag (*s*)|Ag^+ (*aq*) half-cell and a Zn (*s*)|Zn^{2+} (*aq*) half-cell connected by a salt bridge. If oxidation occurs in the zinc half-cell, which of the following represents the cell in standard notation?

A. Ag^+ (*aq*)|Ag (*s*) || Zn (*s*)|Zn^{2+} (*aq*)
B. Zn^{2+} (*aq*)|Zn (*s*) || Ag (*s*)|Ag^+ (*aq*)
C. Zn (*s*)|Zn^{2+} (*aq*) || Ag^+ (*aq*)|Ag (*s*)
D. Zn (*s*)|Zn^{2+} (*aq*) || Ag (*s*)|Ag^+ (*aq*)

10. What is the maximum number of electrons in the n = 3 shell?

A. 32
B. 18
C. 10
D. 8

11. What is the molecular geometry of Cl_2CO?

A. tetrahedral
B. trigonal planar
C. trigonal pyramidal
D. linear

12. Which of the following statements best describes a gas?

A. Definite shape, but indefinite volume
B. Indefinite shape, but definite volume
C. Indefinite shape and volume
D. Definite shape and volume

13. Which of the following is NOT an electrolyte?

A. NaBr
B. Ne
C. KOH
D. HCl

14. A process or reaction that releases heat into the surroundings is:

A. isothermal
B. exothermic
C. endothermic
D. conservative

15. As the temperature of a reaction increases, what happens to the reaction rate and rate constant?

A. Increases, but the rate constant remains the same
B. Increases, in proportion with the rate constant
C. Remains constant, but the rate constant increases
D. Remains constant, as does the rate constant

16. Would a precipitate form, if 1 liter of 0.04 M $Mg(NO_3)_2$ was mixed with 3 liters of 0.08 M K_2SO_4? (Assume complete dissociation of solutions and use the K_{sp} of $MgSO_4 = 4 \times 10^{-5}$)

A. Yes
B. No, because the K_{sp} for $MgSO_4$ is not exceeded
C. No, because the solution does not contain $MgSO_4$
D. No, because the K_{sp} for $MgSO_4$ is exceeded

17. Which of the following is a strong base?

A. $Ba(OH)_2$
B. CH_3CH_2COOH
C. NH_3
D. CH_3CH_2OH

18. Uranium exists in nature in the form of several isotopes that have different:

A. numbers of electrons
B. numbers of protons
C. atomic numbers
D. numbers of neutrons

19. Which of the following pairs is NOT correctly matched?

	Formula	Molecular polarity
A.	SiF_4	nonpolar
B.	H_2O	polar
C.	HCN	nonpolar
D.	H_2CO	polar

20. If the pressure of a gas sample is doubled, according to Gay–Lussac's law, the sample's:

A. temperature is doubled
B. temperature decreases by a factor of 2
C. volume is doubled
D. volume decreases by a factor of 2

21. What is the coefficient (n) of O_2 gas for the balanced equation?

$$nP\ (s) + nO_2\ (g) \rightarrow nP_2O_5\ (s)$$

A. 1 **B.** 2 **C.** 4 **D.** 5

22. Consider the enthalpy of the following balanced reaction:

$$C_3H_8 + 5\ O_2 \rightarrow 3\ CO_2 + 4\ H_2O = +488\ \text{kcal}$$

The reaction is [ ] and ΔH is [ ].

A. exothermic … negative
B. endothermic … negative
C. exothermic … positive
D. endothermic … zero

23. Which of the following causes the reaction to shift towards yielding more carbon dioxide gas in the following balanced endothermic reaction?

$$CaCO_3\ (s) \leftrightarrow CaO\ (s) + CO_2\ (g)$$

A. Increasing the pressure and decreasing the temperature of the system
B. Increasing the pressure of the system
C. Increasing the temperature of the reaction
D. Increasing the pressure and increasing the temperature of the system

24. Assuming the same solute and solvent, which of the following solutions is the least concentrated?

A. 2.8 g solute in 2 mL solution
B. 2.8 g solute in 5 mL solution
C. 25 g solute in 50 mL solution
D. 40 g solute in 160 mL solution

25. What is the pH of a solution that has a $[H_3O^+] = 1.4 \times 10^{-3}$?

A. 1.67
B. 2.86
C. 3.42
D. 4.68

26. In an acidic media, how many electrons are needed to balance the following half-reaction?

$NO_3^- \rightarrow NH_4^+$

A. 8 e^- on the reactant side
B. 4 e^- on the reactant side
C. 3 e^- on the product side
D. 2 e^- on the reactant side

27. Which of the following elements is a metalloid?

A. antimony (Sb)
B. uranium (U)
C. iodine (I)
D. zinc (Zn)

28. Which of the following is the strongest intermolecular force between molecules?

A. hydrophobic
B. van der Waals
C. hydrogen bonding
D. coordinate covalent bonds

29. As the strength of the attractive intermolecular force increases, [ ] decreases.

A. melting point
B. vapor pressure of a liquid
C. viscosity
D. normal boiling temperature

30. What is the oxidation number of Cl in $HClO_4$?

A. –1
B. +5
C. +4
D. +7

31. What is the ΔH for the following reaction?

$$CH_4\,(g) + 2\,O_2\,(g) \rightarrow CO_2\,(g) + 2\,H_2O\,(l)$$

(Use the overall ΔH = –806 kJ/mol and ΔH for $2\,H_2O\,(g) \rightarrow 2\,H_2O\,(l)$ = –86 kJ/mol)

A. –892 kJ/mol
B. –720 kJ/mol
C. 720 kJ/mol
D. 892 kJ/mol

32. Which of the changes listed below shifts the equilibrium to the left for the following reversible reaction?

$PbI_2\,(s) \leftrightarrow Pb^{2+}\,(aq) + 2\,I^{-}\,(aq)$

A. Add $Pb(NO_3)_2\,(s)$

B. Add NaCl (*s*)

C. Decrease $[Pb^{2+}]$

D. Decrease $[I^{-}]$

33. If the chemical formula of the solute is known, which additional information is necessary to determine the molarity of a solution?

A. Mass of the solute dissolved and the volume of the solvent added

B. Mass of the solute dissolved and the final volume of the solution

C. Volume of the solvent used

D. Mass of the solute dissolved

34. Which of the following is a general property of an acidic solution?

A. pH greater than 7

B. Turns litmus paper blue

C. Feels slippery

D. Tastes sour

35. An element can be determined by:

A. protons plus neutrons

B. neutrons only

C. protons plus electrons

D. atomic number

36. Which of the following is the strongest form of an intramolecular attraction?

A. ion-dipole interaction

B. dipole-dipole interaction

C. dipole-induced dipole interaction

D. covalent bond

37. Which is an assumption of the kinetic molecular theory of gases?

A. Nonelastic collisions

B. Constant interaction of molecules

C. Elastic collisions

D. Gas particles take up space

38. Which of the following reaction is NOT classified correctly?

A. $BaCl_2 + H_2SO_4 \rightarrow BaSO_4 + 2\,HCl$ (single-replacement)

B. $F_2 + 2\,NaCl \rightarrow Cl_2 + 2\,NaF$ (single-replacement)

C. $Fe + CuSO_4 \rightarrow Cu + FeSO_4$ (single-replacement)

D. $2\,NO_2 + H_2O_2 \rightarrow 2\,HNO_3$ (synthesis)

39. Which statement is true about the formation of the stable ionic compound NaCl?

A. NaCl is stable because it forms from the combination of isolated gaseous ions
B. The lattice energy provides the necessary stabilization force for the formation of NaCl
C. The net absorption of 147 kJ/mol of energy for the formation of the Na^+ (*g*) and Cl^- (*g*) drives the formation of the stable ionic compound
D. The release of energy as NaCl (*s*) forms leads to an overall increase in the potential energy

40. What is the order of B in a reactant with the rate law of $k[A]^2$?

A. 0th order
B. 1st order
C. 2nd order
D. ½th order

41. Which of the following illustrates the *like dissolves like* rule for a solid solute in a liquid solvent?

I. A nonpolar compound is soluble in a polar solvent
II. A polar compound is soluble in a polar solvent
III. An ionic compound is soluble in a polar solvent

A. I only
B. II only
C. I and II only
D. II and III only

42. If an unknown solution is a good conductor of electricity, which of the following statements is true?

A. The solution is highly reactive
B. The solution is slightly reactive
C. The solution is highly ionized
D. The solution is slightly ionized

43. Which of the following is the balanced chemical reaction for the electrolysis of brine (i.e., concentrated salt water) as a major source of chlorine gas?

A. 2 NaCl (*aq*) + H_2O → 2 NaH (*aq*) + Cl_2O^- (*aq*)
B. 2 NaCl (*aq*) + 2 H_2O → 2 NaOH (*aq*) + Cl_2 (*g*) + 2 H_2 (*g*)
C. 2 NaCl (*aq*) + H_2O → Na_2O (*aq*) + 2 HCl (*aq*)
D. 2 NaOH (*aq*) + Cl_2 (*g*) + 2 H_2 (*g*) → 2 NaCl (*aq*) + H_2O

44. Which element has the electron configuration of $1s^22s^22p^63s^2$ in a neutral state?

A. Na
B. Si
C. Ca
D. Mg

45. The valence shell is the:

A. innermost shell that is complete with electrons
B. last partially filled orbital of an atom
C. shell of electrons in the least reactive atom
D. outermost shell of electrons around an atom

46. What is the new internal pressure of a given mass of nitrogen gas in a 350-ml vessel at 24 °C and 1.2 atm, when heated to 64 °C and compressed to 300 ml?

A. (1.2)·(0.300)·(333) / (0.350)·(297)
B. (0.350)·(333) / (1.2)·(0.300)·(297)
C. (1.2)·(0.350)·(337) / (0.300)·(297)
D. (1.2)·(0.350)·(0.300) / (333)·(278)

47. Which response represents the balanced oxidation half-reaction for the following reaction?

$HCl\ (aq) + Fe\ (s) \rightarrow FeCl_3\ (aq) + H_2\ (g)$

A. $2\ Fe \rightarrow 2\ Fe^{3+} + 6\ e^-$
B. $3\ Fe \rightarrow Fe^{3+} + 3\ e^-$
C. $Fe + 3\ e^- \rightarrow Fe^{3+}$
D. $6\ e^- + 6\ H^+ \rightarrow 3\ H_2$

48. Which component of an insulated vessel design minimizes heat radiation?

A. Tight-fitting screw-on lid
B. Heavy-duty aluminum material
C. Double-walled material
D. Reflective interior coating

49. Which of the following changes the value of the equilibrium constant?

A. Changing the initial concentrations of reactants
B. Changing the initial concentrations of products
C. Changing temperature
D. Adding a catalyst at the onset of the reaction

50. What is the molar concentration of Ca^{2+} *(aq)* in a solution that is prepared by mixing 20 mL of a 0.03 M $CaCl_2$ *(aq)* solution with a 15 mL of a 0.06 M $CaSO_4$ *(aq)* solution?

A. 1.84 M
B. 0.012 M
C. 0.043 M
D. 1.22 M

51. Which of the following compounds is the strongest weak acid?

A. H_2CO_3 ($K_a = 4.3 \times 10^{-7}$)
B. HF ($K_a = 7.2 \times 10^{-4}$)
C. HCN ($K_a = 6.2 \times 10^{-10}$)
D. HClO ($K_a = 3.5 \times 10^{-8}$)

52. Which of the following particles would NOT be deflected by charged plates?

A. alpha particles
B. protons
C. hydrogen atoms
D. cathode rays

53. Which of the general statements regarding covalent bond characteristics is NOT correct?

A. Double bonding can occur with Group VIIA elements
B. Triple bonds are possible when 3 or more electrons are needed to complete an octet
C. Triple bonds are stronger than double bonds
D. Double bonds are stronger than single bonds

54. What are the conditions for a real gas to behave most like an ideal gas?

A. Low temperature, high-pressure
B. Low temperature, low-pressure
C. High temperature, high-pressure
D. High temperature, low-pressure

55. If *A* of element Y equals 14, then 28 grams of element Y represents approximately:

A. 28 moles of atoms
B. 2 moles of atoms
C. 2 atoms
D. ½ of an atom

56. How many grams of Ag can be heated from 21.4 °C to 34 °C by the heat released when 26 g of Au cools from 97.2 °C to 29.4 °C? (Use the specific heat *c* of Ag = 0.240 J/g·°C and the specific heat *c* of Au = 0.130 J/g·°C)

A. 75.8 g
B. 2.7×10^3 g
C. 4.4×10^3 g
D. 22.3 g

57. Which of the following influences the rate of a chemical reaction?

I. Collision orientation
II. Collision kinetic energy
III. Collision frequency

A. II only
B. III only
C. II and III only
D. I, II and III

58. What is the %v/v concentration of red wine that contains 25 mL of ethyl alcohol in 225 mL?

A. 0.110 %v/v
B. 11.1 %v/v
C. 0.011 %v/v
D. 5.5 %v/v

59. Which statement most specifically describes the following reaction?

$$HCl\ (aq) + KOH\ (aq) \rightarrow KCl\ (aq) + H_2O\ (l)$$

A. HCl and potassium hydroxide solutions produce potassium chloride and H_2O
B. HCl and potassium hydroxide solutions produce potassium chloride solution and H_2O
C. Aqueous HCl and potassium hydroxide produce aqueous potassium chloride and H_2O
D. HCl and potassium hydroxide produce potassium chloride and H_2O

60. A galvanic cell is constructed with the following two elements and their ions:

$Mg\ (s) \rightarrow Mg^{2+} + 2\ e^-$ $\qquad$ $E° = 2.35\ V$

$Pb\ (s) \rightarrow Pb^{2+} + 2\ e^-$ $\qquad$ $E° = 0.13\ V$

What is the E° for the net reaction of the oxidation of Mg (*s*) and the reduction of Pb (*s*)?

A. –2.48 V
B. 2.22 V
C. –2.22 V
D. 0.31 V

Use the answer key to check your answers. Review the explanations in detail, focusing on questions you didn't answer correctly. Note the topic of those questions. Complete this BEFORE taking the next Diagnostic Test.

Notes for active learning

Diagnostic Test 3

This Diagnostic Test is designed to assess your proficiency on each topic and NOT to mimic the test. Use your test results and identify areas of strength and weakness to adjust your study plan and enhance your fundamental knowledge. The length of the Diagnostic Tests is optimal for a single study session.

#	Answer:				Review	#	Answer:				Review
1:	A	B	C	D	___	31:	A	B	C	D	___
2:	A	B	C	D	___	32:	A	B	C	D	___
3:	A	B	C	D	___	33:	A	B	C	D	___
4:	A	B	C	D	___	34:	A	B	C	D	___
5:	A	B	C	D	___	35:	A	B	C	D	___
6:	A	B	C	D	___	36:	A	B	C	D	___
7:	A	B	C	D	___	37:	A	B	C	D	___
8:	A	B	C	D	___	38:	A	B	C	D	___
9:	A	B	C	D	___	39:	A	B	C	D	___
10:	A	B	C	D	___	40:	A	B	C	D	___
11:	A	B	C	D	___	41:	A	B	C	D	___
12:	A	B	C	D	___	42:	A	B	C	D	___
13:	A	B	C	D	___	43:	A	B	C	D	___
14:	A	B	C	D	___	44:	A	B	C	D	___
15:	A	B	C	D	___	45:	A	B	C	D	___
16:	A	B	C	D	___	46:	A	B	C	D	___
17:	A	B	C	D	___	47:	A	B	C	D	___
18:	A	B	C	D	___	48:	A	B	C	D	___
19:	A	B	C	D	___	49:	A	B	C	D	___
20:	A	B	C	D	___	50:	A	B	C	D	___
21:	A	B	C	D	___	51:	A	B	C	D	___
22:	A	B	C	D	___	52:	A	B	C	D	___
23:	A	B	C	D	___	53:	A	B	C	D	___
24:	A	B	C	D	___	54:	A	B	C	D	___
25:	A	B	C	D	___	55:	A	B	C	D	___
26:	A	B	C	D	___	56:	A	B	C	D	___
27:	A	B	C	D	___	57:	A	B	C	D	___
28:	A	B	C	D	___	58:	A	B	C	D	___
29:	A	B	C	D	___	59:	A	B	C	D	___
30:	A	B	C	D	___	60:	A	B	C	D	___

This page is intentionally left blank

1. Which characteristics describe a neutron's mass, charge and location, respectively?

A. Approximate mass 1 amu; charge 0; inside nucleus
B. Approximate mass 5×10^{-4} amu; charge 0; inside nucleus
C. Approximate mass 1 amu; charge +1; inside nucleus
D. Approximate mass 1 amu; charge –1; inside nucleus

2. Which bond is formed by the equal sharing of electrons?

A. covalent **B.** dipole **C.** ionic **D.** London

3. At constant pressure, what effect does decreasing the temperature of a liquid by 20 °C have on the magnitude of its vapor pressure?

A. Inversely proportional
B. Requires more information
C. Decrease
D. No effect

4. Which compound contains the lowest number of moles?

A. 15 g CH_4
B. 15 g AlH_3
C. 15 g CO
D. 15 g N_2

5. Which is/are a general guideline for balancing an equation?

I. Balance polyatomic ions as a single unit
II. Begin balancing with the most complex formula
III. Write correct formulas for reactants and products

A. I only
B. II only
C. III only
D. I, II and III

6. In the reaction energy diagrams below, the reaction on the right is [ ] and ΔG is [ ] than for the reaction on the left. (Assume the reactions are drawn to scale)

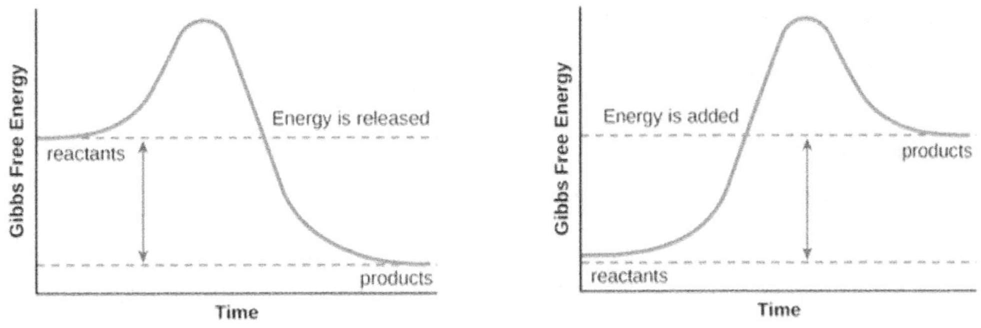

A. exergonic … less than
B. endergonic … greater than
C. exergonic … greater than
D. endergonic … less than

7. Which of the following methods does NOT extract colloidal particles?

A. distillation
B. extraction
C. evaporation
D. simple filtration

8. Which of the following indicators is/are colorless in an acidic solution and pink in a basic solution?

I. phenolphthalein II. bromothymol blue III. methyl red

A. I only
B. II only
C. III only
D. I and II only

9. How many grams of chromium would be electroplated by passing a constant current of 4.8 amps through a solution containing chromium (III) sulfate for 50.0 minutes? (Use the molecular mass for Cr = 52.00 g/mol and 1 Faraday = 96,500 C/mol)

A. 2.27×10^3 g
B. 2.58 g
C. 0.095 g
D. 8.43×10^{-4} g

10. Which electron configuration represents an excited state of potassium which has an atomic number of 19?

A. $1s^22s^22p^63s^23p^6$
B. $1s^22s^22p^63s^23p^64s^2$
C. $1s^22s^22p^63s^23p^54s^2$
D. $1s^22s^22p^63s^23p^7$

11. Which formula for an ionic compound is NOT correct as written?

A. NH_4ClO_4
B. MgS
C. KOH
D. $Ca(SO_4)_2$

12. According to Boyle's law, what happens to a gas as the volume increases?

A. The temperature increases
B. The temperature decreases
C. The pressure increases
D. The pressure decreases

13. By definition, which of the following is true for a strong electrolyte?

A. Must be highly soluble in water
B. Contains both metal and nonmetal atoms
C. Is an ionic compound
D. Dissociates almost completely into its ions in solution

14. What describes a reaction with no change in entropy, if ΔG is negative?

A. Spontaneous and endothermic
B. Spontaneous and exothermic
C. Nonspontaneous and exothermic
D. Nonspontaneous and endothermic

15. For the following reaction of A + B → C, which proceeds at the fastest rate at STP? (Assume that each graph is drawn to proportion)

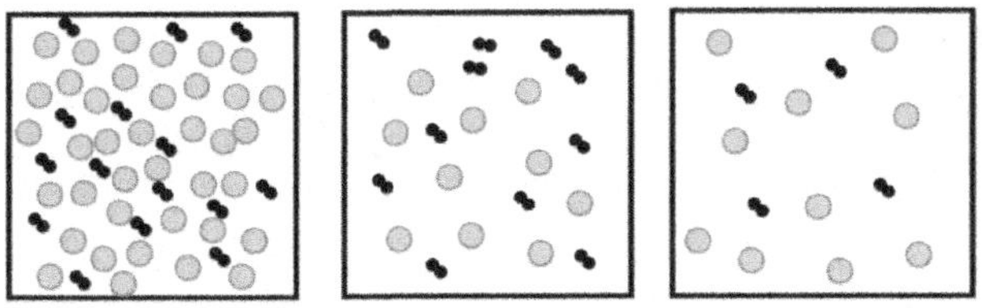

a b c

A. a

B. b

C. c

D. All proceed at the same rate

16. Which ions are the spectator ions in the reaction below?

$$K_2SO_4\,(aq) + Ba(NO_3)_2\,(aq) \rightarrow BaSO_4\,(s) + 2\ KNO_3\,(aq)$$

A. K^+ and SO_4^{2-}

B. Ba^{2+} and NO_3^-

C. K^+ and NO_3^-

D. Ba^{2+} and SO_4^{2-}

17. The neutralization of $Cr(OH)_3$ with H_2SO_4 produces which product?

A. $Cr_2(SO_4)_3$

B. SO_2

C. ^{-}OH

D. H_3O^+

18. Which of the following is/are a general characteristic of a nonmetal?

I. low density II. heat conduction III. brittle

A. I only

B. II only

C. III only

D. I and III only

19. Which of the following statements about electronegativity is NOT correct?

A. Electronegativity increases from left to right within a row

B. Electronegativity increases from bottom to top within a group

C. Metals are generally more electronegative than nonmetals

D. Francium is the least electronegative element

20. When NaCl dissolves in water, what is the force of attraction between Na^+ and H_2O?

A. van der Waals

B. ion-dipole

C. ion-ion

D. dipole-dipole

21. Which of the following has the greatest mass?

A. 35 protons, 35 neutrons and 33 electrons
B. 34 protons, 35 neutrons and 37 electrons
C. 35 protons, 34 neutrons and 37 electrons
D. 34 protons, 35 neutrons and 34 electrons

22. What is the final temperature after 340 J of heat energy is removed from 30.0 g of H_2O at 19.8 °C? (Use the specific heat *c* of H_2O = 4.184 J/g·°C)

A. 28.4 °C
B. 23.2 °C
C. 12.4 °C
D. 17.1 °C

23. Which statement describes how a catalyst affects the rate of a reaction?

A. increases ΔG
B. increases E_{act}
C. decreases E_{act}
D. increases ΔH

24. What is the concentration in mass-volume percent for 14.6 g of CaF_2 in 260 mL of solution?

A. 5.62%
B. 9.95%
C. 0.180%
D. 2.73%

25. Which of the following is/are an example of an Arrhenius acid?

I. HCl (*aq*)   II. H_2SO_4 (*aq*)   III. HNO_3 (*aq*)

A. I only
B. II only
C. III only
D. I, II and III

26. Which of the statements is true for the following two half-reactions occurring in a voltaic cell?

Reaction I: $Cr_2O_7^{2-}(aq) + 14\ H^+(aq) + 6\ e^- \rightarrow 2\ Cr^{3+}(aq) + 7\ H_2O\ (l)$

Reaction II: $6\ I^-(aq) \rightarrow 3\ I_2(s) + 6\ e^-$

A. Reaction II is oxidation and occurs at the cathode
B. Reaction II is oxidation and occurs at the anode
C. Reaction I is reduction and occurs at the anode
D. Reaction I is oxidation and occurs at the anode

27. The *l* quantum number refers to the electron's:

A. angular momentum
B. magnetic orientation
C. spin
D. energy level

28. Which statement is true about the valence shell?

A. The valence shell determines the electron dot structure
B. The electron dot structure is made up of each of the valence shells
C. The valence shell is the innermost shell
D. The valence shell is usually the most unreactive shell

29. What is the relationship between temperature and volume of a fixed amount of gas at constant pressure?

A. Equal
B. Indirectly proportional
C. Directly proportional
D. Decreased by a factor of 2

30. What is the empirical formula of the compound with the mass percent of 71.7% Cl, 24.3% C and 4.0% H?

A. ClC_2H_5
B. Cl_2CH_2
C. $ClCH_3$
D. $ClCH_2$

31. What is the term given for a reaction when the bonds formed during the reaction are stronger than the bonds that are broken?

A. endergonic
B. exergonic
C. endothermic
D. exothermic

32. Which statement about chemical equilibrium is NOT true?

A. At equilibrium, the forward reaction rate equals the reverse reaction rate
B. The same equilibrium state can be attained starting either from the reactant or product side of the equation
C. Chemical equilibrium can only be attained by starting with reagents from the reactant side of the equation
D. At equilibrium, the concentration of reactants and products are constant over time

33. What is the solubility of CaF_2 when it is added to a 0.03 M solution of NaF?

A. Low because $[F^-]$ present in solution inhibits dissociation of CaF_2
B. High because of the common ion effect
C. Unaffected by the presence of NaF
D. Low because less water is available for solvation due to the presence of NaF

34. Which of the following properties is NOT characteristic of a base?

A. Has a slippery feel
B. Is neutralized by an acid
C. Has a bitter taste
D. Produces H^+ in water

35. The element calcium belongs to which family?

A. representative elements
B. alkaline earth metals
C. alkali metals
D. lanthanides

36. Which of the following statements is true?

A. Two electrons in a single covalent bond are paired according to the Pauli exclusion principle
B. Bond energy is the minimum energy required to bring about the pairing of electrons in a covalent bond
C. As the distance between the nuclei decreases for covalent bond formation, there is a corresponding decreased probability of finding both electrons near either nucleus
D. One mole of hydrogen atoms is more stable than one mole of hydrogen molecules

37. Which substance would be expected to have the highest boiling point?

A. Nonvolatile liquid
B. Nonpolar liquid with van der Waals interactions
C. Polar liquid with hydrogen bonding
D. Nonpolar liquid

38. Ethane (C_2H_6) gives off carbon dioxide and water when burning. What is the coefficient of oxygen in the balanced equation for the reaction?

$$\_\_C_2H_6\,(g) + \_\_O_2\,(g) \xrightarrow{\text{Spark}} \_\_CO_2\,(g) + \_\_H_2O\,(g)$$

A. 5
B. 7
C. 10
D. 14

39. An ideal gas fills a closed rigid container. As the number of moles of gas in the chamber is increased at a constant temperature:

A. pressure increases
B. pressure remains constant
C. pressure decreases
D. the effect on pressure cannot be determined

40. Which of the following drives this exothermic reaction towards products?

$$CH_4\,(g) + 2\,O_2\,(g) \leftrightarrow CO_2\,(g) + 2\,H_2O\,(g)$$

A. Addition of CO_2
B. Removal of CH_4
C. Increase in temperature
D. Decrease in temperature

41. What mass of H_2O is needed to prepare 160 grams of 14.0% (m/m) $KHCO_3$ solution?

A. 22.4 g
B. 138 g
C. 1.9 g
D. 11.4 g

42. A 40.0 mL sample of aqueous sulfuric acid was titrated with 0.4 M NaOH (*aq*) until neutralized. What is the molarity of the sulfuric acid solution, if the residue was dried and has a mass of 840 mg?

A. 0.034 M
B. 0.148 M
C. 0.623 M
D. 0.745 M

43. How many electrons are needed to balance the charge for the following half-reaction in acidic solution?

$$C_2H_6O \rightarrow HC_2H_3O_2$$

A. 6 electrons to the right side
B. 4 electrons to the right side
C. 3 electrons to the left side
D. 2 electrons to the left side

44. The three isotopes of hydrogen have different numbers of:

A. neutrons
B. electrons
C. protons
D. charges

45. Which statement describes the lengths of the sulfur-oxygen bonds in SO_3^{2-}?

A. All bonds are of different lengths
B. All bonds are of the same length
C. Two are the same length, while the other is longer
D. Two are the same length, while the other is shorter

46. Which of the following would have the highest boiling point?

A. F_2
B. Br_2
C. Cl_2
D. I_2

47. What substance is the oxidizing agent in the following redox reaction?

$$Co\ (s) + 2\ HCl\ (aq) \rightarrow CoCl_2\ (aq) + H_2\ (g)$$

A. H_2
B. $CoCl_2$
C. HCl
D. Co

48. Which component of an insulated vessel design minimizes heat loss by convection?

A. Heavy-duty aluminum construction
B. Reflective interior coating
C. Tight-fitting, screw-on lid
D. Double-walled construction

49. Dinitrogen tetraoxide decomposes to produce nitrogen dioxide according to the reaction below. Calculate the equilibrium constant (K_{eq}) at 100 °C. (Use the equilibrium concentrations $[N_2O_4] = 0.800$ and $[NO_2] = 0.400$)

$N_2O_4\,(g) \leftrightarrow 2\,NO_2\,(g)$

A. $K_{eq} = 0.725$
B. $K_{eq} = 2.50$
C. $K_{eq} = 0.200$
D. $K_{eq} = 0.500$

50. Which solid compound is most soluble in water?

A. CuS
B. $Ba(OH)_2$
C. $PbBr_2$
D. $NiCO_3$

51. What is the pH of an aqueous solution if $[H^+] = 0.000001$ M?

A. 1
B. 4
C. 6
D. 7

52. Which of the following statements is/are true?

I. There are ten *d* orbitals in the *d* subshell
II. The third energy shell (n=3) has *d* orbitals
III. The *p* subshell can accommodate a maximum of 6 electrons

A. II only
B. III only
C. I and III only
D. II and III only

53. Which of these compounds has a trigonal pyramidal molecular geometry?

A. SO_3
B. XeO_3
C. AlF_3
D. BF_3

54. What is the partial pressure due to CO_2 of a mixture of gases at 700 torr that contains 40% CO_2, 40% O_2 and 20% H_2 by pressure?

A. 760 torr
B. 280 torr
C. 420 torr
D. 560 torr

55. What is the mass of 8.50 moles of glucose ($C_6H_{12}O_6$)?

A. 1.54×10^6 g
B. 21 g
C. 964 g
D. 1,532 g

56. Which statement describes a reaction that is spontaneous?

A. Releases heat to the surroundings
B. Proceeds without external influence once it has begun
C. Proceeds in both the forward and reverse directions
D. Has the same rate in both the forward and reverse direction

57. Which of the following stresses would shift the equilibrium to the left for the following chemical system at equilibrium?

$$6\ H_2O\ (g) + 2\ N_2\ (g) + \textit{heat} \leftrightarrow 4\ NH_3\ (g) + 3\ O_2\ (g)$$

A. Increasing the concentration of H_2O
B. Decreasing the concentration of NH_3
C. Increasing the concentration of O_2
D. Increasing the reaction temperature

58. What is the molarity of a solution formed by dissolving 30.0 g of NaCl in water to make 675 mL of solution? (Use the molecular mass of NaCl = 58.45 g/mol)

A. 0.49 M
B. 0.76 M
C. 0.32 M
D. 0.53 M

59. Which of these salts is basic in an aqueous solution?

A. NaF
B. $CrCl_3$
C. KBr
D. NH_4ClO_4

60. Which of the statements is NOT true regarding the following redox reaction occurring in a nonspontaneous electrochemical cell?

$$3\ C\ (s) + 4\ AlCl_3\ (l) \xrightarrow{\text{Electricity}} 4\ Al\ (l) + 3\ CCl_4\ (g)$$

A. Al metal is produced at the cathode
B. Oxidation half-reaction: $C + 4\ Cl^- \rightarrow CCl_4 + 4\ e^-$
C. Reduction half-reaction: $Al^{3+} + 3\ e^- \rightarrow Al$
D. CCl_4 (*g*) is produced at the cathode

Use the answer key to check your answers. Review the explanations in detail, focusing on questions you didn't answer correctly. Note the topic of those questions.

Notes for active learning

Topical Practice Questions

Electronic and Atomic Structure; Periodic Table

Practice Set 1: Questions 1–20

1. The property defined as the energy required to remove one electron from an atom in the gaseous state is:

A. electronegativity
B. ionization energy
C. electron affinity
D. hyperconjugation

2. Which of the following elements most easily accepts an extra electron?

A. He **B.** Ca **C.** Cl **D.** Fr

3. Which of the following pairs has one metalloid element and one nonmetal element?

A. ^{32}Ge and ^{9}F
B. ^{19}K and ^{9}F
C. ^{51}Sb and ^{20}Ca
D. ^{33}As and ^{14}Si

4. Periods on the periodic table represent elements:

A. in the same group
B. with consecutive atomic numbers
C. known as isotopes
D. with similar chemical properties

5. How many electrons can occupy the $n = 2$ shell?

A. 2 **B.** 6 **C.** 8 **D.** 18

6. Ignoring hydrogen and helium, which area(s) of the periodic table contain(s) metals and nonmetals?

I. *s* area II. *p* area III. *d* area

A. I only
B. II only
C. III only
D. I and II only

7. What is the number of known nonmetals relative to the number of metals?

A. About two times greater
B. About fifty percent
C. About five times less
D. About twenty-five percent greater

8. Based on experimental evidence, Dalton postulated that:

A. atoms of different elements have the same mass
B. not all atoms of a given element are identical
C. atoms can be created and destroyed in chemical reactions
D. each element consists of indivisible minute particles called atoms

9. Which statement is true regarding the average mass of a naturally occurring isotope of iron with an atomic mass equal to 55.91 amu?

A. 55.91 / 1.0078 times greater than a ^{1}H atom
B. 55.91 / 12.000 times greater than a ^{12}C atom
C. 55.91 times greater than a ^{12}C atom
D. 55.91 times greater than a ^{1}H atom

10. Which is the principal quantum number?

A. n **B.** m **C.** l **D.** s

11. How many electrons can occupy the 4*s* subshell?

A. 1 **B.** 2 **C.** 6 **D.** 8

12. An excited hydrogen atom emits a light spectrum of specific, characteristic wavelengths. The light spectrum is a result of:

A. energy released as H atoms form H_2 molecules
B. the light wavelengths, not absorbed by valence electrons when white light passes through the sample
C. particles being emitted as the hydrogen nuclei decay
D. excited electrons dropping to lower energy levels

13. Which has the largest radius?

A. Br^- **B.** K^+ **C.** Ar **D.** Ca^{2+}

14. In its ground state, how many unpaired electrons does a sulfur atom have?

A. 0 **B.** 1 **C.** 2 **D.** 3

15. Which elements would be in the same group as the element whose electronic configuration is $1s^22s^22p^63s^23p^64s^1$?

A. Li **B.** Se **C.** Mg **D.** P

16. When an atom is most stable, how many electrons does it contain in its valence shell?

A. 4 **B.** 6 **C.** 8 **D.** 10

17. Which characteristic(s) is/are responsible for the changes seen in the first ionization energy when moving down a column?

I. Increased shielding of electrons
II. Larger atomic radii
III. Increasing nuclear attraction for electrons

A. I only
B. II only
C. III only
D. I and II only

18. What is the maximum number of electrons that can occupy the $4f$ subshell?

A. 6 **B.** 8 **C.** 14 **D.** 16

19. The property that describes the energy released by a gas phase atom from adding an electron is:

A. ionization energy
B. electronegativity
C. electron affinity
D. hyperconjugation

20. The attraction of the nucleus on the outermost electron in an atom tends to:

A. decrease from right to left and bottom to top on the periodic table
B. decrease from right to left and top to bottom on the periodic table
C. decrease from left to right and top to bottom of the periodic table
D. increase from right to left and top to bottom on the periodic table

---

Practice Set 2: Questions 21–40

---

21. How many protons and neutrons are in ^{35}Cl, respectively?

A. 35; 18
B. 18; 17
C. 17; 18
D. 17; 17

22. What must be the same if two atoms represent the same element?

A. number of neutrons
B. atomic mass
C. number of electron shells
D. atomic number

23. Referring to the periodic table, which of the following is NOT a solid metal under normal conditions?

A. Hg **B.** Os **C.** Ba **D.** Cr

24. Which of the following describe electron affinity?

I. The ability of an atom to attract electrons when it bonds with another atom
II. The energy needed to remove an electron from a neutral atom of the element in the gas phase
III. The energy liberated when an electron is added to a gaseous neutral atom converting it to an anion

A. I only
B. II only
C. III only
D. I and II only

25. Metalloids are elements:

A. larger than nonmetals
B. found in asteroids
C. smaller than metals
D. that have some properties like metals and some like nonmetals

26. Which of the following represent(s) halogens?

I. Br II. H III. I

A. I only
B. II only
C. III only
D. I and III

27. According to John Dalton, atoms of an element:

A. are divisible
B. have the same shape
C. are identical
D. have different masses

28. Which of the following represents a pair of isotopes?

A. $^{32}_{16}S$, $^{32}_{16}S^{2-}$

B. O_2, O_3

C. $^{14}_{6}C$, $^{14}_{7}N$

D. $^{1}_{1}H$, $^{2}_{1}H$

29. Early investigators proposed that the ray of the cathode tube was a negatively charged particle because the ray was:

A. attracted to positively charged electric plates
B. diverted by a magnetic field
C. observed in the presence or absence of a gas
D. able to change colors depending on which gas was within the tube

30. What is the value of quantum numbers n and l in the highest occupied orbital for the element carbon with an atomic number of 6?

A. $n = 1, l = 1$
B. $n = 2, l = 1$
C. $n = 1, l = 2$
D. $n = 2, l = 2$

31. Which type of subshell is filled by the distinguishing electron in an alkaline earth metal?

A. *s* **B.** *p* **C.** *f* **D.** *d*

32. If an element has an electron configuration ending in $3p^4$, which statements about the element's electron configuration is NOT correct?

A. There are six electrons in the 3rd shell
B. Five different subshells contain electrons
C. There are eight electrons in the 2nd shell
D. The 3rd shell needs two more electrons to be filled

33. Halogens form anions by:

A. gaining two electrons
B. gaining one electron
C. gaining one neutron
D. losing one electron

34. What is the term for a broad uninterrupted band of radiant energy?

A. Ultraviolet spectrum
B. Visible spectrum
C. Continuous spectrum
D. Radiant energy spectrum

35. Which of the following is the correct order of increasing atomic radius?

A. Te < Sb < In < Sr < Rb
B. In < Sb < Te < Sr < Rb
C. Te < Sb < In < Rb < Sr
D. Rb < Sr < In < Sb < Te

36. Which of the following electron configurations represents an excited state of an atom?

A. $1s^22s^22p^63s^23p^3$
B. $1s^22s^22p^63s^23p^64s^1$
C. $1s^22s^22p^63d^1$
D. $1s^22s^22p^63s^23p^64s^23d^1$

37. Which of the following elements has the greatest ionization energy?

A. rubidium
B. neon
C. potassium
D. calcium

38. Which of the following elements is a nonmetal?

A. Sodium
B. Chlorine
C. Aluminum
D. Magnesium

39. An atom that contains 47 protons, 47 electrons, and 60 neutrons is an isotope of:

A. Nd **B.** Bh **C.** Ag **D.** Al

40. The shell level of an electron is defined by which quantum number?

A. electron spin quantum number
B. magnetic quantum number
C. azimuthal quantum number
D. principal quantum number

Practice Set 3: Questions 41–60

41. How many neutrons are in a Beryllium atom with an atomic number of 4 and an atomic mass of 9?

A. 4 **B.** 5 **C.** 9 **D.** 13

42. Which characteristics describe the mass, charge, and location of an electron, respectively?

A. approximate mass 5×10^{-4} amu; charge –1; outside nucleus
B. approximate mass 5×10^{-4} amu; charge 0; inside nucleus
C. approximate mass 1 amu; charge –1; inside nucleus
D. approximate mass 1 amu; charge +1; outside nucleus

43. What is the name of the compound $CaCl_2$?

A. dichloromethane
B. dichlorocalcium
C. carbon chloride
D. calcium chloride

44. What is the name for elements in the same column of the periodic table with similar chemical properties?

A. congeners
B. stereoisomers
C. diastereomers
D. epimers

45. Which of the following elements consist of members of the same group on the periodic table?

A. ^{14}Si, ^{15}P, and ^{16}S
B. ^{20}Ca, ^{26}Fe, and ^{34}Se
C. ^{9}F, ^{10}Ne, and ^{11}Na
D. ^{31}Ga, ^{49}In, and ^{81}Tl

46. Which element would most likely be a metal with a low melting point?

A. K **B.** B **C.** N **D.** C

47. Mg is an example of a(n):

A. transition metal
B. noble gas
C. alkaline earth metal
D. halogen

48. Metalloids:

I. have some metallic and some nonmetallic properties
II. may have low electrical conductivities
III. contain elements in group IIIB

A. I and III only
B. II only
C. II and III only
D. I and II only

49. Which element is a halogen?

A. Os
B. I
C. O
D. Te

50. Silicon exists as three isotopes ^{28}Si, ^{29}Si, and ^{30}Si, with atomic masses of 27.98 amu, 28.98 amu, and 29.97 amu, respectively. Which isotope is the most abundant in nature?

A. ^{28}Si
B. ^{29}Si
C. ^{30}Si
D. All are equally abundant

51. Which statement supports why early investigators proposed that the ray of the cathode ray tube was due to the cathode?

A. The ray was diverted by a magnetic field
B. The ray was not seen from the positively charged anode
C. The ray was observed in the presence or absence of a gas
D. The ray changed color depending on the gas used within the tube

52. Which of the following is implied by the spin quantum number?

I. The two spinning electrons generate magnetic fields
II. The values are $+\frac{1}{2}$ or $-\frac{1}{2}$
III. Orbital electrons have opposite spins

A. II only
B. I and II only
C. I and III only
D. I, II and III

53. How many quantum numbers are needed to describe a single electron in an atom?

A. 1
B. 2
C. 3
D. 4

54. Which of the following has the correct order of increasing energy in a subshell?

A. 3*s*, 3*p*, 4*s*, 3*d*, 4*p*, 5*s*, 4*d*
B. 3*s*, 3*p*, 4*s*, 3*d*, 4*p*, 4*d*, 5*s*
C. 3*s*, 3*p*, 3*d*, 4*s*, 4*p*, 4*d*, 5*s*
D. 3*s*, 3*p*, 3*d*, 4*s*, 4*p*, 5*s*, 4*d*

55. Rank the elements below in the order of decreasing atomic radius.

A. Al > P > Cl > Na > Mg
B. Cl > Al > P > Na > Mg
C. Mg > Na > P > Al > Cl
D. Na > Mg > Al > P > Cl

56. Which of the following is the electron configuration of a boron atom?

A. $1s^22s^12p^2$
B. $1s^22p^3$
C. $1s^22s^22p^2$
D. $1s^22s^22p^1$

57. Which element has the electron configuration $1s^22s^22p^63s^23p^64s^23d^{10}4p^65s^24d^{10}5p^2$?

A. Sn **B.** As **C.** Pb **D.** Sb

58. Which of the following elements has the greatest ionization energy?

A. Ar **B.** Sr **C.** Br **D.** In

59. Which element has the lowest electronegativity?

A. Mg **B.** Al **C.** Cl **D.** Br

60. Refer to the periodic table and predict which is a solid nonmetal under normal conditions.

A. Cl **B.** F **C.** Se **D.** As

Practice Set 4: Questions 61–80

---

61. Lines were observed in the spectrum of uranium ore (i.e., naturally occurring solid material) identical to helium in the Sun spectrum. Which of the following produced the lines in the helium spectrum?

A. Excited protons jumping to a higher energy level
B. Excited protons dropping to a lower energy level
C. Excited electrons jumping to a higher energy level
D. Excited electrons dropping to a lower energy level

62. Which set of quantum numbers is possible?

A. $n = 1$; $l = 2$; $m_l = 3$; $m_s = -½$
B. $n = 4$; $l = 2$; $m_l = 2$; $m_s = -½$
C. $n = 2$; $l = 1$; $m_l = 2$; $m_s = -½$
D. $n = 3$; $l = 3$; $m_l = 2$; $m_s = -½$

63. The number of neutrons in an atom is equal to:

A. the mass number
B. the atomic number
C. mass number minus the atomic number
D. atomic number minus the mass number

64. Which of the following represent(s) a compound rather than an element?

I. O_3 II. CCl_4 III. S_8

A. I and III only
B. II only
C. I and II only
D. III only

65. Which of the following elements is NOT correctly classified?

A. Po – halogen
B. Sr – alkaline earth metal
C. K – representative element
D. Ar – noble gas

66. Which is the name of the elements that have properties of metals and nonmetals?

A. alkaline earth metals
B. metalloids
C. nonmetals
D. metals

67. Which of the following is the correct sequence of atomic radii from smallest to largest?

A. $Al < S < Al^{3+} < S^{2-}$
B. $S < S^{2-} < Al < Al^{3+}$
C. $Al^{3+} < Al < S^{2-} < S$
D. $Al^{3+} < S < S^{2-} < Al$

68. Which of the following elements contains 6 valence electrons?

A. S **B.** Cl **C.** Si **D.** P

69. From the periodic table, which of the following elements is a semimetal?

A. Ar **B.** As **C.** Al **D.** Ac

70. Which statement does NOT describe the noble gases?

A. The more massive noble gases react with other elements
B. They belong to group VIIIA (or 18)
C. They contain at least one metalloid
D. He, Ne, Ar, Kr, Xe and Rn are included in the group

71. Which is a good experimental method to distinguish between ordinary hydrogen and deuterium, the rare isotope of hydrogen?

I. Measure the density of the gas at STP
II. Measure the rate at which the gas effuses
III. Infrared spectroscopy

A. I only
B. II only
C. I and II only
D. I, II and III

72. Which statement is true regarding the relative abundances of the ^{6}lithium or ^{7}lithium isotopes?

A. The relative proportions change as neutrons move between the nuclei
B. ^{7}Lithium is much more abundant
C. The relative ratio depends on the temperature of the element
D. ^{6}Lithium is much more abundant

73. Which represents the charge on 1 mole of electrons?

A. 96,485 C
B. 6.02×10^{23} C
C. 6.02×10^{23} grams
D. 1 C

74. Which of the following statement(s) is/are true?

I. The *f* subshell contains 7 orbitals
II. The *d* subshell contains 5 orbitals
III. The third energy shell ($n = 3$) has no *f* orbitals

A. I only
B. II only
C. I and II only
D. I, II and III

75. Which element has the greatest ionization energy?

A. Fr **B.** Cl **C.** Ga **D.** I

76. Electrons fill up subshells in order of:

I. decreasing distance from the nucleus
II. increasing distance from the nucleus
III. increasing energy

A. I only
B. II only
C. I and II only
D. II and III only

77. Which of the following produces the "atomic fingerprint" of an element?

A. Excited protons dropping to a lower energy level
B. Excited protons jumping to a higher energy level
C. Excited electrons dropping to a lower energy level
D. Excited electrons jumping to a higher energy level

78. Which of the following is the electron configuration for manganese (Mn)?

A. $1s^22s^22p^63s^23p^64s^23d^5$
B. $1s^22s^22p^63s^23p^64s^23d^{10}4p^1$
C. $1s^22s^22p^63s^23p^64s^23d^6$
D. $1s^22s^22p^63s^23p^64s^23d^8$

79. Which element listed below has the greatest electronegativity?

A. I **B.** F **C.** H **D.** He

80. Which of the following is NOT an alkali metal?

A. Fr **B.** Cs **C.** Ca **D.** Na

Notes for active learning

Notes for active learning

Chemical Bonding

Practice Set 1: Questions 1–20

1. What is the number of valence electrons in tin (Sn)?

A. 14 **B.** 8 **C.** 2 **D.** 4

2. Unhybridized *p* orbitals participate in π bonds as double and triple bonds. How many distinct and degenerate *p* orbitals exist in the second electron shell, where n = 2?

A. 3 **B.** 2 **C.** 1 **D.** 0

3. What is the number of valence electrons in a sulfite ion, SO_3^{2-}?

A. 22 **B.** 24 **C.** 26 **D.** 34

4. Which type of attractive forces occurs in molecules regardless of the atoms they possess?

A. Dipole–ion interactions
B. Ion–ion interactions
C. Dipole–dipole attractions
D. London dispersion forces

5. Given the structure of glucose, which statement explains the hydrogen bonding between glucose and water?

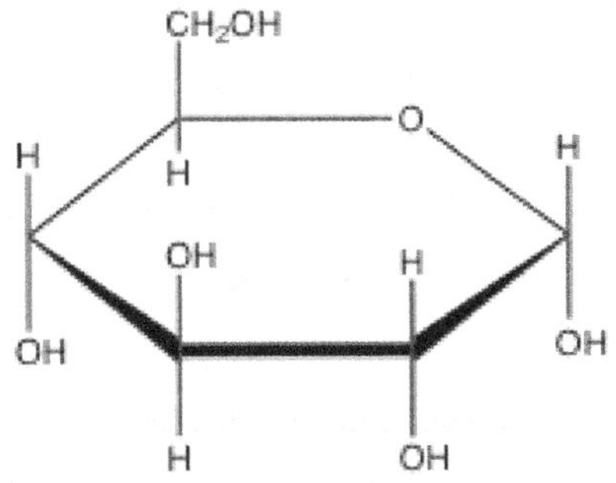

A. Due to the cyclic structure of glucose, there is no H–bonding with water
B. Each glucose molecule could H–bond with up to 17 water molecules
C. H–bonds form, with water being the H–bond donor
D. H–bonds form, with glucose being the H–bond donor

6. How many valence electrons are in the Lewis dot structure of C_2H_6?

A. 2 **B.** 14 **C.** 6 **D.** 8

7. The term for a bond where the electrons are shared unequally is:

A. polar covalent
B. nonpolar covalent
C. coordinate covalent
D. nonpolar ionic

8. What is the name for the weak forces of attraction between nonpolar molecules due to temporary dipoles between adjacent nonpolar molecules?

A. van der Waals forces
B. hydrophobic forces
C. hydrogen bonding forces
D. nonpolar covalent forces

9. Nitrogen has five valence electrons; which of the following types of bonding is/are possible?

I. one single and one double bond
II. three single bonds
III. one triple bond

A. I only
B. II only
C. I and III only
D. I, II and III

10. What is the number of valence electrons in antimony (Sb)?

A. 1 **B.** 2 **C.** 5 **D.** 4

11. How many resonance structures, if any, can be drawn for a nitrite ion?

A. 1 **B.** 2 **C.** 3 **D.** 4

12. What is the formula of the ammonium ion?

A. NH_4^-
B. N_4H^+
C. NH_4^{2-}
D. NH_4^+

13. What is the type of bond that forms between oppositely charged ions?

A. dipole
B. covalent
C. ionic
D. induced dipole

14. Which of the following series of elements are arranged in the order of increasing electronegativity?

A. Fr, Mg, Si, O
B. Br, Cl, S, P
C. F, B, O, Li
D. Cl, S, Se, Te

15. The attraction due to London dispersion forces between molecules depends on what two factors?

A. Volatility and shape
B. Molar mass and volatility
C. Vapor pressure and size
D. Molar mass and shape

16. Which of the substances below would have the largest dipole?

A. CO_2
B. SO_2
C. H_2O
D. CCl_4

17. What is the name for the attraction between H_2O molecules?

A. adhesion
B. polarity
C. cohesion
D. van der Waals

18. Which compound contains only covalent bonds?

A. $HC_2H_3O_2$
B. NaCl
C. NH_4OH
D. $Ca_3(PO_4)_2$

19. The distance between two atomic nuclei in a chemical bond is determined by the:

A. size of the valence electrons
B. size of the nucleus
C. size of the protons
D. balance between the repulsion of the nuclei and the attraction of the nuclei for the bonding electrons

20. Carbonic acid has the chemical formula of H_2CO_3. The carbonate ion has the molecular formula of CO_3^{2-}. From the Lewis structure for the CO_3^{2-} ion, what is the number of reasonable resonance structures for the anion?

A. original structure only
B. 2
C. 3
D. 4

Practice Set 2: Questions 21–40

---

21. Which of the following molecules would contain a dipole?

A. F–F
B. H–H
C. Cl–Cl
D. H–F

22. Which is the correct formula for the ionic compound formed between Ca and I?

A. Ca_3I_2
B. Ca_2I_3
C. CaI_2
D. Ca_2I

23. Which bonding is NOT possible for a carbon atom that has four valence electrons?

A. 1 single and 1 triple bond
B. 1 double and 1 triple bond
C. 4 single bonds
D. 2 single and 1 double bond

24. During strenuous exercise, why does perspiration on a person's skin form droplets?

A. Ability of H_2O to dissipate heat
B. High specific heat of H_2O
C. Adhesive properties of H_2O
D. Cohesive properties of H_2O

25. If an ionic bond is stronger than a dipole–dipole interaction, why does water dissolve an ionic compound?

A. Ions do not overcome their interatomic attraction and therefore are not soluble
B. Ion–dipole interaction causes the ions to heat up and vibrate free of the crystal
C. Ionic bond is weakened by the ion–dipole interactions, and ionic repulsion ejects the ions from the crystal
D. Ion-dipole interactions of several water molecules aggregate with the ionic bond and dissociate it into the solution

26. Which one of these molecules can act as a hydrogen bond acceptor but not a donor?

A. CH_3NH_2
B. CH_3–O–CH_3
C. H_2O
D. C_2H_5OH

27. When NaCl dissolves in water, what is the force of attraction between Na^+ and H_2O?

A. ion–dipole
B. hydrogen bonding
C. ion–ion
D. dipole–dipole

28. Based on the Lewis structure, how many polar and nonpolar bonds are present in H_2CO?

A. 3 polar bonds and 0 nonpolar bonds
B. 2 polar bonds and 1 nonpolar bond
C. 1 polar bond and 2 nonpolar bonds
D. 0 polar bonds and 3 nonpolar bonds

29. How many more electrons can fit within the valence shell of a hydrogen atom?

A. 1 **B.** 2 **C.** 7 **D.** 0

30. Which element likely forms a cation with a +2 charge?

A. Na **B.** S **C.** Si **D.** Mg

31. Which of the statements is an accurate description of the structure of the ionic compound NaCl?

A. Alternating rows of Na^+ and Cl^- ions are present
B. Each ion present is surrounded by six ions of opposite charge
C. Alternating layers of Na and Cl atoms are present
D. Alternating layers of Na^+ and Cl^- ions are present

32. What is the name for the force holding two atoms in a chemical bond?

A. gravitational force
B. strong nuclear force
C. electrostatic force
D. weak nuclear force

33. Which of the following diatomic molecules contains the bond of greatest polarity?

A. CH_4 **B.** Te–F **C.** Cl–F **D.** P_4

34. Why does H_2O have an unusually high boiling point compared to H_2S?

A. Hydrogen bonding
B. Van der Waals forces
C. H_2O molecules pack more closely than H_2S
D. Covalent bonds are stronger in H_2O

35. What is the shape of a molecule in which the central atom has 2 bonding electron pairs and 2 nonbonding electron pairs?

A. trigonal planar
B. trigonal pyramidal
C. linear
D. bent

36. Which of the following represents the breaking of a noncovalent interaction?

A. Ionization of water
B. Decomposition of hydrogen peroxide
C. Hydrolysis of an ester
D. Dissolving of salt crystals

37. Which pair of elements is most likely to form an ionic compound when reacted?

A. C and Cl
B. K and I
C. Ga and Si
D. Fe and Mg

38. Which of the following statements concerning coordinate covalent bonds is correct?

A. Once formed, they are indistinguishable from other covalent bonds
B. They are always single bonds
C. One of the atoms involved must be a metal and the other a nonmetal
D. Both atoms involved in the bond contribute an equal number of electrons to the bond

39. The greatest dipole moment within a bond is when:

A. both bonding elements have a low electronegativity
B. one bonding element has a high electronegativity, and the other has a low electronegativity
C. both bonding elements have a high electronegativity
D. one bonding element has a high electronegativity, and the other has a moderate electronegativity

40. Which of the following must occur for an atom to obtain the noble gas configuration?

A. lose, gain or share an electron
B. loss or gain an electron
C. lose an electron
D. share an electron

Practice Set 3: Questions 41–60

---

41. Based on the Lewis structure for hydrogen peroxide, H_2O_2, how many polar and nonpolar bonds are present?

A. 3 polar and 0 nonpolar bonds
B. 2 polar and 2 nonpolar bonds
C. 2 polar and 1 nonpolar bond
D. 0 polar and 3 nonpolar bonds

42. To form an octet, an atom of selenium must:

A. gain 2 electrons
B. lose 2 electrons
C. gain 6 electrons
D. lose 6 electrons

43. Which of the following is an example of a chemical reaction?

A. two solids mix to form a heterogeneous mixture
B. two liquids mix to form a homogeneous mixture
C. one or more new compounds are formed by rearranging atoms
D. a new element is formed by rearranging nucleons

44. Which compound is NOT correctly matched with the predominant intermolecular force associated with that compound in the liquid state?

Compound	Intermolecular force
A. CH_3OH	hydrogen bonding
B. HF	hydrogen bonding
C. Cl_2O	dipole–dipole interactions
D. HBr	van der Waals interactions

45. Which element forms an ion with the greatest positive charge?

A. Mg **B.** Ca **C.** Al **D.** Na

46. What is the difference between a dipole–dipole and an ion–dipole interaction?

A. One interaction involves dipole attraction between neutral molecules, while the other involves dipole interactions with ions
B. One interaction involves ionic molecules interacting with other ionic molecules, while the other deals with polar molecules
C. One interaction involves salts and water, while the other does not involve water
D. One interaction involves hydrogen bonding, while the others do not

47. Which is a true statement about H_2O as it begins to freeze?

A. Hydrogen bonds break
B. The number of hydrogen bonds increases
C. Covalent bond strength increases
D. Molecules move closer

48. In the process of forming sodium nitride (Na_3N) from its elements, what happens to the electrons of each sodium and the electrons of each nitrogen, respectively?

A. one lost; three gained
B. three lost; three gained
C. three lost; one gained
D. one lost; two gained

49. Which species below has the least number of valence electrons in its Lewis symbol?

A. S^{2-} **B.** Ga^{+} **C.** Ar^{+} **D.** Mg^{2+}

50. Which of the following occur(s) naturally as nonpolar diatomic molecules?

I. sulfur II. chlorine III. argon

A. I only
B. II only
C. I and III only
D. I and II only

51. In a chemical reaction, the bonds being formed are:

A. more energetic than the ones broken
B. less energetic than the ones broken
C. the same as the bonds broken
D. different from the ones broken

52. The charge on a sulfide ion is:

A. +1 **B.** +2 **C.** 0 **D.** –2

53. Which formula for an ionic compound is NOT correct?

A. $Al_2(CO_3)_3$
B. Li_2SO_4
C. Na_2S
D. $MgHCO_3$

54. What term best describes the smallest whole number repeating ratio of ions in an ionic compound?

A. lattice
B. unit cell
C. formula unit
D. covalent unit

55. In the nitrogen monoxide molecule, the dipole moment is 0.16 D, and the bond length is 115 pm. What is the sign and magnitude of the charge on the oxygen atom? (Use the conversion factor of 1 D = 3.34×10^{-30} C·m and the charge of 1 electron = 1.602×10^{-19} C)

A. –0.098 *e*
B. –0.71 *e*
C. –1.3 *e*
D. –0.029 *e*

56. From the electronegativity below, which single covalent bond is the most polar?

Element:	H	C	N	O
Electronegativity	2.1	2.5	3.0	3.5

A. O–C
B. O–N
C. N–C
D. C–H

57. Which of the following pairs is NOT correctly matched?

Formula	Molecular Geometry
A. CH_4	tetrahedral
B. OF_2	bent
C. PCl_3	trigonal planar
D. Cl_2CO	trigonal planar

58. Why are adjacent water molecules attracted to each other?

A. Ionic bonding between the hydrogens of H_2O
B. Covalent bonding between adjacent oxygens
C. Electrostatic attraction between the H of one H_2O and the O of another
D. Covalent bonding between the H of one H_2O and the O of another

59. What is the chemical formula for a compound that contains K^+ and CO_3^{2-} ions?

A. $K(CO_3)_3$
B. K_2CO_3
C. $K_3(CO_3)_3$
D. KCO_3

60. Under what conditions is graphite converted to diamond?

A. low temperature, high-pressure
B. high temperature, low-pressure
C. high temperature, high-pressure
D. low temperature, low-pressure

Practice Set 4: Questions 60–80

---

61. Which of the following molecules is a Lewis acid?

A. BH_3
B. NH_3
C. NH_4^+
D. CH_3COOH

62. Which molecule(s) is/are most likely to show a dipole-dipole interaction?

I. H–C≡C–H II. CH_4 III. CH_3SH IV. CH_3CH_2OH

A. I only
B. II only
C. III only
D. III and IV only

63. C=C, C=O, C=N and N=N bonds are observed in many organic compounds. However, C=S, C=P, C=Si, and other similar bonds are not often found. What is the most probable explanation for this observation?

A. The comparative sizes of 3*p* atomic orbitals make effective overlap between them less likely than between two 2*p* orbitals
B. S, P, and Si do not undergo hybridization of orbitals
C. S, P, and Si do not form π bonds due to the lack of occupied *p* orbitals in their ground state electron configurations
D. Carbon does not combine with elements found below the second row of the periodic table

64. Write the formula for the ionic compound formed from magnesium and sulfur:

A. Mg_2S
B. MgS
C. MgS_2
D. MgS_3

65. Explain why chlorine, Cl_2, is a gas at room temperature, while bromine, Br_2, is a liquid.

A. Bromine ions are held by ionic bonds
B. Chlorine molecules are smaller and, therefore, pack tighter in their physical orientation
C. Bromine atoms are larger, which forms of a stronger induced dipole–induced dipole attraction
D. Chlorine atoms are larger, which forms of a stronger induced dipole–induced dipole attraction

66. What is the major intermolecular force in $(CH_3)_2NH$?

A. hydrogen bonding
B. dipole–dipole attractions
C. London dispersion forces
D. ion–dipole attractions

67. Which of the following correctly describes the molecule potassium oxide and its bond?

A. It is a weak electrolyte with an ionic bond
B. It is a strong electrolyte with an ionic bond
C. It is a non-electrolyte with a covalent bond
D. It is a strong electrolyte with a covalent bond

68. An ion with an atomic number of 34 and 36 electrons has what charge?

A. +2 **B.** –36 **C.** +34 **D.** –2

69. Based on the Lewis structure for $H_3C–NH_2$, the formal charge on N is:

A. –1 **B.** 0 **C.** +2 **D.** +1

70. The name of S^{2-} is:

A. sulfite ion
B. sulfide ion
C. sulfur
D. sulfate ion

71. An ionic bond forms between two atoms when:

A. protons are transferred from the nucleus of the nonmetal to the nucleus of the metal
B. each atom acquires a negative charge
C. electrons are transferred from metallic to nonmetallic atoms
D. four electrons are shared

72. Which of the following pairings of ions is NOT consistent with the formula?

A. Co_2S_3 (Co^{3+} and S^{2-})
B. K_2O (K^+ and O^-)
C. Na_3P (Na^+ and P^{3-})
D. BaF_2 (Ba^{2+} and F^-)

73. Which of the following solids is likely to have the smallest exothermic lattice energy?

A. Al_2O_3 **B.** KF **C.** NaCl **D.** LiF

74. In chlorine monoxide, chlorine has a charge of +0.167 *e*. If the bond length is 154 pm, what is the dipole moment of the molecule? (Use the conversion 1 meter = 1×10^{-12} m/pm and 1 electron = 1.602×10^{-19} C)

A. 2.30 D
B. 0.167 D
C. 1.24 D
D. 3.11 D

75. All of the following are examples of polar molecules, EXCEPT:

A. H_2O
B. CCl_4
C. CH_2Cl_2
D. HF

76. Which is the most likely noncovalent interaction between an alcohol and a carboxylic acid at pH = 10?

A. formation of an anhydride bond
B. dipole–dipole interaction
C. dipole–charge interaction
D. charge–charge interaction

77. Which electron geometry is characteristic of an sp^2 hybridized atom?

A. trigonal planar
B. tetrahedral
C. trigonal bipyramidal
D. bent

78. Which of the following statements about noble gases is NOT correct?

A. They have very stable electron arrangements
B. They are the most reactive of all gases
C. They exist in nature as individual atoms rather than in the molecular form
D. They have 8 valence electrons

79. How are intermolecular forces and solubility related?

A. Solubility is a measure of how weak the intermolecular forces in the solute are
B. Solubility is a measure of how strong a solvent's intermolecular forces are
C. Solubility depends on the solute's ability to overcome the intermolecular forces in the solvent
D. Solubility depends on the solvent's ability to overcome the intermolecular forces in a solute

80. In a bond between two of the following atoms, the bonding electrons would be most strongly attracted to:

A. I
B. He
C. Cs
D. Cl

Notes for active learning

Notes for active learning

Phases and Phase Equilibria

Practice Set 1: Questions 1–20

---

1. Which statement is NOT true regarding vapor pressure?

I. Solids do not have a vapor pressure
II. Vapor pressure of a pure liquid does not depend on the amount of vapor present
III. Vapor pressure of a pure liquid does not depend on the amount of liquid present

A. I only
B. II only
C. III only
D. I and II only

2. How does the volume of a fixed sample of gas change if the temperature is doubled at constant pressure?

A. Decreases by a factor of 2
B. Increases by a factor of 4
C. Doubles
D. Remains the same

3. When a solute is added to a pure solvent, the boiling point [ ] and freezing point [ ]?

A. decreases … decreases
B. decreases … increases
C. increases … decreases
D. increases … increases

4. Consider the phase diagram for H_2O. The termination of the gas-liquid transition at which distinct gas or liquid phases do NOT exist is the:

A. critical point
B. endpoint
C. triple point
D. condensation point

5. The van der Waals equation of state for a real gas is expressed as $[P + n^2a / V^2]\cdot(V - nb) = nRT$. The van der Waals constant, *a*, represents a correction for:

A. negative deviation in the measured value of P from that of an ideal gas due to the attractive forces between the molecules of a real gas
B. positive deviation in the measured value of P from that of an ideal gas due to the attractive forces between the molecules of a real gas
C. negative deviation in the measured value of P from that of an ideal gas due to the finite volume of space occupied by molecules of a real gas
D. positive deviation in the measured value of P from that of an ideal gas due to the finite volume of space occupied by molecules of a real gas

6. What is the value of the ideal gas constant, expressed in units (torr × mL) / mole × K?

A. 62.4
B. 62,400
C. 0.0821
D. 1 / 0.0821

7. If the pressure and the temperature of a gas are halved, the volume is:

A. halved
B. the same
C. doubled
D. quadrupled

8. If container X is occupied by 2.0 moles of O_2 gas, while container Y is occupied by 10.0 grams of N_2 gas, and containers are maintained at 5.0 °C and 760 torrs, then:

A. container X must have a volume of 22.4 L
B. the average kinetic energy of the molecules in X is equal to the average kinetic energy of the molecules in Y
C. container Y must be larger than container X
D. the average speed of the molecules in container X is greater than that of the molecules in container Y

9. Among the following choices, how tall should a properly designed Torricelli mercury barometer be?

A. 100 in
B. 380 mm
C. 76 mm
D. 800 mm

10. When volatile solvents X and Y are mixed in equal proportions, heat is released to the surroundings. If pure X has a higher boiling point than pure Y, which of the following statements is NOT true?

A. The vapor pressure of the mixture is lower than that of pure Y
B. The vapor pressure of the mixture is lower than that of pure X
C. The boiling point of the mixture is lower than that of pure X
D. The boiling point of the mixture is lower than that of pure Y

11. For the balanced reaction 2 Na + Cl_2 → 2 NaCl, which of the following is a gas?

I. Na II. Cl_2 III. NaCl

A. I only
B. II only
C. III only
D. I and II only

12. How does a real gas deviate from an ideal gas?

I. Molecules occupy a significant amount of space
II. Intermolecular forces may exist
III. Pressure is created from molecular collisions with the walls of the container

A. I only
B. II only
C. I and II only
D. II and III only

13. A 2.75 L sample of He gas has a pressure of 0.950 atm. What is the pressure of the gas if the volume is reduced to 0.450 L?

A. 5.80 atm
B. 0.520 atm
C. 0.230 atm
D. 0.960 atm

14. Avogadro's law, in its alternate form, is similar in form to:

I. Boyle's law
II. Charles' law
III. Gay-Lussac's law

A. I only
B. II only
C. III only
D. II and III only

15. What is the relationship between the pressure and volume of a fixed amount of gas at constant temperature?

A. directly proportional
B. equal
C. inversely proportional
D. decreased by a factor of 2

16. What is the term for a change of state from a liquid to a gas?

A. vaporization
B. melting
C. deposition
D. condensing

17. The combined gas law can NOT be written as:

A. $V_2 = V_1 \times P_1 / P_2 \times T_2 / T_1$
B. $P_1 = P_2 \times V_2 / V_1 \times T_1 / T_2$
C. $T_2 = T_1 \times P_1 / P_2 \times V_2 / V_1$
D. $V_1 = V_2 \times P_2 / P_1 \times T_1 / T_2$

18. Which singular molecule is most likely to show a dipole-dipole interaction?

A. CH_4
B. H–C≡C–H
C. SO_2
D. CO_2

19. What happens if the pressure of a gas above liquid increases, such as by pressing a piston above a liquid?

A. Pressure goes down, and the gas moves out of the solvent
B. Pressure goes down, and the gas goes into the solvent
C. The gas is forced into solution, and the solubility increases
D. The solution is compressed, and the gas is forced out of the solvent

20. Which of the following two variables are present in mathematical statements of Avogadro's law?

A. P and V
B. n and V
C. n and T
D. V and T

Practice Set 2: Questions 21–40

21. Which of the following acids has the lowest boiling point elevation?

Monoprotic Acids	*Ka*
Acid I	1.4×10^{-8}
Acid II	1.6×10^{-9}
Acid III	3.9×10^{-10}

A. I
B. II
C. III
D. Requires more information

22. An ideal gas differs from a real gas because the molecules of an ideal gas have:

A. no attraction to each other
B. no kinetic energy
C. molecular weight equal to zero
D. appreciable volumes

23. Which of the following is the definition of standard temperature and pressure?

A. 0 K and 1 atm
B. 298.15 K and 750 mmHg
C. 273.15 K and 10^5 Pa
D. 273 °C and 750 torr

24. At constant volume, as the temperature of a gas sample is decreased, the gas deviates from ideal behavior. Compared to the pressure predicted by the ideal gas law, actual pressure would be:

A. higher because of the volume of the gas molecules
B. higher because of intermolecular attractions between gas molecules
C. lower because of the volume of the gas molecules
D. lower because of intermolecular attractions among gas molecules

25. A flask contains a mixture of O_2, N_2 and CO_2. The pressure exerted by N_2 is 320 torr and by CO_2 is 240 torr. If the total pressure of the gas mixture is 740 torr, what is the percent pressure of O_2?

A. 14%
B. 18%
C. 21%
D. 24%

26. A balloon contains 40 grams of He with a pressure of 1,000 torrs. When He is released from the balloon, the new pressure is 900 torr, and the volume is half the original. If the temperature is the same, how many grams of He remains in the balloon?

A. 18 grams
B. 22 grams
C. 10 grams
D. 40 grams

27. Which of the following is a true statement regarding evaporation?

A. Increasing the surface area of the liquid decreases the rate of evaporation
B. The temperature of the liquid changes during evaporation
C. Decreasing the surface area of the liquid increases the rate of evaporation
D. Molecules with greater kinetic energy escape from the liquid

28. Under which conditions does a real gas behave most nearly like an ideal gas?

A. High temperature and high-pressure
B. High temperature and low-pressure
C. Low temperature and low-pressure
D. Low temperature and high-pressure

29. Which of the following compounds has the highest boiling point?

A. CH_3OH
B. $CH_3CH_2CH_2CH_2CH_2OH$
C. $CH_3CH_2CH_2C(OH)HOH$
D. $CH_3CH_2OCH_2CH_2CH_3$

30. According to the kinetic theory of gases, which of the following is the average kinetic energy of the gas particles directly proportional to?

A. temperature
B. molar mass
C. volume
D. pressure

31. Under ideal conditions, which of the following gases is least likely to behave as an ideal gas?

A. CF_4
B. CH_3OH
C. N_2
D. O_3

32. Which statement is true regarding gases when compared to liquids?

A. Gases have lower compressibility and higher density
B. Gases have lower compressibility and lower density
C. Gases have higher compressibility and higher density
D. Gases have higher compressibility and lower density

33. When nonvolatile solute molecules are added to a solution, the vapor pressure of the solution:

A. stays the same
B. increases
C. decreases
D. is directly proportional to the second power of the amount added

34. Which of the following laws states that the pressure exerted by a mixture of gases equals the sum of the individual gas pressures?

A. Gay-Lussac's law
B. Dalton's law
C. Charles's law
D. Boyle's law

35. Which characteristics best describe a solid?

A. Definite volume; the shape of the container; no intermolecular attractions
B. Volume and shape of the container; no intermolecular attractions
C. Definite shape and volume; strong intermolecular attractions
D. Definite volume; the shape of the container; moderate intermolecular attractions

36. Which of the following statements is true if three 2.0 L flasks are filled with H_2, O_2 and He at STP?

A. There are twice as many He atoms as H_2 or O_2 molecules
B. There are four times as many H_2 or O_2 molecules as He atoms
C. Each flask contains the same number of atoms
D. The number of H_2 or O_2 molecules is the same as the number of He atoms

37. Which of the following atoms could interact through a hydrogen bond?

A. The hydrogen of an amine and the oxygen of an alcohol
B. The hydrogen on an aromatic ring and the oxygen of carbon dioxide
C. The oxygen of a ketone and the hydrogen of an aldehyde
D. The oxygen of methanol and hydrogen on the methyl carbon of methanol

38. Which of the following laws states that the pressure and volume are inversely proportional for gas at a constant temperature?

A. Dalton's law
B. Gay-Lussac's law
C. Boyle's law
D. Charles's law

39. As an automobile travels the highway, the temperature of the air inside the tires [ ] and the pressure [ ]?

A. increases … decreases

B. increases … increases

C. decreases … decreases

D. decreases … increases

40. Using the following unbalanced chemical reaction, what volume of H_2 gas at 780 mmHg and 23 °C is required to produce 12.5 L of NH_3 gas at the same temperature and pressure?

$$N_2\ (g) + H_2\ (g) \rightarrow NH_3\ (g)$$

A. 21.4 L

B. 15.0 L

C. 18.8 L

D. 13.0 L

Practice Set 3: Questions 41–60

41. A mixture of gases containing 16 g of O_2, 14 g of N_2 and 88 g of CO_2 is collected above water at a temperature of 23 °C. The total pressure is 1 atm, and the vapor pressure of water is 38 torr. What is the partial pressure exerted by CO_2?

A. 283 torr **B.** 367 torr **C.** 481 torr **D.** 549 torr

42. Which of the following demonstrate colligative properties?

I. Freezing point II. Boiling point III. Vapor pressure

A. I only
B. II only
C. III only
D. I, II and III

43. What is the proportionality relationship between the pressure of a gas and its volume?

A. directly
B. inversely
C. pressure is raised to the 2nd power
D. pressure raised to the $\sqrt{2}$ power

44. Which of the following statements about gases is correct?

A. Formation of homogeneous mixtures, regardless of the nature of non-reacting gas components
B. Relatively long distances between molecules
C. High compressibility
D. All of the above

45. Which of the following describes a substance in the solid physical state?

I. It compresses negligibly
II. It has a fixed volume
III. It has a fixed shape

A. I only
B. II only
C. I and III only
D. I, II and III

46. Which of the following is NOT a unit used in measuring pressure?

A. kilometers Hg
B. millimeters Hg
C. atmosphere
D. Pascal

47. Which of the following compounds has the highest boiling point?

A. CH_4
B. $CHCl_3$
C. CH_3COOH
D. NH_3

48. Identify the decreasing ordering of attractions among particles in the three states of matter.

A. gas > liquid > solid
B. gas > solid > liquid
C. solid > liquid > gas
D. liquid > solid > gas

49. A sample of N_2 gas occupies a volume of 190 mL at STP. What volume will it occupy at 660 mmHg and 295 K?

A. 1.15 L
B. 0.760 L
C. 0.214 L
D. 1.84 L

50. A nonvolatile liquid would have:

A. a highly explosive propensity
B. strong attractive forces between molecules
C. weak attractive forces between molecules
D. a high vapor pressure at room temperature

51. A closed-end manometer was constructed from a U-shaped glass tube. It was loaded with mercury so that the closed side was filled to the top, which was 820 mm above the neck, while the open end was 160 mm above the neck. The manometer was taken into a chamber used for training astronauts. What is the highest pressure that can be read with assurance on this manometer?

A. 66.0 torr
B. 660 torr
C. 220 torr
D. 760 torr

52. Vessels X and Y each contain 1.00 L of a gas at STP, but vessel X contains oxygen, while vessel Y contains nitrogen. Assuming the gases behave as ideal, they have the same:

I. number of molecules
II. density
III. kinetic energy

A. II only
B. III only
C. I and III only
D. I, II and III

53. What condition must be satisfied for the noble gas Xe to exist in the liquid phase at 180 K, a temperature significantly greater than its normal boiling point?

A. external pressure > vapor pressure of xenon
B. external pressure < vapor pressure of xenon
C. external pressure = partial pressure of water
D. temperature is increased quickly

54. Matter is nearly incompressible in which of these states?

I. solid   II. Liquid   III. gas

A. I only
B. II only
C. III only
D. I and II only

55. Which of the following laws states that volume and temperature are directly proportional for a gas at constant pressure?

A. Gay-Lussac's law
B. Dalton's law
C. Charles' law
D. Boyle's law

56. Which transformation describes sublimation?

A. solid → liquid
B. solid → gas
C. liquid → solid
D. liquid → gas

57. What is the ratio of the diffusion rate of O_2 molecules to the diffusion rate of H_2 molecules if six moles of O_2 gas and six moles of H_2 gas are placed in a large vessel, and the gases and vessels are at the same temperature?

A. 4:1
B. 1:4
C. 12:1
D. 1:1

58. How many molecules of neon gas are present in 6 liters at 10 °C and 320 mmHg? (Use the ideal gas constant R = 0.0821 L·atm K^{-1} mol^{-1})

A. (320 mmHg / 760 atm)·(0.821)·(6 L) / (6×10^{23})·(283 K)
B. (320 mmHg / 760 atm)·(6 L)·(6×10^{23}) / (0.0821)·(283 K)
C. (320 mmHg)·(6 L)·(283 K)·(6×10^{23})
D. (320 mmHg / 760 atm)·(6 L)·(283 K)·(6×10^{23}) / (0.821)

59. Which gas has the greatest density at STP:

A. CO_2
B. O_2
C. N_2
D. NO

60. A hydrogen bond is a special type of:

A. dipole–dipole attraction involving hydrogen-bonded to another hydrogen atom
B. attraction involving molecules that contain hydrogens
C. dipole-dipole attraction involving hydrogen-bonded to a highly electronegative atom
D. dipole–dipole attraction involving hydrogen-bonded to other atoms

==

Practice Set 4: Questions 61–80

==

61. What is the mole fraction of H_2 in a gaseous mixture that consists of 9.50 g of H_2 and 14.0 g of Ne in a 4.50-liter container maintained at 37.5 °C?

A. 0.13 **B.** 0.43 **C.** 0.67 **D.** 0.87

62. Which of the following is true about Liquid A if the vapor pressure of Liquid A is greater than that of Liquid B?

A. Liquid A boils at a lower temperature
B. Liquid A has a higher heat of vaporization
C. Liquid A forms stronger bonds
D. Liquid A boils at a higher temperature

63. When 25 g of non-ionizable compound X is dissolved in 1 kg of camphor, the freezing point of the camphor falls 2.0 K. What is the approximate molecular weight of compound X? (Use $K_{camphor} = 40$)

A. 50 g/mol
B. 500 g/mol
C. 5,000 g/mol
D. 5,500 g/mol

64. The boiling point of a liquid is the temperature:

A. where sublimation occurs
B. where the vapor pressure of the liquid is less than the atmospheric pressure over the liquid
C. where the vapor pressure of the liquid equals the atmospheric pressure over the liquid
D. where the rate of sublimation equals evaporation

65. The van der Waals equation $[(P + n^2a / v^2)\cdot(V - nb) = nRT]$ is used to describe nonideal gases. The terms n^2a / v^2 and nb stand for, respectively:

A. volume of gas molecules and intermolecular forces
B. nonrandom movement and intermolecular forces between gas molecules
C. nonelastic collisions and volume of gas molecules
D. intermolecular forces and volume of gas molecules

66. What are the units of the gas constant R?

A. atm·K/L·mol
B. atm·L/mol·K
C. mol·L/atm·K
D. mol·K/L·atm

67. A sample of a gas occupying a volume of 120.0 mL at STP was placed in a different vessel with a volume of 155.0 mL, in which the pressure was measured at 0.80 atm. What was its temperature?

A. 9.1 °C **B.** 43.1 °C **C.** 4.1 °C **D.** 93.6 °C

68. Why does a beaker of water begin to boil at 22 °C when placed in a closed chamber, and a vacuum pump is used to evacuate the air from the chamber?

A. The vapor pressure decreases
B. Air is released from the water
C. The atmospheric pressure decreases
D. The vapor pressure increases

69. According to the kinetic theory, what happens to the kinetic energy of gaseous molecules when the temperature of a gas decreases?

A. Increase as does velocity
B. Remains constant, as does velocity
C. Increases and velocity decreases
D. Decreases, as does velocity

70. A sample of SO_3 gas is decomposed to SO_2 and O_2: $2\ SO_3\,(g) \rightarrow 2\ SO_2\,(g) + O_2\,(g)$

If the pressure of SO_2 and O_2 is 1,250 torrs, what is the partial pressure of O_2 in torrs?

A. 417 torrs
B. 1,040 torrs
C. 1,250 torrs
D. 884 torrs

71. For the balanced reaction $2\ Na + Cl_2 \rightarrow 2\ NaCl$, which of the following is a solid?

I. Na II. Cl III. NaCl

A. I only
B. II only
C. III only
D. I and III only

72. According to Charles's law, what happens to gas as temperature increases?

A. volume decreases
B. volume increases
C. pressure decreases
D. pressure increases

73. How does the pressure of a sample of gas change if the moles of gas remain constant while the volume is halved and the temperature is quadrupled?

A. increase by a factor of 8
B. quadruple
C. decrease by a factor of 4
D. decrease by a factor of 2

74. For a fixed quantity of gas, gas laws describe the relationships between pressure and which two variables?

A. chemical identity; mass
B. volume; chemical identity
C. temperature; volume
D. temperature; size

75. What is the term for a direct change of state from a solid to a gas?

A. sublimation
B. vaporization
C. condensation
D. deposition

76. The average speed at which a methane molecule effuses at 28.5 °C is 631 m/s. The average speed at which a krypton molecule effuses at the same temperature is:

A. 123 m/s **B.** 276 m/s **C.** 312 m/s **D.** 421 m/s

77. A chemical reaction A (*s*) → B (*s*) + C (*g*) occurs when substance A is vigorously heated. The molecular mass of the gaseous product was determined from the following experimental data:

Mass of A before reaction: 5.2 g

Mass of A after reaction: 0 g

Mass of residue B after cooling and weighing when no more gas evolved: 3.8 g

When all of the gas C evolved, it was collected and stored in a 668.5 mL glass vessel at 32.0 °C, and the gas exerted a pressure of 745.5 torr

Use the ideal gas constant R equals 0.0821 L·atm K^{-1} mol^{-1}

From this data, determine the apparent molecular mass of *C*, assuming it behaves as an ideal gas:

A. 6.46 g/mol
B. 46.3 g/mol
C. 53.9 g/mol
D. 72.2 g/mol

78. Which of the following describes a substance in the liquid physical state?

I. It has a variable shape
II. It compresses negligibly
III. It has a fixed volume

A. I only
B. II only
C. I and III only
D. I, II and III

79. Which is the strongest form of intermolecular attraction between water molecules?

A. ion-dipole
B. covalent bonding
C. induced dipole-induced dipole
D. hydrogen bonding

80. When a helium balloon is placed in a freezer, the temperature in the balloon [ ] and the volume [ ]?

A. increases … increases
B. increases … decreases
C. decreases … increases
D. decreases … decreases

Notes for active learning

Stoichiometry

Practice Set 1: Questions 1–20

1. What is the volume of three moles of O_2 at STP?

A. 11.20 L
B. 22.71 L
C. 68.13 L
D. 32.00 L

2. In the following reaction, which of the following describes H_2SO_4?

$$H_2SO_4 + HI \rightarrow I_2 + SO_2 + H_2O$$

A. reducing agent and is reduced
B. reducing agent and is oxidized
C. oxidizing agent and is reduced
D. oxidizing agent and is oxidized

3. Select the balanced chemical equation for the reaction: $C_6H_{14} + O_2 \rightarrow CO_2 + H_2O$

A. $3\ C_6H_{14} + O_2 \rightarrow 18\ CO_2 + 22\ H_2O$
B. $2\ C_6H_{14} + 12\ O_2 \rightarrow 12\ CO_2 + 14\ H_2O$
C. $2\ C_6H_{14} + 19\ O_2 \rightarrow 12\ CO_2 + 14\ H_2O$
D. $2\ C_6H_{14} + 9\ O_2 \rightarrow 12\ CO_2 + 7\ H_2O$

4. An oxidation number is the [ ] that an atom [ ] when the electrons in each bond are assigned to the [ ] electronegative of the two atoms involved in the bond.

A. number of protons … definitely has … more
B. charge … definitely has … less
C. number of electrons … definitely has … more
D. charge … appears to have … more

5. What is the empirical formula of acetic acid CH_3COOH?

A. CH_3COOH
B. $C_2H_4O_2$
C. CH_2O
D. CO_2H_2

6. In which of the following compounds does Cl have an oxidation number of +7?

A. $NaClO_2$
B. $Al(ClO_4)_3$
C. $Ca(ClO_3)_2$
D. $LiClO_3$

7. What is the formula mass of a molecule of CO_2?

A. 44 amu
B. 52 amu
C. 56.5 amu
D. 112 amu

8. What is the product of heating cadmium metal and powdered sulfur?

A. CdS_2
B. Cd_2S_3
C. CdS
D. Cd_2S

9. After balancing the following redox reaction, what is the coefficient of NaCl?

$$Cl_2\,(g) + NaI\,(aq) \rightarrow I_2\,(s) + NaCl\,(aq)$$

A. 1
B. 2
C. 3
D. 5

10. Which substance listed below is the strongest reducing agent, given the following *spontaneous* redox reaction?

$$FeCl_3\,(aq) + NaI\,(aq) \rightarrow I_2\,(s) + FeCl_2\,(aq) + NaCl\,(aq)$$

A. $FeCl_2$
B. I_2
C. NaI
D. $FeCl_3$

11. 14.5 moles of N_2 gas is mixed with 34 moles of H_2 gas in the following reaction:

$$N_2\,(g) + 3\,H_2\,(g) \rightarrow 2\,NH_3\,(g)$$

How many moles of N_2 gas remain if the reaction produces 18 moles of NH_3 gas when performed at 600 K?

A. 0.6 moles
B. 1.4 moles
C. 5.5 moles
D. 7.4 moles

12. Which equation is NOT correctly classified by the type of chemical reaction?

A. $AgNO_3 + NaCl \rightarrow AgCl + NaNO_3$ (double-replacement/non-redox)
B. $Cl_2 + F_2 \rightarrow 2\,ClF$ (synthesis/redox)
C. $H_2O + SO_2 \rightarrow H_2SO_3$ (synthesis/non-redox)
D. $CaCO_3 \rightarrow CaO + CO_2$ (decomposition/redox)

13. How many grams of H_2O can be formed from a reaction between 10 grams of oxygen and 1 gram of hydrogen?

A. 11 grams of H_2O are formed since mass must be conserved
B. 10 grams of H_2O are formed since the mass of water produced cannot be greater than the amount of oxygen reacting
C. 9 grams of H_2O are formed because oxygen and hydrogen react in an 8:1 mass ratio
D. No H_2O is formed because there is insufficient hydrogen to react with the oxygen

14. How many grams of Ba^{2+} ions are in an aqueous solution of $BaCl_2$ that contains 6.8×10^{22} Cl ions?

A. 3.2×10^{48} g
B. 12 g
C. 14.5 g
D. 7.8 g

15. What is the oxidation number of Br in $NaBrO_3$?

A. −1 **B.** +1 **C.** +3 **D.** +5

16. When an aluminum metal reacts with ferric oxide (Fe_2O_3), a displacement reaction yields two products, with one being metallic iron. What is the sum of the coefficients of the products of the balanced reaction?

A. 3 **B.** 6 **C.** 2 **D.** 5

17. Which of the following reactions is NOT correctly classified?

A. $AgNO_3$ (*aq*) + KOH (*aq*) → KNO_3 (*aq*) + AgOH (*s*) : non-redox / double-replacement
B. 2 H_2O_2 (*s*) → 2 H_2O (*l*) + O_2 (*g*) : non-redox / decomposition
C. $Pb(NO_3)_2$ (*aq*) + 2 Na (*s*) → Pb (*s*) + 2 $NaNO_3$ (*aq*) : redox / single-replacement
D. HNO_3 (*aq*) + LiOH (*aq*) → $LiNO_3$ (*aq*) + H_2O (*l*) : non-redox / double-replacement

18. Calculate the number of O_2 molecules if a 15.0 L cylinder was filled with O_2 gas at STP. (Use the conversion factor of 1 mole of $O_2 = 6.02 \times 10^{23}$ O_2 molecules)

A. 443 molecules
B. 6.59×10^{24} molecules
C. 4.03×10^{23} molecules
D. 2.77×10^{22} molecules

19. Which of the following is a guideline for balancing redox equations by the oxidation number method?

A. Verify that the number of atoms and the ionic charge is the same for reactants and products
B. In front of the substance reduced, place a coefficient that corresponds to the number of electrons lost by the substance oxidized
C. In front of the substance oxidized, place a coefficient that corresponds to the number of electrons gained by the substance reduced
D. All of the above

20. Which of the following represents the oxidation of Co^{2+}?

A. $Co \rightarrow Co^{2+} + 2\,e^-$
B. $Co^{3+} + e^- \rightarrow Co^{2+}$
C. $Co^{2+} + 2\,e^- \rightarrow Co$
D. $Co^{2+} \rightarrow Co^{3+} + e^-$

Practice Set 2: Questions 21–40

---

21. How many moles of phosphorous trichloride are required to produce 365 grams of HCl when the reaction yields 75%?

$$PCl_3\ (g) + 3\ NH_3\ (g) \rightarrow P(NH_2)_3 + 3\ HCl\ (g)$$

A. 1 mol
B. 2.5 mol
C. 3.5 mol
D. 4.5 mol

22. What is the oxidation number of liquid bromine in the elemental state?

A. 0
B. –1
C. –2
D. –3

23. Propane burners are used by campers for cooking. What volume of H_2O is produced by the complete combustion, as shown in the unbalanced equation, of 2.6 L of propane (C_3H_8) gas when measured at the same temperature and pressure?

$$C_3H_8\ (g) + O_2\ (g) \rightarrow CO_2\ (g) + H_2O\ (g)$$

A. 0.65 L
B. 10.4 L
C. 5.2 L
D. 2.6 L

24. Which substance is reduced in the following redox reaction?

$$HgCl_2\ (aq) + Sn^{2+}\ (aq) \rightarrow Sn^{4+}\ (aq) + Hg_2Cl_2\ (s) + Cl^-\ (aq)$$

A. Sn^{4+}
B. Hg_2Cl_2
C. $HgCl_2$
D. Sn^{2+}

25. What is the coefficient (n) of P for the balanced equation: nP (s) + nO$_2$ (g) $\rightarrow$ nP$_2$O$_5$ (s)?

A. 1
B. 2
C. 4
D. 5

26. In basic solution, which of the following are guidelines for balancing a redox equation using the half-reaction method?

I. Add the two half-reactions and cancel identical species on each side of the equation
II. Multiply each half-reaction by a whole number so that the number of electrons lost by the substance oxidized is equal to the electrons gained by the substance reduced
III. Write a balanced half-reaction for the substance oxidized and the substance reduced

A. II only
B. III only
C. I and III only
D. I, II and III

27. What is the molecular formula of galactose if the empirical formula is CH_2O, and the approximate molar mass is 180 g/mol?

A. $C_6H_{12}O_6$
B. CH_2O_6
C. CH_2O
D. CHO

28. How many formula units of lithium iodide (LiI) have a mass equal to 6.45 g? (Use the molecular mass of LiI = 133.85 g)

A. 3.45×10^{23} formula units
B. 2.90×10^{22} formula units
C. 1.65×10^{24} formula units
D. 7.74×10^{25} formula units

29. What is/are the product(s) of the reaction of N_2 and O_2 gases in a combustion engine?

I. NO   II. NO_2 III. N_2O

A. I only
B. II only
C. III only
D. I, II and III

30. What is the coefficient (n) of O_2 gas for the balanced equation?

$$nP\ (s) + nO_2\ (g) \rightarrow nP_2O_3\ (s)$$

A. 1   **B.** 2   **C.** 3   **D.** 5

31. Which substance is the weakest reducing agent given the spontaneous redox reaction?

$$Mg\ (s) + Sn^{2+}\ (aq) \rightarrow Mg^{2+}\ (aq) + Sn\ (s)$$

A. Sn
B. Mg^{2+}
C. Sn^{2+}
D. Mg

32. After balancing the following redox reaction in acidic solution, what is the coefficient of H^+?

$$Mg\ (s) + NO_3^-\ (aq) \rightarrow Mg^{2+}\ (aq) + NO_2\ (aq)$$

A. 1   **B.** 2   **C.** 4   **D.** 6

33. Which substance contains the greatest number of moles in a 10 g sample?

A. SiO_2
B. CH_4
C. CBr_4
D. CO_2

34. Which chemistry law is illustrated when ethyl alcohol contains 52% carbon, 13% hydrogen, and 35% oxygen by mass?

A. law of constant composition
B. law of constant percentages
C. law of multiple proportions
D. law of conservation of mass

35. Which could NOT be true for the following reaction: $N_2(g) + 3\ H_2(g) \rightarrow 2\ NH_3(g)$?

A. 25 grams of N_2 gas reacts with 75 grams of H_2 gas to form 50 grams of NH_3 gas
B. 28 grams of N_2 gas reacts with 6 grams of H_2 gas to form 34 grams of NH_3 gas
C. 15 moles of N_2 gas reacts with 45 moles of H_2 gas to form 30 moles of NH_3 gas
D. 5 molecules of N_2 gas reacts with 15 molecules of H_2 gas to form 10 molecules of NH_3 gas

36. From the following reaction, if 0.2 moles of Al are allowed to react with 0.4 moles of Fe_2O_3, how many grams of aluminum oxide are produced?

$$2\ Al + Fe_2O_3 \rightarrow 2\ Fe + Al_2O_3$$

A. 2.8 g
B. 5.1 g
C. 10.2 g
D. 14.2 g

37. In the following compounds, the oxidation number of hydrogen is +1, EXCEPT:

A. NH_3 **B.** $HClO_2$ **C.** H_2SO_4 **D.** NaH

38. Which equation is NOT correctly classified by the type of chemical reaction?

A. $PbO + C \rightarrow Pb + CO$: single-replacement/non-redox
B. $2\ Na + 2\ HCl \rightarrow 2\ NaCl + H_2$: single-replacement/redox
C. $NaHCO_3 + HCl \rightarrow NaCl + H_2O + CO_2$: double-replacement/non-redox
D. $2\ Na + H_2 \rightarrow 2\ NaH$: synthesis/redox

39. What are the oxidation states of sulfur in H_2SO_4 and H_2SO_3, respectively?

A. +4 and +4
B. +6 and +4
C. +2 and +4
D. +4 and +2

40. A latex balloon has a volume of 500 mL when filled with gas at a pressure of 780 torrs and a temperature of 320 K. How many moles of gas does the balloon contain? (Use the ideal gas constant $R = 0.08206\ L{\cdot}atm\ K^{-1}\ mol^{-1}$)

A. 0.0195
B. 0.822
C. 3.156
D. 18.87

Practice Set 3: Questions 41–60

---

41. What is the term for a substance that causes the reduction of another substance in a redox reaction?

A. oxidizing agent
B. reducing agent
C. anode
D. cathode

42. Which of the following is a method for balancing a redox equation in acidic solution by the half-reaction method?

A. Multiply each half-reaction by a whole number so that the number of electrons lost by the substance oxidized is equal to the electrons gained by the substance reduced
B. Add the two half-reactions and cancel identical species from each side of the equation
C. Write a half-reaction for the substance oxidized and the substance reduced
D. All of the above

43. What is the formula mass of a molecule of $C_6H_{12}O_6$?

A. 148 amu
B. 27 amu
C. 91 amu
D. 180 amu

44. Is it possible to have a macroscopic sample of oxygen that has a mass of 12 atomic mass units?

A. No, because this is less than the mass of a single oxygen atom
B. Yes, because it would have the same density as nitrogen
C. No, because this is less than a macroscopic quantity
D. Yes, but it would need to be made of isotopes of oxygen atoms

45. What is the coefficient for O_2 when balanced with the lowest whole number coefficients?

$$C_2H_6 + O_2 \rightarrow CO_2 + H_2O$$

A. 3 **B.** 4 **C.** 6 **D.** 7

46. Which element is reduced in the following redox reaction?

$$BaSO_4 + 4\ C \rightarrow BaS + 4\ CO$$

A. O in CO
B. S in $BaSO_4$
C. S in BaS
D. C in CO

47. Ethanol (C_2H_5OH) is blended with gasoline as a fuel additive. If the combustion of ethanol produces carbon dioxide and water, what is the coefficient of oxygen in the balanced equation?

$$\_\_C_2H_5OH\ (g) + \_\_O_2\ (g) \xrightarrow{\text{Spark}} \_\_CO_2\ (g) + \_\_H_2O\ (g)$$

A. 1 **B.** 2 **C.** 3 **D.** 6

48. How many moles of C are in a 4.50 g sample if the atomic mass of C is 12.011 amu?

A. 0.375 moles
B. 2.67 moles
C. 1.00 moles
D. 0.54 moles

49. Upon combustion analysis, a 6.84 g sample of a hydrocarbon yielded 8.98 grams of carbon dioxide. The percent by mass of carbon in the hydrocarbon is:

A. 18.6%
B. 23.7%
C. 35.8%
D. 11.4%

50. Select the balanced chemical equation:

A. $2\ C_2H_5OH + 2\ Na_2Cr_2O_7 + 8\ H_2SO_4 \rightarrow 2\ HC_2H_3O_2 + 2\ Cr_2(SO_4)_3 + 4\ Na_2SO_4 + 11\ H_2O$
B. $3\ C_2H_5OH + 2\ Na_2Cr_2O_7 + 8\ H_2SO_4 \rightarrow 3\ HC_2H_3O_2 + 2\ Cr_2(SO_4)_3 + 2\ Na_2SO_4 + 11\ H_2O$
C. $C_2H_5OH + 2\ Na_2Cr_2O_7 + 8\ H_2SO_4 \rightarrow HC_2H_3O_2 + 2\ Cr_2(SO_4)_3 + 2\ Na_2SO_4 + 11\ H_2O$
D. $C_2H_5OH + Na_2Cr_2O_7 + 2\ H_2SO_4 \rightarrow HC_2H_3O_2 + Cr_2(SO_4)_3 + 2\ Na\ _2SO_4 + 11\ H_2O$

51. Under acidic conditions, what is the sum of the coefficients in the balanced reaction below?

$$Fe^{2+} + Cr_2O_7^{2-} \rightarrow Fe^{3+} + Cr^{3+}$$

A. 8 **B.** 14 **C.** 17 **D.** 36

52. Assuming STP, if 49 g of H_2SO_4 are produced in the following reaction, what volume of O_2 must be used in the reaction?

$$RuS\ (s) + O_2 + H_2O \rightarrow Ru_2O_3\ (s) + H_2SO_4$$

A. 25.2 liters
B. 31.2 liters
C. 28.3 liters
D. 29.1 liters

53. From the following reaction, if 0.20 mole of Al is allowed to react with 0.40 mole of Fe_2O_3, how many moles of iron are produced?

$$2\ Al + Fe_2O_3 \rightarrow 2\ Fe + Al_2O_3$$

A. 0.05 mole
B. 0.075 mole
C. 0.20 mole
D. 0.10 mole

54. What is the oxidation number of sulfur in the $S_2O_8^{2-}$ ion?

A. −1 **B.** +7 **C.** +2 **D.** +6

55. What are the oxidation numbers for the elements in Na_2CrO_4?

A. +2 for Na, +5 for Cr and −6 for O
B. +2 for Na, +3 for Cr and −2 for O
C. +1 for Na, +4 for Cr and −6 for O
D. +1 for Na, +6 for Cr and −2 for O

56. What is the oxidation state of sulfur in sulfuric acid?

A. +8 **B.** +6 **C.** −2 **D.** −6

57. Which substance is oxidized in the following redox reaction?

$$HgCl_2\ (aq) + Sn^{2+}\ (aq) \rightarrow Sn^{4+}\ (aq) + Hg_2Cl_2\ (s) + Cl^-\ (aq)$$

A. Hg_2Cl_2
B. Sn^{4+}
C. Sn^{2+}
D. $HgCl_2$

58. Which of the following statements is NOT true with respect to the balanced equation?

$$Na_2SO_4\ (aq) + BaCl_2\ (aq) \rightarrow 2\ NaCl\ (aq) + BaSO_4\ (s)$$

A. Barium sulfate and sodium chloride are products
B. Barium chloride is dissolved in water
C. Barium sulfate is a solid
D. 2 NaCl (*aq*) could be written as Na_2Cl_2 (*aq*)

59. Which of the following represents 1 mol of phosphine gas (PH_3)?

I. 22.71 L phosphine gas at STP
II. 34.00 g phosphine gas
III. 6.02×10^{23} phosphine molecules

A. I only
B. II only
C. III only
D. II and III only

60. Which of the following represents the oxidation of Co^{2+}?

A. $Co \rightarrow Co^{2+} + 2\ e^-$
B. $Co^{3+} + e^- \rightarrow Co^{2+}$
C. $Co^{2+} + 2\ e^- \rightarrow Co$
D. $Co^{2+} \rightarrow Co^{3+} + e^-$

==

Practice Set 4: Questions 61–80

==

61. What is the term for a substance that causes oxidation in a redox reaction?

A. oxidized
B. oxidizing agent
C. anode
D. cathode

62. Carbon tetrachloride (CCl_4) is a potent hepatotoxin (toxic to the liver) commonly used in the past in fire extinguishers and as a refrigerant. What is the percent by mass of Cl in carbon tetrachloride?

A. 25% **B.** 66% **C.** 78% **D.** 92%

63. For a redox reaction to be balanced, which of the following is true?

I. Total ionic charge of reactants must equal ionic charge of the products
II. Atoms of each reactant must equal atoms of the product
III. Electron gain must equal electron loss

A. I only
B. II only
C. I and II only
D. I, II and III

64. The mass percent of a compound is approximately 71.8% Cl, 24.2% C, and 4.0% H. If the molecular weight of the compound is 99 g/mol, what is the molecular formula of the compound?

A. $Cl_2C_3H_6$
B. $Cl_2C_2H_4$
C. $ClCH_3$
D. ClC_2H_2

65. In the early 1980s, benzene was used as a solvent for waxes and oils until listed as a carcinogen by the EPA. What is the molecular formula of benzene if the empirical formula is C_1H_1 and the approximate molar mass is 78 g/mol?

A. CH_{12}
B. $C_{12}H_{12}$
C. C_6H_6
D. CH_6

66. How many O atoms are in the formula unit $GaO(NO_3)_2$?

A. 3 **B.** 4 **C.** 5 **D.** 7

67. Which reaction represents the balanced reaction for the combustion of ethanol?

A. $4\ C_2H_5OH + 13\ O_2 \rightarrow 8\ CO_2 + 10\ H_2$
B. $C_2H_5OH + 3\ O_2 \rightarrow 2\ CO_2 + 3\ H_2O$
C. $C_2H_5OH + 2\ O_2 \rightarrow 2\ CO_2 + 2\ H_2O$
D. $C_2H_5OH + O_2 \rightarrow CO_2 + H_2O$

68. What is the coefficient for O_2 when the following equation is balanced with the lowest whole number coefficients?

$$\_\_C_3H_7OH + \_\_O_2 \rightarrow \_\_CO_2 + \_\_H_2O$$

A. 3 **B.** 6 **C.** 9 **D.** 13/2

69. After balancing the following redox reaction, what is the coefficient of CO_2?

$$\_\_Co_2O_3\ (s) + \_\_CO\ (g) \rightarrow \_\_Co\ (s) + \_\_CO_2\ (g)$$

A. 1 **B.** 2 **C.** 3 **D.** 4

70. What is the term for the volume occupied by 1 mol of gas at STP?

A. STP volume
B. molar volume
C. standard volume
D. Avogadro's volume

71. Which substance listed below is the weakest oxidizing agent given the following spontaneous redox reaction?

$$Mg\ (s) + Sn^{2+}\ (aq) \rightarrow Mg^{2+}\ (aq) + Sn\ (s)$$

A. Mg^{2+}
B. Sn
C. Mg
D. Sn^{2+}

72. In the following reaction performed at 500 K, 18.0 moles of N_2 gas is mixed with 24.0 moles of H_2 gas. What is the percent yield of NH_3, if the reaction produces 13.5 moles of NH_3?

$$N_2\ (g) + 3\ H_2\ (g) \rightarrow 2\ NH_3\ (g)$$

A. 16%
B. 66%
C. 72%
D. 84%

73. How many grams of H_2O can be produced from the reaction of 25.0 grams of H_2 and 225 grams of O_2?

A. 266 grams
B. 223 grams
C. 184 grams
D. 27 grams

74. What is the oxidation number of Cl in $LiClO_2$?

A. –1 **B.** +1 **C.** +3 **D.** +5

75. In acidic conditions, what is the sum of the coefficients in the products of the balanced reaction?

$MnO_4^- + C_3H_7OH \rightarrow Mn^{2+} + C_2H_5COOH$

A. 12 **B.** 16 **C.** 18 **D.** 20

76. Which reaction is NOT correctly classified?

A. $PbO\ (s) + C\ (s) \rightarrow Pb\ (s) + CO\ (g)$: (double-replacement)
B. $CaO\ (s) + H_2O\ (l) \rightarrow Ca(OH)_2\ (aq)$: (synthesis)
C. $Pb(NO_3)_2\ (aq) + 2\ LiCl\ (aq) \rightarrow 2\ LiNO_3\ (aq) + PbCl_2\ (s)$: (double-replacement)
D. $Mg\ (s) + 2\ HCl\ (aq) \rightarrow MgCl_2\ (aq) + H_2\ (g)$: (single-replacement)

77. The reactants for this chemical reaction are:

$C_6H_{12}O_6 + 6\ H_2O + 6\ O_2 \rightarrow 6\ CO_2 + 12\ H_2O$

A. $C_6H_{12}O_6$, H_2O, O_2 and CO_2
B. $C_6H_{12}O_6$ and H_2O
C. $C_6H_{12}O_6$
D. $C_6H_{12}O_6$, H_2O and O_2

78. What is the charge of the electrons in 4 grams of He? (Use the Faraday constant $F = 96{,}500$ C/mol)

A. 48,250 C
B. 96,500 C
C. 193,000 C
D. 386,000 C

79. What is the oxidation number of iron in the compound $FeBr_3$?

A. –2 **B.** +1 **C.** +2 **D.** +3

80. What is the term for the amount of substance that contains 6.02×10^{23} particles?

A. molar mass
B. mole
C. Avogadro's number
D. formula mass

Notes for active learning

Notes for active learning

Thermochemistry

Practice Set 1: Questions 1–20

1. Which of the following statement regarding the symbol ΔG is NOT true?

A. Specifies the enthalpy of the reaction
B. Refers to the free energy of the reaction
C. Predicts the spontaneity of a reaction
D. Describes the effect of enthalpy and entropy on a reaction

2. What happens to the kinetic energy of a gas molecule when the gas is heated?

A. Depends on the gas
B. Kinetic energy increases
C. Kinetic energy decreases
D. Kinetic energy remains constant

3. How much heat energy (in Joules) is required to heat 21.0 g of copper from 21.0 °C to 68.5 °C? (Use specific heat *c* of Cu = 0.382 J/g·°C)

A. 462 J **B.** 188 J **C.** 522 J **D.** 381 J

4. For *n* moles of gas, which term expresses the kinetic energy?

A. nPA, where n = number of moles of gas, P = total pressure, and A = surface area of container walls
B. ½nPA, where n = number of moles of gas, P = total pressure, and A = surface area of container walls
C. 3/2 nRT, where n = number of moles of gas, R = ideal gas constant, and T = absolute temperature
D. MV^2, where M = molar mass of the gas and V = volume of the container

5. Which of the following is NOT an endothermic process?

A. Condensation of water vapor
B. Boiling liquid
C. Water evaporating
D. Ice melting

6. What is true of an endothermic reaction if it causes a decrease in the entropy (S) of the system?

A. Only occurs at low temperatures when ΔS is insignificant
B. Occurs if coupled to an endergonic reaction
C. Never occurs because it decreases ΔS of the system
D. Never occurs because ΔG is positive

7. Which of the following terms describe(s) energy contained in an object or transferred to an object?

I. chemical   II. electrical   III. heat

A. I only
B. II only
C. I and II only
D. I, II and III

8. The greatest entropy is observed for which 10 g sample of CO_2?

A. CO_2 (*g*)
B. CO_2 (*aq*)
C. CO_2 (*s*)
D. CO_2 (*l*)

9. What is the term for a reaction that proceeds by absorbing heat energy?

A. Isothermal reaction
B. Exothermic reaction
C. Endothermic reaction
D. Spontaneous

10. If a chemical reaction has $\Delta H = X$, $\Delta S = Y$, $\Delta G = X - RY$ and occurs at R K, the reaction is:

A. spontaneous
B. at equilibrium
C. nonspontaneous
D. cannot be determined

11. The thermodynamic systems that have high stability tend to demonstrate:

A. maximum ΔH and maximum ΔS
B. maximum ΔH and minimum ΔS
C. minimum ΔH and maximum ΔS
D. minimum ΔH and minimum ΔS

12. Whether a reaction is endothermic or exothermic is determined by:

A. energy balance between bond breaking and bond forming, resulting in net loss or gain of energy
B. the presence of a catalyst
C. the activation energy
D. the physical state of the reaction system

13. What role does entropy play in chemical reactions?

A. The entropy change determines whether the reaction occurs spontaneously
B. The entropy change determines whether the chemical reaction is favorable
C. The entropy determines how much product is actually produced
D. The entropy change determines whether the reaction is exothermic or endothermic

14. Calculate the value of $\Delta H°$ of reaction using the provided bond energies.

$H_2C{=}CH_2\ (g) + H_2\ (g) \rightarrow H_3C{-}CH_3\ (g)$

C–C: 348 kJ

C≡C: 960 kJ

C=C: 612 kJ C–H: 412 kJ

H–H: 436 kJ

A. –348 kJ
B. +134 kJ
C. –546 kJ
D. –124 kJ

15. The bond dissociation energy is:

I. useful in estimating the enthalpy change in a reaction
II. the energy required to break a bond between two gaseous atoms
III. the energy released when a bond between two gaseous atoms is broken

A. I only
B. II only
C. I and II only
D. I and III only

16. Based on the following reaction, which statement is true?

$N_2 + O_2 \rightarrow 2\ NO$ (Use the value for enthalpy, $\Delta H = 43.3$ kcal)

A. 43.3 kcal are consumed when 2.0 mole of O_2 reacts
B. 43.3 kcal are consumed when 2.0 moles of NO are produced
C. 43.3 kcal are produced when 1.0 g of N_2 reacts
D. 43.3 kcal are consumed when 2.0 g of O_2 reacts

17. Which of the following properties of gas is/are a state function?

I. temperature II. heat III. work

A. I only
B. I and II only
C. II and III only
D. I, II and III

18. The following statements concerning temperature change as a substance is heated are correct, EXCEPT:

A. As a liquid is heated, its temperature rises until its boiling point is reached
B. During the time a liquid is changing to the gaseous state, the temperature gradually increases until all the liquid is changed
C. As a solid is heated, its temperature rises until its melting point is reached
D. During the time for a solid to melt to a liquid, the temperature remains constant

19. Calculate the value of ΔH° of reaction for:

$$O{=}C{=}O\ (g) + 3\ H_2\ (g) \rightarrow CH_3{-}O{-}H\ (g) + H{-}O{-}H\ (g)$$

Use the following bond energies, ΔH°:

C–C: 348 kJ	C=C: 612 kJ	C≡C: 960 kJ	C–H: 412 kJ
C–O: 360 kJ	C=O: 743 kJ	H–H: 436 kJ	H–O: 463 kJ

A. –348 kJ
B. +612 kJ
C. –191 kJ
D. –769 kJ

20. Which statement(s) is/are true for ΔS?

I. ΔS of the universe is conserved
II. ΔS of a system is conserved
III. ΔS of the universe increases with each reaction

A. I only
B. II only
C. III only
D. I and II only

Practice Set 2: Questions 21–40

21. Which of the following reaction energies is the most endothermic?

A. 360 kJ/mole
B. –360 kJ/mole
C. 88 kJ/mole
D. –88 kJ/mole

22. A fuel cell contains hydrogen and oxygen gas that react explosively, and the energy converts water to steam which drives a turbine to turn a generator that produces electricity. The fuel cell and the steam represent which forms of energy, respectively?

A. Electrical and heat energy
B. Electrical and chemical energy
C. Chemical and heat energy
D. Chemical and mechanical energy

23. Which of the following is true for the ΔG of formation for N_2 (*g*) at 25 °C?

A. 0 kJ/mol
B. positive
C. negative
D. 1 kJ/mol

24. Which of the statements best describes the following reaction?

$$HC_2H_3O_2\ (aq) + NaOH\ (aq) \rightarrow NaC_2H_3O_2\ (aq) + H_2O\ (l)$$

A. Acetic acid and NaOH solutions produce sodium acetate and H_2O
B. Aqueous solutions of acetic acid and NaOH produce aqueous sodium acetate and H_2O
C. Acetic acid and NaOH solutions produce sodium acetate solution and H_2O
D. Acetic acid and NaOH produce sodium acetate and H_2O

25. If a chemical reaction is spontaneous, which value must be negative?

A. C_p
B. ΔS
C. ΔG
D. ΔH

26. What purpose is the hollow walls in a closed hollow-walled container that effectively maintains the temperature inside?

A. To trap air trying to escape from the container, which minimizes convection
B. To act as an effective insulator, which minimizes convection
C. To act as an effective insulator, which minimizes conduction
D. To provide an additional source of heat for the container

27. If the heat of reaction is exothermic, which of the following is always true?

A. The energy of the reactants is greater than the products
B. The energy of the reactants is less than the products
C. The reaction rate is fast
D. The reaction rate is slow

28. How much heat must be absorbed to evaporate 16 g of NH_3 to its condensation point at −33 °C? (Use heat of condensation for NH_3 = 1,380 J/g)

A. 86.5 J
B. 2,846 J
C. 118 J
D. 22,080 J

29. Which of the reactions is the most exothermic, if the following energy profiles have the same scale? (Use the notation of R = reactants and P = products)

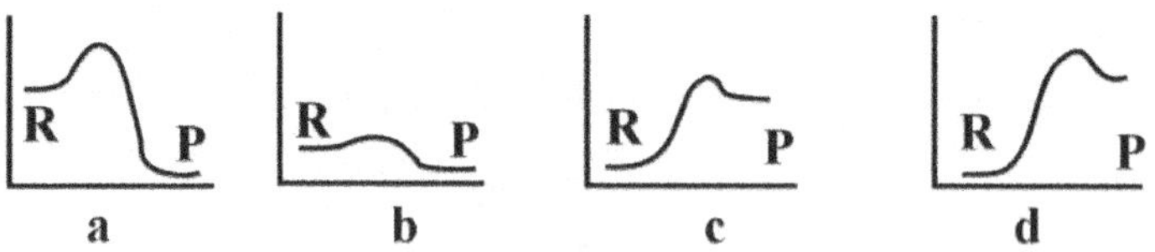

A. a
B. b
C. c
D. d

30. What is the heat of formation of NH_3 (*g*) of the following reaction:

$$2\ NH_3\ (g) \rightarrow N_2\ (g) + 3\ H_2\ (g)$$

(Use $\Delta H°$ = 92.4 kJ/mol)

A. −92.4 kJ/mol
B. −184.4 kJ/mol
C. 46.2 kJ/mol
D. −46.2 kJ/mol

31. If a chemical reaction has a positive ΔH and a negative ΔS, the reaction tends to be:

A. at equilibrium
B. nonspontaneous
C. spontaneous
D. irreversible

32. What happens to the entropy of a system as the components of the system are introduced to a larger number of possible arrangements, such as when liquid water transforms into water vapor?

A. The entropy of a system is solely dependent upon the amount of material undergoing reaction
B. The entropy of a system is independent of introducing the components of the system to a larger number of possible arrangements
C. The entropy increases because there are more ways for the energy to disperse
D. The Entropy decreases because there are fewer ways in which the energy can disperse

33. Which quantities are needed to calculate the heat energy released as water turns to ice at 0 °C?

A. The heat of condensation for water and the mass
B. The heat of vaporization for water and the mass
C. The heat of fusion for water and the mass
D. The heat of solidification for water and the mass

34. A nuclear power plant uses ^{235}U to convert water to steam that drives a turbine which turns a generator to produce electricity. What are the initial and final forms of energy, respectively?

A. Heat energy and electrical energy
B. Nuclear energy and electrical energy
C. Chemical energy and mechanical energy
D. Chemical energy and heat energy

35. Which statement(s) is/are correct for the entropy?

I. Higher for a sample of gas than for the same sample of liquid
II. A measure of the disorder in a system
III. Available energy for conversion into mechanical work

A. I only
B. II only
C. III only
D. I and II only

36. Given the following data, what is the heat of formation for ethanol?

$C_2H_5OH + 3\ O_2 \rightarrow 2\ CO_2 + 3\ H_2O : \Delta H = 327.0$ kcal/mole

$H_2O \rightarrow H_2 + \frac{1}{2}\ O_2 : \Delta H = +68.3$ kcal/mole

$C + O_2 \rightarrow CO_2 : \Delta H = -94.1$ kcal/mole

A. –720.1 kcal
B. –327.0 kcal
C. +62.6 kcal
D. +720.1 kcal

37. Based on the reaction shown, which statement is true?

$S + O_2 \rightarrow SO_2 + 69.8$ kcal

A. 69.8 kcal are consumed when 32.1 g of sulfur reacts
B. 69.8 kcal are produced when 32.1 g of sulfur reacts
C. 69.8 kcal are consumed when 1 g of sulfur reacts
D. 69.8 kcal are produced when 1 g of sulfur reacts

38. In the reaction, $2\ H_2\ (g) + O_2\ (g) \rightarrow 2\ H_2O\ (g)$, entropy is:

A. increasing
B. the same
C. inversely proportional
D. decreasing

39. For an isolated system, which of the following is NOT exchanged between the system and its surroundings?

I. Temperature II. Matter III. Energy

A. I only
B. II only
C. II and III only
D. I, II and III

40. Which is a state function?

I. ΔG II. ΔH III. ΔS

A. I only
B. II only
C. III only
D. I, II and III

Practice Set 3: Questions 41–60

41. To simplify comparisons, the energy value of fuels is expressed in units of:

A. kcal/g
B. kcal/L
C. J/kcal
D. kcal/mol

42. Which process is slowed down when an office worker places a lid on a hot cup of coffee?

I. Radiation II. Conduction III. Convection

A. I only
B. II only
C. III only
D. I and III only

43. Which statement below is true for a spontaneous chemical reaction?

A. $\Delta S_{sys} - \Delta S_{surr} = 0$
B. $\Delta S_{sys} + \Delta S_{surr} > 0$
C. $\Delta S_{sys} + \Delta S_{surr} < 0$
D. $\Delta S_{sys} + \Delta S_{surr} = 0$

44. A fuel cell contains hydrogen and oxygen gas that react explosively, and the energy converts water to steam, which drives a turbine to turn a generator that produces electricity. What energy changes are employed in the process?

I. Mechanical → electrical energy
II. Heat → mechanical energy
III. Chemical → heat energy

A. I only
B. II only
C. I and II only
D. I, II and III

45. Determine the value of $\Delta E°_{rxn}$ for this reaction, whereby the standard enthalpy of reaction ($\Delta H°_{rxn}$) is –311.5 kJ mol^{-1}:

$C_2H_2\ (g) + 2\ H_2\ (g) \rightarrow C_2H_6\ (g)$

A. –306.5 kJ mol^{-1}
B. –318.0 kJ mol^{-1}
C. +346.0 kJ mol^{-1}
D. +306.5 kJ mol^{-1}

46. For a closed system, what can be exchanged between the system and its surroundings?

I. Heat II. Matter III. Energy

A. I only
B. II only
C. III only
D. I and III only

47. Which reaction is accompanied by an increase in entropy?

A. Na_2CO_3 (*s*) + CO_2 (*g*) + H_2O (*g*) → 2 $NaHCO_3$ (*s*)
B. BaO (*s*) + CO_2 (*g*) → $BaCO_3$ (*s*)
C. CH_4 (*g*) + H_2O (*g*) → CO (*g*) + 3 H_2 (*g*)
D. ZnS (*s*) + 3/2 O_2 (*g*) → ZnO (*s*) + SO_2 (*g*)

48. Under standard conditions, which reaction has the largest difference between the energy of the reaction and enthalpy?

A. C (*graphite*) → C (*diamond*)
B. C (*graphite*) + O_2 (*g*) → C (*diamond*)
C. 2 C (*graphite*) + O_2 (*g*) → 2 CO (*g*)
D. C (*graphite*) + O_2 (*g*) → CO_2 (*g*)

49. Which type of reaction tends to be the most stable?

I. Isothermic II. Exergonic III. Endergonic

A. I only
B. II only
C. III only
D. I and II only

50. Which is true for the thermodynamic functions G, H, and S in $\Delta G = \Delta H - T\Delta S$?

A. G refers to the universe, H to the surroundings, and S to the system
B. G, H, and S refer to the system
C. G and H refers to the surroundings and S to the system
D. G and H refer to the system and S to the surroundings

51. Which of the following statements is true for the following reaction? (Use the change in enthalpy, $\Delta H° = -113.4$ kJ/mol and the change in entropy, $\Delta S° = -145.7$ J/K mol)

2 NO (*g*) + O_2 (*g*) → 2 NO_2 (*g*)

A. The reaction is at equilibrium at 25 °C under standard conditions
B. The reaction is spontaneous at only high temperatures
C. The reaction is spontaneous only at low temperatures
D. The reaction is spontaneous at all temperatures

52. Which of the following expressions defines enthalpy? (Use the conventions: q = heat, U = internal energy, P = pressure and V = volume)

A. $q - \Delta U$
B. $U + q$
C. U + PV
D. ΔU

53. Which statement is true regarding entropy?

I. It is a state function
II. It is an extensive property
III. It has an absolute zero value

A. I only
B. III only
C. I and II only
D. I and III only

54. Where does the energy released during an exothermic reaction originate from?

A. The kinetic energy of the surrounding
B. The kinetic energy of the reacting molecules
C. The potential energy of the reacting molecules
D. The thermal energy of the reactants

55. The species in the reaction $KClO_3$ (*s*) → KCl (*s*) + 3/2 O_2 (*g*) have the values for standard enthalpies of formation at 25 °C. At constant physical states, assume that the values of ΔH° and ΔS° are constant throughout a broad temperature range. Which of the following conditions may apply to the reaction? (Use $KClO_3$ (*s*) with $\Delta H_f° = -391.2$ kJ mol^{-1} and KCl (*s*) with $\Delta H_f° = -436.8$ kJ mol^{-1})

A. Nonspontaneous at low temperatures, but spontaneous at high temperatures
B. Spontaneous at low temperatures but nonspontaneous at high temperatures
C. Nonspontaneous at temperatures over a broad temperature range
D. Spontaneous at temperatures over a broad temperature range

56. What is the standard enthalpy change for the reaction?

P_4 (*s*) + 6 Cl_2 (*g*) → 4 PCl_3 (*l*)   ΔH° = –1,289 kJ

3 P_4 (*s*) + 18 Cl_2 (*g*) → 12 PCl_3 (*l*)

A. –3,867 kJ
B. –1,345 kJ
C. –366 kJ
D. 1,289 kJ

57. Which of the following reactions is endothermic?

A. $PCl_3 + Cl_2 \rightarrow PCl_5$ + *heat*
B. 2 $NO_2 \rightarrow N_2 + 2\ O_2$ + *heat*
C. $CH_4 + NH_3$ + *heat* → HCN + 3 H_2
D. NH_3 + HBr → NH_4Br

58. When the system undergoes a spontaneous reaction, is it possible for entropy of a system to decrease?

A. No, because this violates the second law of thermodynamics
B. No, because this violates the first law of thermodynamics
C. Yes, but only if the reaction is endothermic
D. Yes, but only if entropy gain of the environment is greater than entropy loss in the system

59. A 500 ml beaker of distilled water is placed under a bell jar, then covered by a layer of opaque insulation. After several days, some of the water evaporated. The contents of the bell jar are what kind of system?

A. endothermic
B. isolated
C. closed
D. open

60. Which law explains the observation that the amount of heat transfer accompanying a change in one direction is equal in magnitude but opposite in sign to the amount of heat transfer in the opposite direction?

A. Law of Conservation of Energy
B. Law of Definite Proportions
C. Avogadro's law
D. Boyle's law

Practice Set 4: Questions 61–80

61. What is the term for a reaction that proceeds by releasing heat energy?

A. Endothermic reaction
B. Isothermal reaction
C. Exothermic reaction
D. Nonspontaneous

62. If a stationary gas has a kinetic energy of 500 J at 25 °C, what is its kinetic energy at 50 °C?

A. 125 J
B. 450 J
C. 540 J
D. 1,120 J

63. Which is NOT true for entropy in a closed system according to the equation $\Delta S = Q / T$?

A. Entropy is a measure of energy dispersal of the system
B. The equation is only valid for a reversible process
C. Changes due to heat transfer are greater at low temperatures
D. Disorder of the system decreases as heat is transferred out of the system

64. In which of the following pairs of physical changes are both processes exothermic?

A. Melting and condensation
B. Freezing and condensation
C. Sublimation and evaporation
D. Freezing and sublimation

65. A fuel cell contains hydrogen and oxygen gas that react explosively, and the energy converts water to steam which drives a turbine to turn a generator that produces electricity. What are the initial and final forms of energy, respectively?

A. Chemical and electrical energy
B. Nuclear and electrical energy
C. Chemical and mechanical energy
D. Chemical and heat energy

66. Which must be true concerning a solution that reached an equilibrium where chemicals are mixed in a redox reaction?

A. $\Delta G° = \Delta G$
B. $E = 0$
C. $\Delta G° < 1$
D. $K = 1$

67. A solid sample at room temperature spontaneously sublimes, forming a gas. This change in state is accompanied by which of the changes in the sample?

A. Entropy decreases and energy increases
B. Entropy increases and energy decreases
C. Entropy and energy decrease
D. Entropy and energy increase

68. At constant temperature and pressure, a negative ΔG indicates that the:

A. reaction is nonspontaneous
B. reaction is fast
C. reaction is spontaneous
D. reaction is endothermic

69. The process of H_2O (*g*) → H_2O (*l*) is nonspontaneous under the pressure of 760 torr and temperatures of 378 K because:

A. $\Delta H > T\Delta S$
B. $\Delta G < 0$
C. $\Delta H > 0$
D. $\Delta H < T\Delta S$

70. Consider the contribution of entropy to the spontaneity of the reaction. As written, the reaction is [ ] and the entropy of the system [ ].

$$2\ Al_2O_3\ (s) \rightarrow 4\ Al\ (s) + 3\ O_2\ (g),\ \Delta G = +138\ \text{kcal}$$

A. non-spontaneous … decreases
B. non-spontaneous … increases
C. spontaneous … decreases
D. spontaneous … increases

71. The ΔG of a reaction is the maximum energy that the reaction releases to do:

A. P–V work only
B. work and release heat
C. any type of work
D. non P–V work only

72. Which of the following represent forms of internal energy?

I. bond energy II. thermal energy III. gravitational energy

A. I only
B. II only
C. I and II only
D. I and III only

73. If it takes energy to break bonds and energy is gained when forming bonds, how can some reactions be exothermic while others are endothermic?

A. Some products have more energy than others and require energy to be formed
B. Some reactants have more energetic bonds than others and release energy
C. It is the number of bonds that is determinative. Since all bonds have the same amount of energy, the net gain or net loss of energy depends on the number of bonds
D. It is the amount of energy that is determinative. Some bonds are stronger than others, so there is a net gain or net loss of energy when formed

74. Which constant is represented by A in the following calculation to determine how much heat is required to convert 60 g of ice at −25 °C to steam at 320 °C?

$$\text{Total heat} = [(A)\cdot(60\text{ g})\cdot(25\text{ °C})] + [(\text{heat of fusion})\cdot(60\text{ g}]) + [(4.18\text{ J/g·°C})\cdot(60\text{ g})\cdot(100\text{ °C})] + [(B)\cdot(60\text{ g})] + [(C)\cdot(60\text{ g})\cdot(220\text{ °C})]$$

A. Specific heat of ice
B. Heat of vaporization of water
C. Heat of condensation
D. Heat capacity of steam

75. The heat of formation of water vapor is:

A. positive, but greater than the heat of formation for H_2O (*l*)
B. positive and smaller than the heat of formation for H_2O (*l*)
C. negative, but greater than the heat of formation for H_2O (*l*)
D. negative and smaller than the heat of formation for H_2O (*l*)

76. Entropy can be defined as the amount of:

A. equilibrium in a system
B. chemical bonds changed during a reaction
C. energy required to initiate a reaction
D. disorder in a system

77. Which is true of an atomic fission bomb according to the conservation of mass and energy law?

A. The mass of the bomb and the fission products are identical
B. A small amount of mass is converted into energy
C. The energy of the bomb and the fission products are identical
D. The mass of the fission bomb is greater than the mass of the products

78. Once an object enters a black hole, astronomers consider it to have left the universe, which means the universe is:

A. entropic
B. isolated
C. closed
D. open

79. Which ranking from lowest to highest entropy per gram of NaCl is correct?

A. NaCl (*s*) < NaCl (*l*) < NaCl (*aq*) < NaCl (*g*)
B. NaCl (*s*) < NaCl (*l*) < NaCl (*g*) < NaCl (*aq*)
C. NaCl (*g*) < NaCl (*aq*) < NaCl (*l*) < NaCl (*s*)
D. NaCl (*s*), NaCl (*aq*), NaCl (*l*), NaCl (*g*)

80. From the given bond energies, how many kJ of energy are released or absorbed from the reaction of one mole of N_2 with three moles of H_2 to form two moles of NH_3?

$$N{\equiv}N + H{-}H + H{-}H + H{-}H \rightarrow NH_3 + NH_3$$

H–N: 389 kJ/mol H–H: 436 kJ/mol N≡N: 946 kJ/mol

A. –80 kJ/mol released

B. +89.5 kJ/mol absorbed

C. –946 kJ/mol released

D. +895 kJ/mol absorbed

Notes for active learning

Notes for active learning

Kinetics and Equilibrium

Practice Set 1: Questions 1–20

1. What is the general equilibrium constant (K_{eq}) expression for the following reversible reaction?

$2\,A + 3\,B \leftrightarrow C$

A. $K_{eq} = [C] / [A]^2 \cdot [B]^3$

B. $K_{eq} = [C] / [A] \cdot [B]$

C. $K_{eq} = [A] \cdot [B] / [C]$

D. $K_{eq} = [A]^2 \cdot [B]^3 / [C]$

2. What are gases A and B likely to be if, in a mixture of these two gases, gas A has twice the average velocity of gas B?

A. Ar and Kr

B. N and Fe

C. Ne and Ar

D. Mg and K

3. For a hypothetical reaction, A + B → C, predict which reaction occurs at the slowest rate from the following reaction conditions.

Reaction	Activation energy	Temperature
1	103 kJ/mol	15 °C
2	46 kJ/mol	22 °C
3	103 kJ/mol	24 °C
4	46 kJ/mol	30 °C

A. 1 **B.** 2 **C.** 3 **D.** 4

4. From the data below, what is the order of the reaction with respect to reactant A?

Determining Rate Law from Experimental Data: A + B → Products

Exp.	Initial [A]	Initial [B]	Initial Rate M/s
1	0.015	0.022	0.125
2	0.030	0.044	0.500
3	0.060	0.044	0.500
4	0.060	0.066	1.125
5	0.085	0.088	?

A. Zero **B.** First **C.** Second **D.** Third

5. For the combustion of ethanol (C_2H_6O) to form carbon dioxide and water, what is the rate at which carbon dioxide is produced if the ethanol is consumed at 4.0 M s^{-1}?

A. 1.5 M s^{-1}
B. 12.0 M s^{-1}
C. 8.0 M s^{-1}
D. 9.0 M s^{-1}

6. Which is the correct equilibrium constant (K_{eq}) expression for the following reaction?

$$2\ Ag\ (s) + Cl_2\ (g) \leftrightarrow 2\ AgCl\ (s)$$

A. $K_{eq} = [AgCl] / [Ag]^2 \times [Cl_2]$
B. $K_{eq} = [2\ AgCl] / [2\ Ag] \times [Cl_2]$
C. $K_{eq} = [AgCl]^2 / [Ag]^2 \times [Cl_2]$
D. $K_{eq} = 1 / [Cl_2]$

7. The equilibrium position for a system where $K_{eq} = 6.3 \times 10^{-14}$ can be described as being favored for [ ], and the product concentration is relatively [ ].

A. the left; large
B. the left; small
C. the right; large
D. the right; small

8. What can be deduced about the activation energy of a reaction that takes billions of years to go to completion and a reaction that takes only a fraction of a millisecond?

A. The slow reaction has high activation energy, while the fast reaction has a low activation energy
B. The slow reaction has low activation energy, while the fast reaction has a high activation energy
C. The activation energy of both reactions is very low
D. The activation energy of both reactions is very high

9. Which influences the rate of a first-order chemical reaction?

I. catalyst II. temperature III. concentration

A. I only
B. I and II only
C. I and III only
D. I, II and III

Questions **10** through **14** are based on the following:

Energy profiles for four reactions (with the same scale).

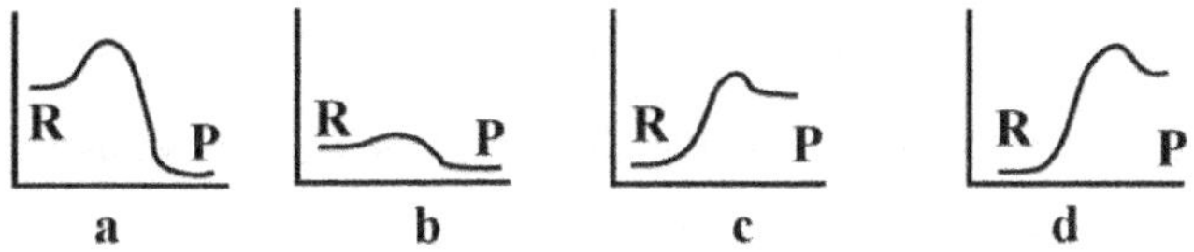

R = reactants   P = products

10. Which reaction requires the most energy?

A. a **B.** b **C.** c **D.** d

11. Which reaction has the highest activation energy?

A. a **B.** b **C.** c **D.** d

12. Which reaction has the lowest activation energy?

A. a **B.** b **C.** c **D.** d

13. Which reaction proceeds the slowest?

A. a **B.** b **C.** c **D.** d

14. If the graphs are for the same reaction, which most likely has a catalyst?

A. a **B.** b **C.** c **D.** d

15. What is an explanation for the observation that the reaction stops before all reactants are converted to products in the following reaction? (Use the notation of [ ] for concentration)

$$NH_3\ (aq) + HC_2H_3O_2\ (aq) \rightarrow NH_4^+\ (aq) + C_2H_3O_2^-\ (aq)$$

A. The catalyst is depleted
B. The reverse rate increases, while the forward rate decreases until they are equal
C. As [products] increases, the acetic acid begins to dissociate, stopping the reaction
D. As [reactants] decreases, NH_3 and $HC_2H_3O_2$ molecules stop colliding

16. What is the term for the principle that the rate of reaction is regulated by the frequency, energy, and orientation of molecules striking each other?

A. orientation theory
B. frequency theory
C. energy theory
D. collision theory

17. What is the effect on the energy of the activated complex and on the rate of the reaction when a catalyst is added to a chemical reaction?

A. The energy of the activated complex increases, and the reaction rate decreases
B. The energy of the activated complex decreases, and the reaction rate increases
C. The energy of the activated complex and the reaction rate increase
D. The energy of the activated complex and the reaction rate decrease

18. Which change shifts the equilibrium to the right for the reversible reaction in an aqueous solution?

$HNO_2\,(aq) \leftrightarrow H^+\,(aq) + NO_2^-\,(aq)$

I. Add solid NaOH
II. Decrease $[NO_2^-]$
III. Decrease $[H^+]$
IV. Increase $[HNO_2]$

A. II and III only
B. I and IV only
C. II, III and IV only
D. I, II, III and IV

19. Which conditions would favor driving the reaction to completion?

$2\,N_2\,(g) + 6\,H_2O\,(g) + heat \leftrightarrow 4\,NH_3\,(g) + 3\,O_2\,(g)$

A. Increasing the reaction temperature
B. Continual addition of NH_3 gas to the reaction mixture
C. Decreasing the pressure on the reaction vessel
D. Continual removal of N_2 gas

20. The system, $H_2\,(g) + X_2\,(g) \leftrightarrow 2\,HX\,(g)$ has a value of 24.4 for K_c. A catalyst was introduced into a reaction within a 4.0-liters reactor containing 0.20 moles of H_2, 0.20 moles of X_2 and 0.800 moles of HX. The reaction proceeds in which direction?

A. to the right, $Q > K_c$
B. to the left, $Q > K_c$
C. to the right, $Q < K_c$
D. to the left, $Q < K_c$

Practice Set 2: Questions 21–40

---

21. Which is the K_c equilibrium expression for the following reaction?

$$4\ CuO\ (s) + CH_4\ (g) \leftrightarrow CO_2\ (g) + 4\ Cu\ (s) + 2\ H_2O\ (g)$$

A. $[Cu]^4 / [CuO]^4$

B. $[CO_2]\cdot[H_2O]^2 / [CH_4]$

C. $[CH_4] / [CO_2]\cdot[H_2O]^2$

D. $[CuO]^4 / [Cu]^4$

22. Which of the following concentrations of CH_2Cl_2 should be used in the rate law for Step 2 if CH_2Cl_2 is a product of the fast (first) step and a reactant of the slow (second) step?

A. $[CH_2Cl_2]$ at equilibrium

B. $[CH_2Cl_2]$ in Step 2 cannot be predicted because Step 1 is the fast step

C. Zero moles per liter

D. $[CH_2Cl_2]$ after Step 1 is completed

23. What is the term for a substance that allows a reaction to proceed faster by lowering activation energy?

A. rate barrier

B. energy barrier

C. collision energy

D. catalyst

24. Which statement is true for the grams of products present after a chemical reaction reaches equilibrium?

A. Must equal the grams of the initial reactants

B. May be less than, equal to, or greater than the grams of reactants present, depending upon the chemical reaction

C. Must be greater than the grams of the initial reactants

D. Must be less than the grams of the initial reactants

25. What is the rate law when rates were measured at different concentrations for the dissociation of hydrogen gas: $H_2\ (g) \rightarrow 2\ H\ (g)$?

$[H_2]$	Rate M/s s^{-1}
1.0	1.3×10^5
1.5	2.6×10^5
2.0	5.2×10^5

A. rate = $k^2[H] / [H_2]$

B. rate = $k[H]^2 / [H_2]$

C. rate = $k[H_2]^2$

D. rate = $k[H_2] / [H]^2$

26. Which change to this reaction system causes the equilibrium to shift to the right?

$N_2\ (g) + 3\ H_2\ (g) \leftrightarrow 2\ NH_3\ (g) + \textit{heat}$

A. Heating the system
B. Removal of $H_2\ (g)$
C. Addition of $NH_3\ (g)$
D. Lowering the temperature

27. Which statement is NOT correct for $aA + bB \rightarrow dD + eE$ whereby rate = $k[A]^{q} \cdot [B]^{r}$?

A. The overall order of the reaction is $q + r$
B. The exponents q and r are equal to the coefficients a and b, respectively
C. The exponents q and r must be determined experimentally
D. The exponents q and r are often integers

28. Which is the correct equilibrium constant (K_{eq}) expression for the following reaction?

$CO\ (g) + 2\ H_2\ (g) \leftrightarrow CH_3OH\ (l)$

A. $K_{eq} = 1 / [CO]\cdot[H_2]^2$
B. $K_{eq} = [CO]\cdot[H_2]^2$
C. $K_{eq} = [CH_3OH] / [CO]\cdot[H_2]^2$
D. $K_{eq} = [CH_3OH] / [CO]\cdot[H_2]$

Questions **29-32** are based on the following graph and net reaction:

The reaction proceeds in two consecutive steps.

$XY + Z \leftrightarrow XYZ \leftrightarrow X + YZ$

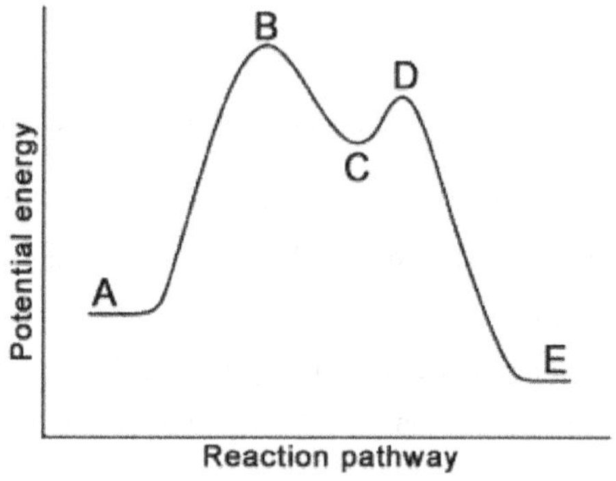

29. Where is the activated complex for this reaction?

A. A
B. A and C
C. C
D. B and D

30. The activation energy of the slow step for the forward reaction is given by:

A. A → B
B. A → C
C. B → C
D. C → E

31. The activation energy of the slow step for the reverse reaction is given by:

A. C → A
B. E → C
C. C → B
D. E → D

32. The change in energy (ΔE) of the overall reaction is given by the difference between:

A. A and B
B. A and E
C. A and C
D. B and D

33. Which of the following increases the collision energy of gaseous molecules?

I. Increasing the temperature
II. Adding a catalyst
III. Increasing the concentration

A. I only
B. II only
C. III only
D. I and II only

34. Which of the following conditions favors the formation of NO (*g*) in a closed container?

N_2 (*g*) + O_2 (*g*) ↔ 2 NO (*g*) (Use $\Delta H = +181$ kJ/mol)

A. Increasing the temperature
B. Decreasing the temperature
C. Increasing the pressure
D. Decreasing the pressure

35. A chemical system is considered to have reached dynamic equilibrium when the:

A. activation energy of the forward reaction equals the activation energy of the reverse reaction
B. rate of production of each of the products equals the rate of their consumption by the reverse reaction
C. frequency of collisions between the reactant molecules equals the frequency of collisions between the product molecules
D. sum of the concentrations of each of the reactant species equals the sum of the concentrations of each of the product species

36. Chemical equilibrium is reached in a system when:

A. complete conversion of reactants to products has occurred
B. product molecules begin reacting with each other
C. reactant concentrations steadily decrease
D. product and reactant concentrations remain constant

37. At a given temperature, $K = 46.0$ for the reaction:

4 HCl (*g*) + O_2 (*g*) ↔ 2 H_2O (*g*) + 2 Cl_2 (*g*)

At equilibrium, [HCl] = 0.150, [O_2] = 0.395 and [H_2O] = 0.625. What is the concentration of Cl_2 at equilibrium?

A. 0.153 M
B. 0.444 M
C. 1.14 M
D. 0.00547 M

38. If at equilibrium, reactant concentrations are slightly smaller than product concentrations, the equilibrium constant would be:

A. slightly greater than 1
B. slightly lower than 1
C. much lower than 1
D. much greater than 1

39. Hydrogen gas reacts with iron (III) oxide to form iron metal (which produces steel), as shown in the reaction below. Which statement is NOT correct concerning the equilibrium system?

$$Fe_2O_3\ (s) + 3\ H_2\ (g) + \textit{heat} \leftrightarrow 2\ Fe\ (s) + 3\ H_2O\ (g)$$

A. Continually removing water from the reaction chamber increases the yield of iron
B. Decreasing the volume of hydrogen gas reduces the yield of iron
C. Lowering the reaction temperature increases the concentration of hydrogen gas
D. Increasing the pressure on the reaction chamber increases the formation of products

40. Assuming vessels a, b and c are drawn to relative proportions, which of the following reactions proceeds the fastest? (Assume equal temperatures)

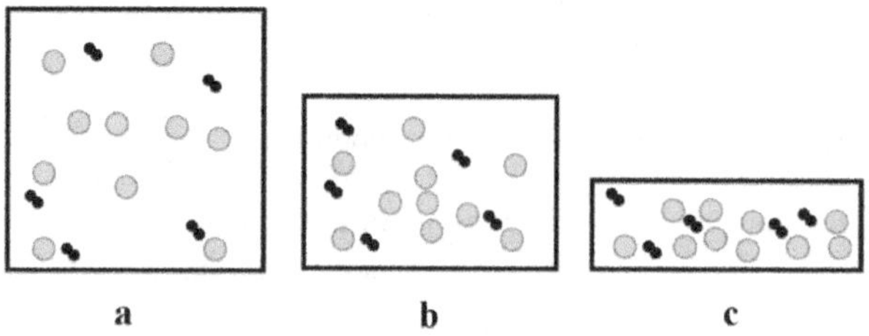

$$A + B \rightarrow C$$

A. a
B. b
C. c
D. All proceed at same rate

Practice Set 3: Questions 41–60

41. Which is the correct equilibrium constant (K_{eq}) expression for the following reaction?

$$4\ NH_3\ (g) + 5\ O_2\ (g) \leftrightarrow 4\ NO\ (g) + 6\ H_2O\ (g)$$

A. $K_{eq} = [NO]^4 \times [H_2O]^6 / [NH_3]^4 \times [O_2]^5$
B. $K_{eq} = [NH_3]^4 \times [O_2]^5 / [NO]^4 \times [H_2O]^6$
C. $K_{eq} = [NO] \times [H_2O] / [NH_3] \times [O_2]$
D. $K_{eq} = [NH_3] \times [O_2] / [NO] \times [H_2O]$

42. Carbonic acid equilibrium in blood:

$$CO_2\ (g) + H_2O\ (l) \leftrightarrow H_2CO_3\ (aq) \leftrightarrow H^+\ (aq) + HCO_3^-\ (aq)$$

If a person hyperventilates, rapid breathing expels carbon dioxide gas. Which of the following decreases when a person hyperventilates?

I. $[HCO_3^-]$ II. $[H^+]$ III. $[H_2CO_3]$

A. I only
B. II only
C. III only
D. I, II and III

43. What is the overall order of the reaction if the units of the rate constant for a particular reaction are min^{-1}?

A. Zero
B. First
C. Second
D. Third

44. Heat is often added to chemical reactions performed in the laboratory to:

A. compensate for the natural tendency of energy to disperse
B. increase the rate at which reactants collide
C. allow a greater number of reactants to overcome the barrier of the activation energy
D. all of the above

45. Predict which reaction occurs faster for a hypothetical reaction X + Y → W + Z.

Reaction	Activation energy	Temperature
1	low	low
2	low	high
3	high	high
4	high	low

A. 1 **B.** 2 **C.** 3 **D.** 4

46. For the reaction, 2 $XO + O_2 \rightarrow 2\ XO_2$, data obtained from measurement of the initial rate of reaction at varying concentrations are:

Experiment	[XO]	[O_2]	Rate (mmol L^{-1} s^{-1})
1	0.010	0.010	2.5
2	0.010	0.020	5.0
3	0.030	0.020	45.0

What is the expression for the rate law?

A. rate = $k[XO]\cdot[O_2]$
B. rate = $k[XO]^2\cdot[O_2]^2$
C. rate = $k[XO]^2\cdot[O_2]$
D. rate = $k[XO]\cdot[O_2]^2$

47. What is the equilibrium (K_{eq}) expression for the following reaction?

$CaO\ (s) + CO_2\ (g) \leftrightarrow CaCO_3\ (s)$

A. $K_{eq} = [CaCO_3]\ /\ [CaO]$
B. $K_{eq} = 1\ /\ [CO_2]$
C. $K_{eq} = [CaCO_3]\ /\ [CaO]\cdot[CO_2]$
D. $K_{eq} = [CO_2]$

48. If a reaction does not occur extensively and gives a low concentration of products at equilibrium, which of the following is true?

A. The rate of the forward reaction is greater than the reverse reaction
B. The rate of the reverse reaction is greater than the forward reaction
C. The equilibrium constant is greater than one; that is, K_{eq} is larger than 1
D. The equilibrium constant is less than one; that is, K_{eq} is smaller than 1

49. Which of the following changes most likely decreases the rate of a reaction?

A. Increasing the reaction temperature
B. Increasing the concentration of a reactant
C. Increasing the activation energy for the reaction
D. Decreasing the activation energy for the reaction

50. Which factors would increase the rate of a reversible chemical reaction?

I. Increasing the temperature of the reaction
II. Removing products as they form
III. Adding a catalyst to the reaction vessel

A. I only
B. II only
C. I and III only
D. I, II and III

51. What is the ionization equilibrium constant (K_i) expression for the following weak acid?

$H_2S\ (aq) \leftrightarrow H^+\ (aq) + HS^-\ (aq)$

A. $K_i = [H^+]^2 \cdot [S^{2-}] / [H_2S]$

B. $K_i = [H_2S] / [H^+] \cdot [HS^-]$

C. $K_i = [H^+] \cdot [HS^-] / [H_2S]$

D. $K_i = [H^+]^2 \cdot [HS^-] / [H_2S]$

52. Increasing the temperature of a chemical reaction:

A. increases the reaction rate by lowering the activation energy
B. increases the reaction rate by increasing reactant collisions per unit time
C. increases the activation energy, thus increasing the reaction rate
D. raises the activation energy, thus decreasing the reaction rate

53. Reaction rates are determined by the following factors, EXCEPT:

A. orientation of collisions between molecules
B. spontaneity of the reaction
C. force of collisions between molecules
D. number of collisions between molecules

54. Coal burning plants release sulfur dioxide into the atmosphere, while nitrogen monoxide is released into the atmosphere via industrial processes and combustion engines. Sulfur dioxide can also be produced in the atmosphere by the following equilibrium reaction:

$SO_3\ (g) + NO\ (g) + \textit{heat} \leftrightarrow SO_2\ (g) + NO_2\ (g)$

Which of the following does NOT shift the equilibrium to the right?

A. $[NO_2]$ decrease

B. [NO] increase

C. Decrease the reaction chamber volume

D. Temperature increase

55. Which of the following statements can be assumed to be true about how reactions occur?

A. Reactant particles must collide with each other
B. Energy must be released as the reaction proceeds
C. Catalysts must be present in the reaction
D. Energy must be absorbed as the reaction proceeds

56. At equilibrium, increasing the temperature of an exothermic reaction likely:

A. increases the heat of reaction

B. decreases the heat of reaction

C. increases the forward reaction

D. decreases the forward reaction

57. Which shifts the equilibrium to the left for the reversible reaction in an aqueous solution?

$HC_2H_3O_2\ (aq) \leftrightarrow H^+\ (aq) + C_2H_3O_2^-\ (aq)$

I. increase pH

II. increase $[HC_2H_3O_2]$

III. add solid $KC_2H_3O_2$

A. I only

B. I and III only

C. III only

D. II and III only

58. Which of the following statements is true concerning the equilibrium system, whereby S combines with H_2 to form hydrogen sulfide, toxic gas from the decay of organic material?
(Use the equilibrium constant, $K_{eq} = 2.8 \times 10^{-21}$)

$S\ (g) + H_2\ (g) \leftrightarrow H_2S\ (g)$

A. Almost all the starting molecules are converted to product

B. Little hydrogen sulfide gas is present in the equilibrium

C. Decreasing $[H_2]$ shifts the equilibrium to the right

D. Increasing the volume of the sealed reaction container shifts the equilibrium to the right

59. Which of the following conditions characterizes a system in a state of chemical equilibrium?

A. Product concentrations are greater than reactant concentrations

B. Reactant molecules no longer react with each other

C. Concentrations of reactants and products are equal

D. Reactants are being consumed at the same rate they are being produced

60. For the reaction A + B → C, which proceeds the slowest? (Assume equal temperatures)

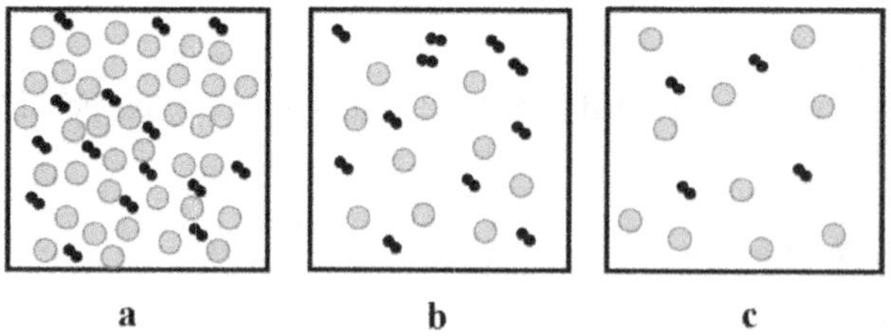

A. a

B. b

C. c

D. All proceed at the same rate

Practice Set 4: Questions 61–80

Questions **61** through **63** refer to the rate data for the conversion of reactants W, X, and Y to product Z.

Trial number	Concentration (moles/L)			Rate of formation of Z
	W	X	Y	(moles/l·s)
1	0.01	0.05	0.04	0.04
2	0.015	0.07	0.06	0.08
3	0.01	0.15	0.04	0.36
4	0.03	0.07	0.06	0.08
5	0.01	0.05	0.16	0.08

61. From the above data, what is the overall order of the reaction?

A. 2½ **B.** 4 **C.** 3 **D.** 2

62. From the above data, the order with respect to W suggests that the rate of formation of Z is:

A. dependent on [W]
B. independent of [W]
C. semi-dependent on [W]
D. unable to be determined

63. From the above data, the magnitude of *k* for trial 1 is:

A. 20 **B.** 40 **C.** 60 **D.** 80

64. Which of the following changes shifts the equilibrium to the left for the given reversible reaction?

$SO_3\,(g) + NO\,(g) + heat \leftrightarrow SO_2\,(g) + NO_2\,(g)$

A. Decrease temperature
B. Decrease volume
C. Increase [NO]
D. Decrease [SO_2]

65. What is the ionization equilibrium constant (K_i) expression for the following weak acid?

$H_3PO_4\,(aq) \leftrightarrow H^+\,(aq) + H_2PO_4^-\,(aq)$

A. $K_i = [H_3PO_4] / [H^+]\cdot[H_2PO_4^-]$
B. $K_i = [H^+]^3\cdot[PO_4^{3-}] / [H_3PO_4]$
C. $K_i = [H^+]^3\cdot[H_2PO_4^-] / [H_3PO_4]$
D. $K_i = [H^+]\cdot[H_2PO_4^-] / [H_3PO_4]$

66. What effect does a catalyst have on equilibrium?

A. It increases the rate of the forward reaction
B. It shifts the reaction to the right
C. It increases the rate at which equilibrium is reached without changing ΔG
D. It increases the rate at which equilibrium is reached and lowers ΔG

67. What is the correct ionization equilibrium constant (K_i) expression for the following weak acid?

$H_2SO_3\,(aq) \leftrightarrow H^+\,(aq) + HSO_3^-\,(aq)$

A. $K_i = [H_2SO_3] / [H^+]\cdot[HSO_3^-]$
B. $K_i = [H^+]^2\cdot[SO_3^{2-}] / [H_2SO_3]$
C. $K_i = [H^+]^2\cdot[HSO_3^-] / [H_2SO_3]$
D. $K_i = [H^+]\cdot[HSO_3^-] / [H_2SO_3]$

68. Which of the following is true if a reaction occurs extensively and yields a high concentration of products at equilibrium?

A. The rate of the reverse reaction is greater than the forward reaction
B. The rate of the forward reaction is greater than the reverse reaction
C. The equilibrium constant is less than one; K_{eq} is much smaller than 1
D. The equilibrium constant is greater than one; K_{eq} is much larger than 1

69. The minimum combined kinetic energy reactants must possess for collisions to result in a reaction is:

A. orientation energy
B. activation energy
C. collision energy
D. dissociation energy

70. Which factors decrease the rate of a reaction?

I. Lowering the temperature
II. Increasing the concentration of reactants
III. Adding a catalyst to the reaction vessel

A. I only
B. II only
C. III only
D. I and II only

71. For a collision between molecules to result in a reaction, the molecules must possess a favorable orientation relative to each other and:

A. be in the gaseous state
B. have a certain minimum energy
C. adhere for at least 2 nanoseconds
D. exchange electrons

72. Most reactions are carried out in liquid solution or the gaseous phase because in such situations:

A. kinetic energies of reactants are lower
B. reactant collisions occur more frequently
C. activation energies are higher
D. reactant activation energies are lower

73. Find the reaction rate for A + B → C:

Trial	$[A]_{t=0}$	$[B]_{t=0}$	**Initial rate (M/s)**
1	0.05 M	1.0 M	1.0×10^{-3}
2	0.05 M	4.0 M	16.0×10^{-3}
3	0.15 M	1.0 M	3.0×10^{-3}

A. rate = $k[A]^2 \cdot [B]^2$
B. rate = $k[A] \cdot [B]^2$
C. rate = $k[A]^2 \cdot [B]$
D. rate = $k[A] \cdot [B]$

74. Why does a glowing splint of wood burn only slowly in the air but rapidly in a burst of flames when placed in pure oxygen?

A. A glowing wood splint is extinguished within pure oxygen because oxygen inhibits the smoke
B. Pure oxygen can absorb carbon dioxide at a faster rate
C. Oxygen is a flammable gas
D. There is an increased number of collisions between the wood and oxygen molecules

75. Which of the changes shifts the equilibrium to the right for the following system at equilibrium?

$N_2\ (g) + 3\ H_2\ (g) \leftrightarrow 2\ NH_3\ (g) + 92.94\ kJ$

I. Removing NH_3
II. Adding NH_3
III. Removing N_2
IV. Adding N_2

A. I and III
B. II and III
C. II and IV
D. I and IV

76. Which of the changes does not affect the equilibrium for the reversible reaction in an aqueous solution?

$HC_2H_3O_2\ (aq) \leftrightarrow H^+\ (aq) + C_2H_3O_2^-\ (aq)$

A. Adding solid $NaC_2H_3O_2$
B. Adding solid $NaNO_3$
C. Increasing $[HC_2H_3O_2]$
D. Increasing $[H^+]$

77. Consider the following reaction: $H_2\ (g) + I_2\ (g) \rightarrow 2\ HI\ (g)$

At 160 K, this reaction has an equilibrium constant of 35. If at 160 K, the concentration of hydrogen gas is 0.4 M, iodine gas is 0.6 M, and hydrogen iodide gas is 3 M:

A. system is at equilibrium
B. [iodine] decreases
C. [hydrogen iodide] increases
D. [hydrogen iodide] decreases

78. If the concentration of reactants decreases, which of the following is true?

I. The amount of products increases
II. The heat of reaction decreases
III. The rate of reaction decreases

A. I only
B. II only
C. III only
D. II and III only

79. What does a chemical equilibrium expression of a reaction depend on?

I. mechanism
II. stoichiometry
III. rate

A. I only
B. II only
C. III only
D. I and II only

80. For the following reaction where $\Delta H < 0$, which factor decreases the magnitude of the equilibrium constant *K*?

$CO\ (g) + 2\ H_2O\ (g) \leftrightarrow CH_3OH\ (g)$

A. Decreasing the temperature of this system
B. Decreasing volume
C. Decreasing the pressure of this system
D. None of the above

Notes for active learning

Notes for active learning

Solution Chemistry

Practice Set 1: Questions 1–20

---

1. If the solubility of nitrogen in blood is 1.90 cc/100 cc at 1.00 atm, what is the solubility of nitrogen in a scuba diver's blood at a depth of 125 feet where the pressure is 4.5 atm?

A. 1.90 cc/100 cc
B. 8.55 cc/100 cc
C. 4.5 cc/100 cc
D. 0.236 cc/100 cc

2. All of the statements about molarity are correct, EXCEPT:

A. volume = moles/molarity
B. moles = molarity × volume
C. molarity equals moles of solute per mole of solvent
D. abbreviation is M

3. Which of the following molecules is expected to be most soluble in water?

A. NaCl
B. $CH_3CH_2CH_2COOH$
C. $CH_3CH_2CH_2$ OH
D. $Al(OH)_3$

4. The equation for the reaction shown below can be written as an ionic equation.

$$BaCl_2\ (aq) + K_2CrO_4\ (aq) \rightarrow BaCrO_4\ (s) + 2\ KCl\ (aq)$$

In the ionic equation, the spectator ions are:

A. K^+ and Cl^-
B. Ba^{2+} and CrO_4^{2-}
C. Ba^{2+} and K^+
D. K^+ and CrO_4^{2-}

5. In commercially prepared soft drinks, carbon dioxide gas is injected into soda. Under what conditions are carbon dioxide gas most soluble?

A. High temperature, high-pressure
B. High temperature, low-pressure
C. Low temperature, low-pressure
D. Low temperature, high-pressure

6. A solute is a:

A. substance that dissolves into a solvent
B. substance containing a solid, liquid or gas
C. solid substance that does not dissolve into water
D. solid substance that does not dissolve at a given temperature

7. How many ions are produced in solution by dissociating one formula unit of $Co(NO_3)_2 \cdot 6H_2O$?

A. 2 **B.** 3 **C.** 4 **D.** 6

8. 15 grams of an unknown substance is dissolved in 60 grams of water. When the solution is transferred to another container, it weighs 78 grams. Which of the following is a possible explanation?

A. The solution reacted with the second container, forming a precipitate
B. Some of the solution remained in the first container
C. The reaction was endothermic, which increased the average molecular speed
D. Solution reacted with the first container, and some byproducts were transferred with the solution

9. Which of the following is NOT soluble in H_2O?

A. Iron (III) hydroxide
B. Iron (III) nitrate
C. Potassium sulfate
D. Ammonium sulfate

10. What is the v/v% concentration of a solution made by adding 25 mL of acetone to 75 mL of water?

A. 33% v/v
B. 0.33% v/v
C. 25% v/v
D. 2.5% v/v

11. Why is octane less soluble in H_2O than in benzene?

A. Bonds between benzene and octane are much stronger than the bonds between H_2O and octane
B. Octane cannot dissociate in the presence of H_2O
C. Bonds between H_2O and octane are weaker than the bonds between H_2O molecules
D. Octane and benzene have similar molecular weights

12. What is the K_{sp} for slightly soluble copper (II) phosphate in an aqueous solution?

$$Cu_3(PO_4)_2\ (s) \leftrightarrow 3\ Cu^{2+}\ (aq) + 2\ PO_4^{3-}\ (aq)$$

A. $K_{sp} = [Cu^{2+}]^3 \cdot [PO_4^{3-}]^2$
B. $K_{sp} = [Cu^{2+}] \cdot [PO_4^{3-}]^2$
C. $K_{sp} = [Cu^{2+}]^3 \cdot [PO_4^{3-}]$
D. $K_{sp} = [Cu^{2+}] \cdot [PO_4^{3-}]$

13. Apply the *like dissolves like* rule to predict which of the following liquids is miscible with water.

I. carbon tetrachloride, CCl_4
II. toluene, C_7H_8
III. ethanol, C_2H_5OH

A. I only
B. II only
C. III only
D. I and II only

14. Which of the following pairs of substances would have both species in the pair be written in the molecular form in a net ionic equation?

I. CO_2 and H_2SO_4 II. LiOH and H_2 III. HF and CO_2

A. I only
B. II only
C. III only
D. I, II and III

15. Which statement best describes a supersaturated solution?

A. It contains dissolved solute in equilibrium with undissolved solid
B. It rapidly precipitates if a seed crystal is added
C. It contains as much solvent as it can accommodate
D. It contains no double bonds

16. What is the mass of a 7.50% urine sample that contains 122 g of dissolved solute?

A. 1,250 g
B. 935 g
C. 49.35 g
D. 1,627 g

17. Calculate the molarity of a solution prepared by dissolving 15.0 g of NH_3 in 250 g of water with a final density of 0.974 g/mL.

A. 36.2 M
B. 3.23 M
C. 0.0462 M
D. 0.664 M

18. Which of the following compounds has the highest boiling point?

A. 0.2 M $Al(NO_3)_3$
B. 0.2 M $MgCl_2$
C. 0.2 M glucose ($C_6H_{12}O_6$)
D. 0.2 M Na_2SO_4

19. Which compound produces four ions per formula unit by dissociation when dissolved in water?

A. Li_3PO_4
B. $Ca(NO_3)_2$
C. $MgSO_4$
D. $(NH_4)_2SO_4$

20. Which of the following would be a weak electrolyte in a solution?

A. HBr (*aq*)
B. KCl
C. KOH
D. $HC_2H_3O_2$

Practice Set 2: Questions 21–40

21. The ions Ca^{2+}, Mg^{2+}, Fe^{2+}, Fe^{3+}, which are present in groundwater, can be removed by pretreating the water with:

A. $PbSO_4$
B. $Na_2CO_3 \cdot 10H_2O$
C. KNO_3
D. $CaCl_2$

22. Choose the spectator ions: $Pb(NO_3)_2$ (*aq*) + H_2SO_4 (*aq*) → ?

A. NO_3^- and H^+
B. H^+ and SO_4^{2-}
C. Pb^{2+} and H^+
D. Pb^{2+} and NO_3^-

23. From the *like dissolves like* rule, predict which of the following vitamins is soluble in water:

A. α-tocopherol ($C_{29}H_{50}O_2$)
B. calciferol ($C_{27}H_{44}O$)
C. ascorbic acid ($C_6H_8O_6$)
D. retinol ($C_{20}H_{30}O$)

24. How much water must be added when 125 mL of a 2.00 M solution of HCl is diluted to a final concentration of 0.400 M?

A. 150 mL
B. 500 mL
C. 625 mL
D. 750 mL

25. Which of the following is the sulfate ion?

A. SO_4^{2-}
B. S^{2-}
C. CO_3^{2-}
D. PO_4^{3-}

26. Which of the following explains why bubbles form inside a pot of water when water is heated?

A. As temperature increases, the vapor pressure increases
B. As temperature increases, the atmospheric pressure decreases
C. As temperature increases, the solubility of air decreases
D. As temperature increases, the kinetic energy decreases

27. A solution in which the rate of crystallization is equal to the rate of dissolution is:

A. saturated
B. supersaturated
C. dilute
D. unsaturated

28. What is the term that refers to liquids that do not dissolve in one another and separate into two layers?

A. Soluble
B. Miscible
C. Insoluble
D. Immiscible

29. Which is a correctly balanced hydration equation for the hydration of Na_2SO_4?

A. $Na_2SO_4\,(s) \xrightarrow{H_2O} Na^+\,(aq) + 2SO_4^{2-}\,(aq)$

B. $Na_2SO_4\,(s) \xrightarrow{H_2O} 2\,Na^{2+}\,(aq) + S^{2-}\,(aq) + O_4^{2-}\,(aq)$

C. $Na_2SO_4\,(s) \xrightarrow{H_2O} Na_2^{2+}\,(aq) + SO_4^{2-}\,(aq)$

D. $Na_2SO_4\,(s) \xrightarrow{H_2O} 2\,Na^+\,(aq) + SO_4^{2-}\,(aq)$

30. Soft drinks are carbonated by injection with carbon dioxide gas. Under what conditions is carbon dioxide gas least soluble?

A. High temperature, low-pressure
B. High temperature, high-pressure
C. Low temperature, high-pressure
D. Low temperature, low-pressure

31. Which type of compound is likely to dissolve in H_2O?

I. One with hydrogen bonds
II. Highly polar compound
III. Salt

A. I only
B. **B.** II only
C. III only
D. I, II and III

32. Which of the following might have the best solubility in water?

A. CH_3CH_3
B. CH_3OH
C. CCl_4
D. O_2

33. What is the K_{sp} for slightly soluble gold (III) chloride in an aqueous solution for the reaction shown?

$AuCl_3\,(s) \leftrightarrow Au^{3+}\,(aq) + 3\,Cl^-\,(aq)$

A. $K_{sp} = [Au^{3+}]^3 \cdot [Cl^-] / [AuCl_3]$
B. $K_{sp} = [Au^{3+}] \cdot [Cl^-]^3$
C. $K_{sp} = [Au^{3+}]^3 \cdot [Cl^-]$
D. $K_{sp} = [Au^{3+}] \cdot [Cl^-]$

34. What is the net ionic equation for the reaction shown?

$$CaCO_3 + 2\ HNO_3 \rightarrow Ca(NO_3)_2 + CO_2 + H_2O$$

A. $CO_3^{2-} + H^+ \rightarrow CO_2$
B. $CaCO_3 + 2\ H^+ \rightarrow Ca^{2+} + CO_2 + H_2O$
C. $Ca^{2+} + 2\ NO_3^- \rightarrow Ca(NO_3)_2$
D. $CaCO_3 + 2\ NO_3^- \rightarrow Ca(NO_3)_2 + CO_3^{2-}$

35. What is the volume of a 0.550 M $Fe(NO_3)_3$ solution needed to supply 0.950 moles of nitrate ions?

A. 265 mL
B. 0.828 mL
C. 22.2 mL
D. 576 mL

36. What is the molarity of a solution that contains 48 mEq Ca^{2+} per liter?

A. 0.024 M
B. 0.048 M
C. 1.8 M
D. 2.4 M

37. If 36.0 g of LiOH is dissolved in water to make 975 mL of solution, what is the molarity of the LiOH solution? (Use molecular mass of LiOH = 24.0 g/mol)

A. 1.54 M
B. 2.48 M
C. 0. 844 M
D. 0.268 M

38. What volume of 8.50% (m/v) solution contains 60.0 grams of glucose?

A. 170 mL
B. 448 mL
C. 706 mL
D. 344 mL

39. In an AgCl solution, if the K_{sp} for AgCl is A, and the concentration Cl^- in a container is B molar, what is the concentration of Ag (in moles/liter)?

I. A moles/liter II. B moles/liter III. A/B moles/liter

A. I only
B. II only
C. III only
D. II and III only

40. Which statement below is generally true?

A. Bases are strong electrolytes and ionize completely when dissolved in water
B. Salts are strong electrolytes and dissociate completely when dissolved in water
C. Acids are strong electrolytes and ionize completely when dissolved in water
D. Bases are weak electrolytes and ionize completely when dissolved in water

---

Practice Set 3: Questions 41–60

---

41. Which of the following intermolecular attractions is/are important for the formation of a solution?

I. solute-solute II. solvent-solute III. solvent-solvent

A. I only
B. III only
C. I and II only
D. I, II and III

42. What is the concentration of I^- ions in a 0.40 M solution of magnesium iodide?

A. 0.05 M
B. 0.80 M
C. 0.60 M
D. 0.20 M

43. Which is most likely soluble in NH_3?

A. CO_2 **B.** SO_2 **C.** CCl_4 **D.** N_2

44. Which of the following represents the symbol for the chlorite ion?

A. ClO_2^-
B. ClO^-
C. ClO_4^-
D. ClO_3^-

45. Which of the following statements best describes what is happening in a water softening unit?

A. Sodium is removed from the water, making the water interact less with the soap molecules
B. Ions in the water softener are softened by chemically bonding with sodium
C. Hard ions are trapped in the softener, which filters out the ions
D. Hard ions in water are exchanged for ions that do not interact as strongly with soaps

46. Which of the following compounds are soluble in water?

I. $Mn(OH)_2$ II. $Cr(NO_3)_3$ III. $Ni_3(PO_4)_2$

A. I only
B. II only
C. III only
D. I and III only

47. The hydration number of an ion is the number of:

A. water molecules bonded to an ion in an aqueous solution
B. water molecules required to dissolve one mole of ions
C. ions bonded to one mole of water molecules
D. ions dissolved in one liter of an aqueous solution

48. When a solid dissolves, each molecule is removed from the crystal by interaction with the solvent. This process of surrounding each ion with solvent molecules is called:

A. hemolysis
B. electrolysis
C. solvation
D. dilution

49. The term *miscible* describes which type of solution?

A. Solid/solid
B. Liquid/gas
C. Liquid/solid
D. Liquid/liquid

50. Apply the *like dissolves like* rule to predict which of the following vitamins is insoluble in water:

A. niacinamide ($C_6H_6N_2O$)
B. pyridoxine ($C_8H_{11}NO_3$)
C. retinol ($C_{20}H_{30}O$)
D. thiamine ($C_{12}H_{17}N_4OS$)

51. Which species is NOT written as its constituent ions when the equation is expanded into the ionic equation?

$$Mg(OH)_2\ (s) + 2\ HCl\ (aq) \rightarrow MgCl_2\ (aq) + 2\ H_2O\ (l)$$

A. $Mg(OH)_2$ only
B. H_2O and $Mg(OH)_2$
C. HCl
D. $MgCl_2$

52. If x moles of $PbCl_2$ fully dissociate in 1 liter of H_2O, the K_{sp} is equivalent to:

A. x^2
B. $2x^4$
C. $4x^3$
D. $2x^3$

53. Which of the following are strong electrolytes?

I. salts
II. strong bases
III. weak acids

A. I only
B. I and II only
C. III only
D. I, II and III

54. What are the spectator ions in the reaction between KOH and HNO_3?

A. K^+ and NO_3^-
B. H^+ and NO_3^-
C. K^+ and H^+
D. H^+ and ^-OH

55. What volume of 14 M acid must be diluted with distilled water to prepare 6.0 L of 0.20 M acid?

A. 86 mL
B. 62 mL
C. 0.94 mL
D. 6.8 mL

56. What is the molarity of the solution obtained by diluting 160 mL of 4.50 M NaOH to 595 mL?

A. 0.242 M
B. 1.21 M
C. 2.42 M
D. 1.72 M

57. Which compound is most likely to be more soluble in the nonpolar solvent of benzene than in water?

A. SO_2
B. CO_2
C. Silver chloride
D. H_2S

58. Which of the following concentrations is dependent on temperature?

A. Mole fraction
B. Molarity
C. Mass percent
D. Molality

59. Which of the following solutions is the most concentrated?

A. One liter of water with 1 gram of sugar
B. One liter of water with 2 grams of sugar
C. One liter of water with 5 grams of sugar
D. One liter of water with 10 grams of sugar

60. What mass of NaOH is contained in 75.0 mL of a 5.0% (w/v) NaOH solution?

A. 6.50 g
B. 15.0 g
C. 3.75 g
D. 0.65 g

Practice Set 4: Questions 61–80

Questions **61** through **63** are based on the following data:

	K_{sp}
$PbCl_2$	1.0×10^{-5}
$AgCl$	1.0×10^{-10}
$PbCO_3$	1.0×10^{-15}

61. Consider a saturated solution of $PbCl_2$. The addition of NaCl would:

I. decrease $[Pb^{2+}]$
II. increase the precipitation of $PbCl_2$
III. not affect the precipitation of $PbCl_2$

A. I only
B. II only
C. III only
D. I and II only

62. What occurs when $AgNO_3$ is added to a saturated solution of $PbCl_2$?

I. AgCl precipitates
II. $Pb(NO_3)_2$ forms a white precipitate
III. More $PbCl_2$ forms

A. I only
B. II only
C. III only
D. I, II and III

63. Comparing equal volumes of saturated solutions for $PbCl_2$ and AgCl, which solution contains a greater concentration of Cl^-?

I. $PbCl_2$ II. AgCl III. Both have the same concentration of Cl^-

A. I only
B. II only
C. III only
D. I and II only

64. Which of the following is the reason why hexane is significantly soluble in octane?

A. Entropy increases for the two substances as the dominant factor in the ΔG when mixed
B. Hexane hydrogen bonds with octane
C. Intermolecular bonds between hexane-octane are much stronger than hexane-hexane or octane-octane molecular bonds
D. ΔH for hexane-octane is greater than hexane-H_2O

65. Which of the following are characteristics of an ideally dilute solution?

I. Solute molecules do not interact with each other
II. Solvent molecules do not interact with each other
III. The mole fraction of the solvent approaches 1

A. I only
B. II only
C. I and III only
D. I, II and III

66. Which principle states that the solubility of a gas in a liquid is proportional to the partial pressure of the gas above the liquid?

A. Solubility principle
B. Tyndall effect
C. Colloid principle
D. Henry's law

67. Water and methanol are two liquids that dissolve in each other. When the two are mixed, they form one layer because the liquids are:

A. unsaturated
B. saturated
C. miscible
D. immiscible

68. What is the molarity of a glucose solution that contains 10.0 g of $C_6H_{12}O_6$ dissolved in 100.0 mL of solution? (Use the molecular mass of $C_6H_{12}O_6$ = 180.0 g/mol)

A. 1.80 M
B. 0.555 M
C. 0.0555 M
D. 0.00555 M

69. Which of the following solid compounds is insoluble in water?

I. $BaSO_4$ II. Hg_2Cl_2 III. $PbCl_2$

A. I only
B. II only
C. III only
D. I, II and III

70. The net ionic equation for the reaction between zinc and a hydrochloric acid solution is:

A. $Zn\ (s) + 2\ H^+\ (aq) + 2\ Cl^-\ (aq) \rightarrow Zn^{2+}\ (aq) + 2\ Cl^-\ (aq) + H_2\ (g)$
B. $ZnCl_2\ (aq) + H_2\ (g) \rightarrow Zn\ (s) + 2\ HCl\ (aq)$
C. $Zn\ (s) + 2\ H^+\ (aq) \rightarrow Zn^{2+}\ (aq) + H_2\ (g)$
D. $Zn\ (s) + 2HCl\ (aq) \rightarrow ZnCl_2\ (aq) + H_2\ (g)$

71. Apply the *like dissolves like* rule to predict which of the following liquids is/are miscible with water:

I. methyl ethyl ketone, C_4H_8O
II. glycerin, $C_3H_5(OH)_3$
III. formic acid, $HCHO_2$

A. I only
B. II only
C. III only
D. I, II and III

72. What is the K_{sp} for calcium fluoride (CaF_2) if the calcium ion concentration in a saturated solution is 0.00021 *M*?

A. $K_{sp} = 3.7 \times 10^{-11}$
B. $K_{sp} = 2.6 \times 10^{-10}$
C. $K_{sp} = 3.6 \times 10^{-9}$
D. $K_{sp} = 8.1 \times 10^{-10}$

73. A 4 M solution of H_3A is completely dissociated in water. How many equivalents of H^+ are in 1/3 liter?

A. ¼ **B.** 4 **C.** 1.5 **D.** 3

74. Which of the following aqueous solutions is a poor conductor of electricity?

I. sucrose, $C_{12}H_{22}O_{11}$
II. barium nitrate, $Ba(NO_3)_2$
III. calcium bromide, $CaBr_2$

A. I only
B. II only
C. III only
D. I and II only

75. Which of the following solid compounds is insoluble in water?

A. $BaSO_4$
B. Na_2S
C. $(NH_4)_2CO_3$
D. K_2CrO_4

76. Which is true if the ion concentration product of a solution of AgCl is less than the K_{sp}?

I. Precipitation occurs
II. The ions are insoluble in water
III. Precipitation does not occur

A. I only
B. II only
C. III only
D. I and II only

77. Under which conditions is the expected solubility of oxygen gas in water the highest?

A. High temperature and high O_2 pressure above the solution
B. Low temperature and low O_2 pressure above the solution
C. Low temperature and high O_2 pressure above the solution
D. High temperature and low O_2 pressure above the solution

78. What is the molarity of an 8.60 molal solution of methanol (CH_3OH) with a density of 0.94 g/mL?

A. 0.155 M
B. 23.5 M
C. 6.34 M
D. 9.68 M

79. Which of the following is NOT a unit factor related to a 15.0% aqueous potassium iodide (KI) solution?

A. 100 g solution / 85.0 g water
B. 15.0 g KI / 100 g water
C. 85.0 g water / 100 g solution
D. 15.0 g KI / 85.0 g water

80. What is the molar concentration of a solution containing 0.75 mol of solute in 75 cm^3 of a solution?

A. 0.1 M
B. 1.5 M
C. 3 M
D. 10 M

Notes for active learning

Acids and Bases

Practice Set 1: Questions 1–20

1. Which is the conjugate acid–base pair in the reaction?

$$CH_3NH_2 + HCl \leftrightarrow CH_3NH_3^+ + Cl^-$$

A. HCl and Cl^-
B. $CH_3NH_3^+$ and Cl^-
C. CH_3NH_2 and Cl^-
D. CH_3NH_2 and HCl

2. What is the pH of an aqueous solution if the $[H^+] = 0.10$ M?

A. 0.0 **B.** 1.0 **C.** 2.0 **D.** 10.0

3. In which of the following pairs of substances are both species salts?

A. NH_4F and KCl
B. $CaCl_2$ and HCN
C. LiOH and K_2CO_3
D. NaOH and $CaCl_2$

4. Which reactant is a Brønsted-Lowry acid?

$$HCl\ (aq) + KHS\ (aq) \rightarrow KCl\ (aq) + H_2S\ (aq)$$

A. KCl
B. H_2S
C. HCl
D. KHS

5. If a light bulb in a conductivity apparatus glows brightly when testing a solution, which of the following must be true about the solution?

A. It is highly reactive
B. It is slightly reactive
C. It is highly ionized
D. It is slightly ionized

6. What is the term for a substance capable of donating or accepting a proton in an acid–base reaction?

I. Nonprotic II. Aprotic III. Amphoteric

A. I only
B. II only
C. III only
D. I and III only

7. Which of the following compounds is a strong acid?

I. $HClO_4$ (*aq*) II. H_2SO_4 (*aq*) III. HNO_3 (*aq*)

A. I only
B. II only
C. II and III only
D. I, II and III

8. What is the approximate pH of a solution of a strong acid where $[H_3O^+] = 8.30 \times 10^{-5}$?

A. 4 **B.** 11 **C.** 3 **D.** 5

9. Which of the following reactions represents the ionization of H_2O?

A. $H_2O + H_2O \rightarrow 2\ H_2 + O_2$
B. $H_2O + H_2O \rightarrow H_3O^+ + {}^-OH$
C. $H_2O + H_3O^+ \rightarrow H_3O^+ + H_2O$
D. $H_3O^+ + {}^-OH \rightarrow H_2O + H_2O$

10. Which set below contains only weak electrolytes?

A. NH_4Cl (*aq*), $HClO_2$ (*aq*), HCN (*aq*)
B. NH_3 (*aq*), HCO_3^- (*aq*), HCN (*aq*)
C. KOH (*aq*), H_3PO_4 (*aq*), $NaClO_4$ (*aq*)
D. HNO_3 (*aq*), H_2SO_4 (*aq*), HCN (*aq*)

11. Which of the following statements describes a neutral solution?

A. $[H_3O^+] / [^-OH] = 1 \times 10^{-14}$
B. $[H_3O^+] / [^-OH] = 1$
C. $[H_3O^+] < [^-OH]$
D. $[H_3O^+] > [^-OH]$

12. Which of the following is an example of an Arrhenius acid?

A. H_2O (*l*)
B. RbOH (*aq*)
C. $Ba(OH)_2$ (*aq*)
D. None of the above

13. In the following reaction, which reactant is a Brønsted-Lowry base?

H_2CO_3 (*aq*) + Na_2HPO_4 (*aq*) $\rightarrow$ $NaHCO_3$ (*aq*) + NaH_2PO_4 (*aq*)

A. $NaHCO_3$
B. NaH_2PO_4
C. Na_2HPO_4
D. H_2CO_3

14. Which of the following is the conjugate base of ^-OH?

A. O_2 **B.** O^{2-} **C.** H_2O **D.** O^-

15. Which of the following describes the solution for a vinegar sample at a pH of 5?

A. Weakly basic
B. Neutral
C. Weakly acidic
D. Strongly acidic

16. What is the pI for glutamic acid that contains two carboxylic acid groups and an amino group? (Use the carboxyl $pK_{a1} = 2.2$, carboxyl $pK_{a2} = 4.2$ and amino $pK_a = 9.7$)

A. 3.2 **B.** 1.0 **C.** 6.4 **D.** 5.4

17. Which of the following compounds cannot act as an acid?

A. NH_3
B. H_2SO_4
C. HSO_4^{1-}
D. SO_4^{2-}

18. A weak acid is titrated with a strong base. When the concentration of the conjugate base is equal to the concentration of the acid, the titration is at the:

A. endpoint
B. equivalence point
C. buffering region
D. diprotic point

19. If $[H_3O^+]$ in an aqueous solution is 7.5×10^{-9} M, what is the $[^-OH]$?

A. 6.4×10^{-5} M
B. $3.8 \times 10^{+8}$ M
C. 7.5×10^{-23} M
D. 1.3×10^{-6} M

20. Which species has a K_a of 5.7×10^{-10} if NH_3 has a K_b of 1.8×10^{-5}?

A. H^+
B. NH_2^-
C. NH_4^+
D. H_2O

Practice Set 2: Questions 21–40

---

21. Which of the following are the conjugate bases of HSO_4^-, CH_3OH and H_3O^+, respectively:

A. SO_4^{2-}, CH_2OH^- and ^-OH
B. CH_3O^-, SO_4^{2-} and H_2O
C. SO_4^{2-}, CH_3O^- and H_2O
D. SO_4^-, CH_2OH^- and H_2O

22. If 30.0 mL of 0.10 M $Ca(OH)_2$ is titrated with 0.20 M HNO_3, what volume of nitric acid is required to neutralize the base according to the following expression?

$$2\ HNO_3\ (aq) + Ca(OH)_2\ (aq) \rightarrow 2\ Ca(NO_3)_2\ (aq) + 2\ H_2O\ (l)$$

A. 30.0 mL
B. 15.0 mL
C. 10.0 mL
D. 20.0 mL

23. Which of the following expressions describes an acidic solution?

A. $[H_3O^+] / [^-OH] = 1 \times 10^{-14}$
B. $[H_3O^+] \times [^-OH] \neq 1 \times 10^{-14}$
C. $[H_3O^+] < [^-OH]$
D. $[H_3O^+] > [^-OH]$

24. Which is incorrectly classified as an acid, a base, or an amphoteric species?

A. LiOH / base
B. H_2O / amphoteric
C. H_2S / acid
D. NH_4^+ / base

25. Which of the following is the strongest weak acid?

A. CH_3COOH; $K_a = 1.8 \times 10^{-5}$
B. HF; $K_a = 6.5 \times 10^{-4}$
C. HCN; $K_a = 6.3 \times 10^{-10}$
D. HClO; $K_a = 3.0 \times 10^{-8}$

26. What are the products from the complete neutralization of phosphoric acid with aqueous lithium hydroxide?

A. $LiHPO_4$ (aq) and H_2O (l)
B. Li_3PO_4 (aq) and H_2O (l)
C. Li_2HPO_4 (aq) and H_2O (l)
D. LiH_2PO_4 (aq) and H_2O (l)

27. Which of the following compounds is NOT a strong base?

A. $Ca(OH)_2$
B. $Fe(OH)_3$
C. KOH
D. NaOH

28. What is the $[H^+]$ in stomach acid that registers a pH of 2.0 on a strip of pH paper?

A. 0.2 M **B.** 0.1 M **C.** 0.02 M **D.** 0.01 M

29. Which statement is true about distinguishing between dissociation and ionization?

A. Ionization is the separation of existing charged particles
B. Dissociation produces new charged particles
C. Ionization involves polar covalent compounds
D. Dissociation involves polar covalent compounds

30. Which of the following is a general property of a basic solution?

I. Turns litmus paper red
II. Tastes sour
III. Causes the skin of the fingers to feel slippery

A. I only
B. II only
C. III only
D. I and II only

31. Which of the following compound-classification pairs is incorrectly matched?

A. $Ca(OH)_2$ – weak base
B. $LiC_2H_3O_2$ –salt
C. NH_3 – weak base
D. HI – strong acid

32. What is the term for a substance that releases H^+ in H_2O?

A. Brønsted-Lowry acid
B. Brønsted-Lowry base
C. Arrhenius acid
D. Arrhenius base

33. Which molecule is acting as a base in the following reaction?

$$^{-}OH + NH_4^+ \rightarrow H_2O + NH_3$$

A. ^{-}OH
B. NH_4^+
C. H_2O
D. NH_3

34. Citric acid is a triprotic acid with three carboxylic acid groups having pK_a values of 3.2, 4.8 and 6.4. At a pH of 5.7, what is the predominant protonation state of citric acid?

A. Three carboxylic acid groups are deprotonated
B. Three carboxylic acid groups are protonated
C. One carboxylic acid group is deprotonated, while two are protonated
D. Two carboxylic acid groups are deprotonated, while one is protonated

35. When fully neutralized by treatment with barium hydroxide, a phosphoric acid yields $Ba_2P_2O_7$ as one of its products. The parent acid for the anion in this compound is:

A. tetraprotic acid
B. diprotic acid
C. triprotic acid
D. hexaprotic acid

36. Does a solution become more or less acidic when a weak acid solution is added to a concentrated HCl solution?

A. Less acidic because the concentration of OH^- increases
B. No change in acidity because [HCl] is too high to be changed by the weak solution
C. Less acidic because the solution becomes more dilute with a less concentrated solution of H_3O^+ being added
D. More acidic because more H_3O^+ is being added to the solution

37. Which of the following is a triprotic acid?

A. HNO_3
B. H_3PO_4
C. H_2SO_3
D. $HC_2H_3O_2$

38. For which of the following pairs of substances do the two pair members NOT react?

A. Na_3PO_4 and HCl
B. KCl and NaI
C. HF and LiOH
D. $PbCl_2$ and H_2SO_4

39. What happens to the pH when sodium acetate is added to a solution of acetic acid?

A. Decreases due to the common ion effect
B. Increases due to the common ion effect
C. Remains constant because sodium acetate is a buffer
D. Remains constant because sodium acetate is neither acidic nor basic

40. Which of the following is the acidic anhydride of phosphoric acid (H_3PO_4)?

A. P_2O
B. P_2O_3
C. PO_3
D. P_4O_{10}

Practice Set 3: Questions 41–60

41. Which of the following does NOT act as a Brønsted-Lowry acid?

A. CO_3^{2-}
B. HS^-
C. HSO_4^-
D. H_2O

42. Which of the following is the strongest weak acid?

A. HF ; $pK_a = 3.17$
B. HCO_3^- ; $pK_a = 10.32$
C. $H_2PO_4^-$; $pK_a = 7.18$
D. NH_4^+ ; $pK_a = 9.20$

43. Why does boiler scale form on the walls of hot water pipes from groundwater?

A. Transformation of $H_2PO_4^-$ ions to PO_4^{3-} ions, which precipitate with the "hardness ions," Ca^{2+}, Mg^{2+}, Fe^{2+}/Fe^{3+}
B. Transformation of HSO_4^- ions to SO_4^{2-} ions, which precipitate with the "hardness ions," Ca^{2+}, Mg^{2+}, Fe^{2+}/Fe^{3+}
C. Transformation of HSO_3^- ions to SO_3^{2-} ions, which precipitate with the "hardness ions," Ca^{2+}, Mg^{2+}, Fe^{2+}/Fe^{3+}
D. Transformation of HCO_3^- ions to CO_3^{2-} ions, which precipitate with the "hardness ions," Ca^{2+}, Mg^{2+}, Fe^{2+}/Fe^{3+}

44. Which of the following substances, when added to a solution of sulfoxylic acid (H_2SO_2), could be used to prepare a buffer solution?

A. H_2O
B. $NaHSO_2$
C. KCl
D. HCl

45. Which of the following statements is NOT correct?

A. Acidic salts are formed by partial neutralization of a diprotic acid by a diprotic base
B. Acidic salts are formed by partial neutralization of a triprotic acid by a diprotic base
C. Acidic salts are formed by partial neutralization of a monoprotic acid by a monoprotic base
D. Acidic salts are formed by partial neutralization of a diprotic acid by a monoprotic base

46. Which of the following is the acid anhydride for $HClO_4$?

A. ClO
B. ClO_2
C. ClO_3
D. Cl_2O_7

47. Identify the acid/base behavior of each substance for the reaction:

$H_3O^+ + Cl^- \rightleftharpoons H_2O + HCl$

A. H_3O^+ acts as an acid, Cl^- acts as a base, H_2O acts as a base, and HCl acts as an acid
B. H_3O^+ acts as a base, Cl^- acts as an acid, H_2O acts as a base, and HCl acts as an acid
C. H_3O^+ acts as an acid, Cl^- acts as a base, H_2O acts as an acid, and HCl acts as a base
D. H_3O^+ acts as a base, Cl^- acts as an acid, H_2O acts as an acid, and HCl acts as a base

48. Given the pK_a values for phosphoric acid of 2.15, 6.87, and 12.35, what is the HPO_4^{2-} / $H_2PO_4^-$ ratio in a typical muscle cell when the pH is 7.35?

A. 6.32×10^{-6}
B. 1.18×10^{5}
C. 0.46
D. 3.02

49. If a light bulb in a conductivity apparatus glows dimly when testing a solution, which of the following must be true about the solution?

I. It is slightly reactive
II. It is slightly ionized
III. It is highly ionized

A. I only
B. II only
C. III only
D. I and II only

50. Which of the following properties is NOT characteristic of an acid?

A. It is neutralized by a base
B. It has a slippery feel
C. It produces H^+ in water
D. It tastes sour

51. What is the term for a substance that releases hydroxide ions in water?

A. Brønsted-Lowry base
B. Brønsted-Lowry acid
C. Arrhenius base
D. Arrhenius acid

52. For the reaction below, which of the following is the conjugate acid of C_5H_5N?

$C_5H_5N + H_2CO_3 \leftrightarrow C_5H_6N^+ + HCO_3^-$

A. $C_5H_6N^+$
B. HCO_3^-
C. C_5H_5N
D. H_2CO_3

53. Which of the following terms applies to Cl^- in the reaction below?

$HCl\ (aq) \rightarrow H^+ + Cl^-$

A. Weak conjugate base
B. Strong conjugate base
C. Weak conjugate acid
D. Strong conjugate acid

54. Which of the following compounds is a diprotic acid?

A. HCl
B. H_3PO_4
C. HNO_3
D. H_2SO_3

55. Lysine contains two amine groups (pK_a = 9.0 and 10.0) and a carboxylic acid group (pK_a = 2.2). In a solution of pH 9.5, which describes the protonation and charge state of lysine?

A. Carboxylic acid is deprotonated and negative; amine (pK_a 9.0) is deprotonated and neutral, whereby the amine (pK_a = 10.0) is protonated and positive
B. Carboxylic acid is deprotonated and negative; amine (pK_a = 9.0) is protonated and positive, whereby the amine (pK_a = 10.0) is deprotonated and neutral
C. Carboxylic acid is deprotonated and neutral; both amines are protonated and positive
D. Carboxylic acid is deprotonated and negative; both amines are deprotonated and neutral

56. Which of the following is the chemical species present in acidic solutions?

A. H_2O^+ (*aq*)
B. H_3O^+ (*l*)
C. H_3O^+ (*aq*)
D. ^-OH (*aq*)

57. Which compound has a value of K_a that is approximately equal to 10^{-5}?

A. $CH_3CH_2CH_2CO_2H$
B. KOH
C. NaBr
D. HNO_3

58. Relative to a pH of 7, a solution with a pH of 4 has:

A. 30 times less [H^+]
B. 300 times less [H^+]
C. 1,000 times greater [H^+]
D. 300 times greater [H^+]

59. What is the pH of this buffer system if the concentration of undissociated weak acid is equal to the concentration of the conjugate base? (Use the K_a of the buffer = 4.6×10^{-4})

A. 1 and 2
B. 3 and 4
C. 5 and 6
D. 7 and 8

60. Which of the following is the ionization constant expression for water?

A. $K_w = [H_2O] / [H^+]\cdot[^-OH]$
B. $K_w = [H^+]\cdot[^-OH] / [H_2O]$
C. $K_w = [H^+]\cdot[^-OH]$
D. $K_w = [H_2O]\cdot[H_2O]$

Practice Set 4: Questions 61–80

---

61. Which of the following statements about strong or weak acids is true?

A. A weak acid reacts with a strong base
B. A strong acid does not react with a strong base
C. A weak acid readily forms ions when dissolved in water
D. A weak acid and a strong acid at the same concentration are equally corrosive

62. What is the value of K_w at 25 °C?

A. 1.0
B. 1.0×10^{-7}
C. 1.0×10^{-14}
D. 1.0×10^{7}

63. Which of the statements below best describes the following reaction?

$$HNO_3\ (aq) + LiOH\ (aq) \rightarrow LiNO_3\ (aq) + H_2O\ (l)$$

A. Nitric acid and lithium hydroxide solutions produce lithium nitrate solution and H_2O
B. Nitric acid and lithium hydroxide solutions produce lithium nitrate and H_2O
C. Nitric acid and lithium hydroxide produce lithium nitrate and H_2O
D. Aqueous solutions of nitric acid and lithium hydroxide produce aqueous lithium nitrate and H_2O

64. The Brønsted-Lowry acid and base in the following reaction are, respectively:

$$NH_4^+ + CN^- \rightarrow NH_3 + HCN$$

A. NH_4^+ and ^-CN
B. ^-CN and HCN
C. NH_4^+ and HCN
D. NH_3 and ^-CN

65. Which would NOT be used to make a buffer solution?

A. H_2SO_4
B. H_2CO_3
C. NH_4OH
D. CH_3COOH

66. Which of the following is a general property of an acidic solution?

A. Turns litmus paper blue
B. Neutralizes acids
C. Tastes bitter
D. None of the above

67. What is the term for a solution that is a good conductor of electricity?

A. Strong electrolyte
B. Weak electrolyte
C. Non-electrolyte
D. Aqueous electrolyte

68. Which of the following compounds is an acid?

A. HBr
B. C_2H_6
C. KOH
D. NaF

69. A metal and a salt solution react only if the metal introduced into the solution is:

A. below the replaced metal in the activity series
B. above the replaced metal in the activity series
C. below hydrogen in the activity series
D. above hydrogen in the activity series

70. Which of the following is an example of an Arrhenius base?

I. NaOH (*aq*) II. $Al(OH)_3$ (*s*) III. $Ca(OH)_2$ (*aq*)

A. I only
B. II only
C. III only
D. I and III only

71. Which of the following is NOT a conjugate acid/base pair?

A. S^{2-} / H_2S
B. HSO_4^- / SO_4^{2-}
C. H_2O / ^-OH
D. PH_4^+ / PH_3

72. If a buffer is made with the pH below the pK_a of the weak acid, the [base] / [acid] ratio is:

A. equal to 0
B. equal to 1
C. greater than 1
D. less than 1

73. Which of the following acids listed below has the strongest conjugate base?

Monoprotic Acids	K_a
Acid I	1.3×10^{-8}
Acid II	2.9×10^{-9}
Acid III	4.2×10^{-10}
Acid IV	3.8×10^{-8}

A. I **B.** II **C.** III **D.** IV

74. Complete neutralization of phosphoric acid with barium hydroxide, when separated and dried, yields $Ba_3(PO_4)_2$ as one of the products. Therefore, which term describes phosphoric acid?

A. Monoprotic acid
B. Diprotic acid
C. Hexaprotic acid
D. Triprotic acid

75. Which is the correct net ionic equation for the hydrolysis reaction of Na_2S?

A. $Na^+ + H_2O \rightarrow NaOH + H_2$
B. $S^{2-} + H_2O \rightarrow OH^- + HS^-$
C. $S^{2-} + H_2O \rightarrow 2\ HS^- + OH^-$
D. $S^{2-} + 2\ H_2O \rightarrow HS^- + H_3O^+$

76. Calculate the pH of 0.0765 M HNO_3.

A. 1.1
B. 3.9
C. 11.7
D. 7.9

77. Which of the following compounds is NOT a strong acid?

A. HBr (*aq*)
B. HNO_3
C. CH_3COOH
D. H_2SO_4

78. Which reaction produces $NiCr_2O_7$ as a product?

A. Nickel (II) hydroxide and dichromic acid
B. Nickel (II) hydroxide and chromic acid
C. Nickelic acid and chromium (II) hydroxide
D. Nickel (II) hydroxide and chromate acid

79. Which of the following statements describes a Brønsted-Lowry base?

A. Donates protons to other substances
B. Accepts protons from other substances
C. Produces hydrogen ions in an aqueous solution
D. Produces hydroxide ions in an aqueous solution

80. When dissolved in water, the Arrhenius acid/bases KOH, H_2SO_4 and HNO_3 are, respectively:

A. base, acid and base
B. base, base and acid
C. base, acid and acid
D. acid, base and base

Notes for active learning

Notes for active learning

Electrochemistry

Practice Set 1: Questions 1–20

1. Which is NOT true regarding the redox reaction occurring in a spontaneous electrochemical cell?

$$Cl_2\ (g) + 2\ Br^-\ (aq) \rightarrow Br_2\ (l) + 2\ Cl^-\ (aq)$$

A. Anions flow towards the anode
B. Electrons flow from the anode to the cathode
C. Cl_2 is reduced at the cathode
D. Br^- is oxidized at the cathode

2. What is the relationship between an element's ionization energy and its ability to function as an oxidizing agent? As ionization energy increases:

A. the ability of an element to function as an oxidizing agent remains the same
B. the ability of an element to function as an oxidizing agent decreases
C. the ability of an element to function as an oxidizing agent increases
D. the ability of an element to function as a reducing agent remains the same

3. How many electrons are needed to balance the following half-reaction $H_2S \rightarrow S_8$ in an acidic solution?

A. 16 electrons to the right side
B. 6 electrons to the right side
C. 12 electrons to the left side
D. 8 electrons to the right side

4. Which is NOT true regarding the following redox reaction occurring in an electrolytic cell?

$$3\ C + 2\ Co_2O_3 \xrightarrow{\text{Electricity}} 4\ Co + 3\ CO_2$$

A. $CO_2\ (g)$ is produced at the anode
B. Co metal is produced at the anode
C. Oxidation half-reaction: $C + 2\ O^{2-} \rightarrow CO_2 + 4\ e^-$
D. Reduction half-reaction: $Co^{3+} + 3\ e^- \rightarrow Co$

5. The anode in a galvanic cell attracts:

A. cations
B. neutral particles
C. anions
D. anions and neutral particles

6. The electrode with the standard reduction potential of 0 V is assigned as the standard reference electrode and uses the half-reaction:

A. $2\ H^+\ (aq) + 2\ e^- \leftrightarrow H_2\ (g)$
B. $Ag^+\ (aq) + e^- \leftrightarrow Ag\ (s)$
C. $Cu^{2+}\ (aq) + 2\ e^- \leftrightarrow Cu\ (s)$
D. $Zn^{2+}\ (aq) + 2\ e^- \leftrightarrow Zn\ (s)$

7. In the reaction for a discharging nickel–cadmium (NiCd) battery, which substance is being oxidized?

$$Cd\ (s) + NiO_2\ (s) + 2\ H_2O\ (l) \rightarrow Cd(OH)_2\ (s) + Ni(OH)_2\ (s)$$

A. H_2O
B. $Cd(OH)_2$
C. Cd
D. NiO_2

8. What happens at the anode if rust forms when Fe is in contact with H_2O?

$$4\ Fe + 3\ O_2 \rightarrow 2\ Fe_2O_3$$

A. Fe is reduced
B. Oxygen is reduced
C. Oxygen is oxidized
D. Fe is oxidized

9. Which is true regarding the redox reaction occurring in the spontaneous electrochemical cell for $Cl_2\ (g) + 2\ Br^-\ (aq) \rightarrow Br_2\ (l) + 2\ Cl^-\ (aq)$?

A. Electrons flow from the cathode to the anode
B. Cl_2 is oxidized at the cathode
C. Br^- is reduced at the anode
D. Cations in the salt bridge flow from the Br_2 half-cell to the Cl_2 half-cell

10. What is the term for a chemical reaction that involves electron transfer between two reacting substances?

A. Reduction reaction
B. Redox reaction
C. Oxidation reaction
D. Half-reaction

11. Using the following metal ion/metal reaction potentials,

$Cu^{2+}\ (aq)\|Cu\ (s)$	$Ag^+\ (aq)\|Ag\ (s)$	$Co^{2+}\ (aq)\|Co\ (s)$	$Zn^{2+}\ (aq)\|Zn\ (s)$
+0.34 V	+0.80 V	–0.28 V	–0.76 V

calculate the standard cell potential for the cell whose reaction is:

$$Co\ (s) + Cu^{2+}\ (aq) \rightarrow Co^{2+}\ (aq) + Cu\ (s)$$

A. +0.62 V
B. –0.62 V
C. +0.48 V
D. –0.48 V

12. How are photovoltaic cells different from many other forms of solar energy?

A. Light is reflected and the coolness of the shade is used to provide a temperature differential
B. Light is passively converted into heat
C. Light is converted directly to electricity
D. Light is converted into heat and then into electricity

13. What is the term for an electrochemical cell that has a single electrode where oxidation or reduction can occur?

A. Half-cell
B. Voltaic cell
C. Dry cell
D. Electrolytic cell

14. By which method could electrolysis be used to raise the hull of a sunken ship?

A. Electrolysis could only be used to raise the hull if the ship is made of iron. If so, the electrolysis of the iron metal might produce sufficient gas to lift the ship
B. The gaseous products of the electrolysis of H_2O are collected with bags attached to the hull of the ship and the inflated bags raise the ship
C. The electrolysis of the H_2O under the ship's hull boils H_2O and creates upward pressure to raise the ship
D. An electric current passed through the hull of the ship produces electrolysis, and the gases trapped in compartments of the vessel would push it upwards

15. What are the products for the single-replacement reaction $Zn\ (s) + CuSO_4\ (aq) \rightarrow$?

A. CuO and $ZnSO_4$
B. CuO and $ZnSO_3$
C. Cu and $ZnSO_4$
D. Cu and $ZnSO_3$

16. Which of the following is a true statement about the electrochemical reaction for an electrochemical cell with a cell potential of +0.36 V?

A. The reaction favors the formation of reactants and would be considered a galvanic cell
B. The reaction favors the formation of reactants and would be considered an electrolytic cell
C. The reaction is at equilibrium and is a galvanic cell
D. The reaction favors the formation of products and would be considered a galvanic cell

17. Which of the following is a unit of electrical energy?

A. Coulomb
B. Joule
C. Watt
D. Ampere

18. Which observation describes the solution near the cathode when an aqueous solution of sodium chloride is electrolyzed, and hydrogen gas is evolved at the cathode?

A. Colored
B. Acidic
C. Basic
D. Frothy

19. How many grams of Ag is deposited in the cathode of an electrolytic cell if a current of 3.50 A is applied to a solution of $AgNO_3$ for 12 minutes? (Use the molecular mass of Ag = 107.86 g/mol and the conversion of 1 mol e^- = 9.65×10^4 C)

A. 0.32 g
B. 0.86 g
C. 2.82 g
D. 3.86 g

20. In an electrochemical cell, which of the following statements is FALSE?

A. The anode is the electrode where oxidation occurs
B. A salt bridge provides electrical contact between the half-cells
C. The cathode is the electrode where reduction occurs
D. All of the above are true

Practice Set 2: Questions 21–40

21. What is the relationship between an element's ionization energy and its ability to function as an oxidizing and reducing agent? Elements with high ionization energy are:

A. strong oxidizing and weak reducing agents
B. weak oxidizing and weak reducing agents
C. strong oxidizing and strong reducing agents
D. weak oxidizing and strong reducing agents

22. In a basic solution, how many electrons are needed to balance the following half-reaction?

$$C_8H_{10} \rightarrow C_8H_4O_4^{2-}$$

A. 8 electrons on the left side
B. 12 electrons on the left side
C. 4 electrons on the left side
D. 12 electrons on the right side

23. Which statement is true regarding the following redox reaction occurring in an electrolytic cell?

$$3\ C\ (s) + 2\ Co_2O_3\ (l) \xrightarrow{\text{Electricity}} 4\ Co\ (l) + 3\ CO_2\ (g)$$

A. CO_2 gas is produced at the anode
B. Co^{3+} is produced at the cathode
C. Oxidation half-reaction: $Co^{3+} + 3\ e^- \rightarrow Co$
D. Reduction half-reaction: $C + 2\ O^{2-} \rightarrow CO_2 + e^-$

24. Which is true regarding the following redox reaction occurring in a spontaneous electrochemical cell?

$$Sn\ (s) + Cu^{2+}\ (aq) \rightarrow Cu\ (s) + Sn^{2+}\ (aq)$$

A. Anions in the salt bridge flow from the Cu half-cell to the Sn half-cell
B. Electrons flow from the Cu electrode to the Sn electrode
C. Cu^{2+} is oxidized at the cathode
D. Sn is reduced at the anode

25. In galvanic cells, reduction occurs at the:

I. salt bridge   II. cathode   III. anode

A. I only
B. II only
C. III only
D. I and II only

26. For a battery, what is undergoing reduction in the following oxidation-reduction reaction?

$Mn_2O_3 + ZnO \rightarrow 2\ MnO_2 + Zn$

A. Mn_2O_3
B. MnO_2
C. ZnO
D. Zn

27. Which of the metals listed below has the greatest tendency to undergo oxidation, given that the following redox reactions go essentially to completion?

$Ni\ (s) + Ag^+\ (aq) \rightarrow Ag\ (s) + Ni^{2+}\ (aq)$

$Al\ (s) + Cd^{2+}\ (aq) \rightarrow Cd\ (s) + Al^{3+}\ (aq)$

$Cd\ (s) + Ni^{2+}\ (aq) \rightarrow Ni\ (s) + Cd^{2+}\ (aq)$

$Ag\ (s) + H^+\ (aq) \rightarrow$ no reaction

A. (H) **B.** Cd **C.** Ni **D.** Al

28. What is the purpose of the salt bridge in a voltaic cell?

A. Allows for a balance of charge between the two chambers
B. Allows the Fe^{2+} and the Cu^{2+} to flow freely between the two chambers
C. Allows for the buildup of positively charged ions in one container and negatively charged ions in the other container
D. Prevents further migration of electrons through the wire

29. If $\Delta G°$ for a cell is positive, the E° is:

A. neutral
B. negative
C. positive
D. unable to be determined

30. In an electrochemical cell, which of the following statements is FALSE?

A. Oxidation occurs at the anode
B. Reduction occurs at the cathode
C. Electrons flow through the salt bridge to complete the cell
D. A spontaneous electrochemical cell is called a voltaic cell

31. Which of the following statements about electrochemistry is NOT true?

A. The study of how protons are transferred from one chemical compound to another
B. The use of electrical current to produce an oxidation-reduction reaction
C. The use of a set of oxidation-reduction reactions to produce electrical current
D. The study of how electrical energy and chemical reactions are related

32. In an operating photovoltaic cell, electrons move through the external circuit to the negatively charged p-type silicon wafer. How can the electrons move to the negatively charged silicon wafer if electrons are negatively charged?

A. The p-type silicon wafer is positively charged
B. The energy of the sunlight moves electrons in a nonspontaneous direction
C. Advancements in photovoltaic technology have solved this technological impediment
D. An electric current occurs because the energy from the sun reverses the charge of the electrons

33. What is the term for the value assigned to an atom in a substance that indicates whether the atom is electron-poor or electron-rich compared to a free atom?

A. Reduction number
B. Oxidation number
C. Cathode number
D. Anode number

34. What is the term for the relative ability of a substance to undergo reduction?

A. Oxidation potential
B. Reduction potential
C. Anode potential
D. Cathode potential

35. What is the primary difference between a fuel cell and a battery?

A. Fuel cells oxidize to supply electricity, while batteries reduce to supply electricity
B. Batteries supply electricity, while fuel cells supply heat
C. Batteries can be recharged, while fuel cells cannot
D. Fuel cells do not run down because they can be refueled, while batteries run down and need to be recharged

36. In balancing the equation for a disproportionation reaction, using the oxidation number method, the substance undergoing disproportionation is:

A. initially written twice on the reactant side of the equation
B. initially written twice on the product side of the equation
C. initially written on the reactant and product sides of the equation
D. always assigned an oxidation number of zero

37. What is the term for an electrochemical cell in which electrical energy is generated from a spontaneous redox reaction?

A. Voltaic cell
B. Photoelectric cell
C. Dry cell
D. Electrolytic cell

38. Which observation describes the solution near the anode when an aqueous solution of sodium sulfate is electrolyzed, and gas is evolved at the anode?

A. Colored
B. Acidic
C. Basic
D. Frothy

39. Which of the statements listed below is true regarding the following redox reaction occurring in a nonspontaneous electrochemical cell?

$$Cd\ (s) + Zn(NO_3)_2\ (aq) + \text{electricity} \rightarrow Zn\ (s) + Cd(NO_3)_2\ (aq)$$

A. Oxidation half-reaction: $Zn^{2+} + 2\ e^- \rightarrow Zn$
B. Reduction half-reaction: $Cd \rightarrow Cd^{2+} + 2\ e^-$
C. Cd metal is produced at the anode
D. Zn metal is produced at the cathode

40. Which statement is true for electrolysis?

A. A spontaneous redox reaction produces electricity
B. A nonspontaneous redox reaction is forced to occur by applying an electric current
C. Only pure, drinkable water is produced
D. There is a cell that reverses the flow of ions

Practice Set 3: Questions 41–60

---

41. Which relationship explains an element's electronegativity and its ability to act as an oxidizing and reducing agent?

A. Atoms with large electronegativity tend to act as strong oxidizing and strong reducing agents
B. Atoms with large electronegativity tend to act as weak oxidizing and strong reducing agents
C. Atoms with large electronegativity tend to act as strong oxidizing and weak reducing agents
D. Atoms with large electronegativity tend to act as weak oxidizing and weak reducing agents

42. Which substance is undergoing reduction for a battery if the following two oxidation-reduction reactions take place?

Reaction I: $Zn + 2\ OH^- \rightarrow ZnO + H_2O + 2\ e^-$

Reaction II: $2\ MnO_2 + H_2O + 2\ e^- \rightarrow Mn_2O_3 + 2\ OH^-$

A. MnO_2 **B.** H_2O **C.** Zn **D.** ZnO

43. What is the term for an electrochemical cell in which a nonspontaneous redox reaction occurs by forcing electricity through the cell?

I. voltaic cell II. dry cell III. electrolytic cell

A. I only
B. II only
C. III only
D. I and II only

44. Which of the statements is true regarding the following redox reaction occurring in a galvanic cell?

$Cl_2\ (g) + 2\ Br^-\ (aq) \rightarrow Br_2\ (l) + 2\ Cl^-\ (aq)$?

A. Cations in the salt bridge flow from the Cl_2 half-cell to the Br_2 half-cell
B. Electrons flow from the anode to the cathode
C. Cl_2 is reduced at the anode
D. Br^- is oxidized at the cathode

45. Which of the statements listed below is true regarding the following redox reaction occurring in a nonspontaneous electrolytic cell?

$$3\ C\ (s) + 4\ AlCl_3\ (l) \xrightarrow{\text{Electricity}} 4\ Al\ (l) + 3\ CCl_4\ (g)$$

A. Cl^- is produced at the anode
B. Al metal is produced at the cathode
C. Oxidation half-reaction is $Al^{3+} + 3\ e^- \rightarrow Al$
D. Reduction half-reaction is $C + 4\ Cl^- \rightarrow CCl_4 + 4\ e^-$

46. Which of the following materials is most likely to undergo oxidation?

I. Cl^- II. Na III. Na^+

A. I only
B. II only
C. III only
D. I and II only

47. Which statement is NOT true regarding the redox reaction occurring in a spontaneous electrochemical cell?

$$Zn\ (s) + Cd^{2+}\ (aq) \rightarrow Cd\ (s) + Zn^{2+}\ (aq)$$

A. Anions in the salt bridge flow from the Zn half-cell to the Cd half-cell
B. Electrons flow from the Zn electrode to the Cd electrode
C. Cd^{2+} is reduced at the cathode
D. Zn is oxidized at the anode

48. Using the provided electrochemical series, determine which is capable of oxidizing Cu (*s*) to Cu^{2+} (*aq*) when added to Cu (*s*) in solution?

A. Al^{3+} (*aq*)
B. Ag^+ (*aq*)
C. K^+ (*aq*)
D. Pb^{2+} (*aq*)

Equilibrium	**E°**
$Li^+\ (aq) + e^- \leftrightarrow Li\ (s)$	–3.03 volts
$K^+\ (aq) + e^- \leftrightarrow K\ (s)$	–2.92
$^*Ca^{2+}\ (aq) + 2\ e^- \leftrightarrow Ca\ (s)$	–2.87
$^*Na^+\ (aq) + e^- \leftrightarrow Na\ (s)$	–2.71
$Mg^{2+}\ (aq) + 2\ e^- \leftrightarrow Mg\ (s)$	–2.37
$Al^{3+}\ (aq) + 3\ e^- \leftrightarrow Al\ (s)$	–1.66
$Zn^{2+}\ (aq) + 2\ e^- \leftrightarrow Zn\ (s)$	–0.76
$Fe^{2+}\ (aq) + 2\ e^- \leftrightarrow Fe\ (s)$	–0.44
$Pb^{2+}\ (aq) + 2\ e^- \leftrightarrow Pb\ (s)$	–0.13
$2\ H^+\ (aq) + 2\ e^- \leftrightarrow H_2\ (g)$	0.0
$Cu^{2+}\ (aq) + 2\ e^- \leftrightarrow Cu\ (s)$	+0.34
$Ag^+\ (aq) + e^- \leftrightarrow Ag\ (s)$	+0.80
$Au^{3+}\ (aq) + 3\ e^- \leftrightarrow Au\ (s)$	+1.50

49. Why is the anode of a battery indicated with a negative (–) sign?

A. Electrons move to the anode to react with NH_4Cl in the battery
B. It indicates the electrode where the chemicals are reduced
C. Electrons are attracted to the negative electrode
D. The electrode is the source of negatively charged electrons

50. What is the term for the conversion of chemical energy to electrical energy from redox reactions?

A. Redox chemistry
B. Electrochemistry
C. Cell chemistry
D. Battery chemistry

51. In electrolysis, E° tends to be:

A. negative
B. neutral
C. positive
D. greater than 1

52. A major source of chlorine gas is the electrolysis of concentrated saltwater, NaCl (*aq*). What is the sign of the electrode where the chlorine gas is formed?

A. Neither, since chlorine gas is a neutral molecule and there is no electrode attraction
B. Negative, since the chlorine gas needs to deposit electrons to form chloride ions
C. Positive, since the chloride ions lose electrons to form chlorine molecules
D. Both, since the chloride ions from NaCl (*aq*) are attracted to the positive electrode to form chlorine molecules, while the produced chlorine gas molecules move to deposit electrons at the negative electrode

53. Which statement is correct for an electrolytic cell that has two electrodes?

A. Oxidation occurs at the anode, which is negatively charged
B. Oxidation occurs at the anode, which is positively charged
C. Oxidation occurs at the cathode, which is positively charged
D. Oxidation occurs at the cathode, which is negatively charged

54. Which of the statements listed below is true regarding the following redox reaction occurring in a nonspontaneous electrochemical cell?

$$Cd\ (s) + Zn(NO_3)_2\ (aq) + \text{electricity} \rightarrow Zn\ (s) + Cd(NO_3)_2\ (aq)$$

A. Oxidation half-reaction: $Zn^{2+} + 2\ e^- \rightarrow Zn$
B. Reduction half-reaction: $Cd \rightarrow Cd^{2+} + 2\ e^-$
C. Cd metal dissolves at the anode
D. Zn metal dissolves at the cathode

55. Which battery system is based on half-reactions involving zinc metal and manganese dioxide?

A. Alkaline batteries
B. Fuel cells
C. Lead-acid storage batteries
D. Dry-cell batteries

56. Which of the following equations is a disproportionation reaction?

A. $2\ H_2O \rightarrow 2\ H_2 + O_2$

B. $H_2SO_3 \rightarrow H_2O + SO_2$

C. $HNO_2 \rightarrow NO + HNO_3$

D. $Mg + H_2SO_4 \rightarrow MgSO_4 + H_2$

57. How is electrolysis different from the chemical process inside a battery?

A. Pure compounds cannot be generated *via* electrolysis

B. Electrolysis only uses electrons from a cathode

C. Electrolysis does not use electrons

D. They are the same process in reverse

58. Which fact about fuel cells is FALSE?

A. Fuel cell automobiles are powered by water and only emit hydrogen

B. Fuel cells are based on the tendency of some elements to gain electrons from other elements

C. Fuel cell automobiles are quiet

D. Fuel cell automobiles are environmentally friendly

59. Which process occurs when copper is refined using the electrolysis technique?

A. Impure copper goes into solution at the anode, and pure copper plates out on the cathode

B. Impure copper goes into solution at the cathode, and pure copper plates out on the anode

C. Pure copper goes into solution from the anode and forms a precipitate at the bottom of the tank

D. Pure copper goes into solution from the cathode and forms a precipitate at the bottom of the tank

60. Which of the following statements describes electrolysis?

A. A chemical reaction which results when electrical energy is passed through a metallic liquid

B. A chemical reaction which results when electrical energy is passed through a liquid electrolyte

C. The splitting of atomic nuclei by electrical energy

D. The splitting of atoms by electrical energy

Practice Set 4: Questions 61–80

61. In a battery, which of the species in the two oxidation-reduction reactions is undergoing oxidation?

Reaction I: $Zn + 2\ OH^- \rightarrow ZnO + H_2O + 2\ e^-$

Reaction II: $2\ MnO_2 + H_2O + 2\ e^- \rightarrow Mn_2O_3 + 2\ OH^-$

A. H_2O
B. MnO_2
C. ZnO
D. Zn

62. What is the term for an electrochemical cell where the anode and cathode reactions do not occur in aqueous solutions?

A. Voltaic cell
B. Electrolytic cell
C. Dry cell
D. Galvanic cell

63. How many electrons are needed to balance the charge for the following half-reaction in an acidic solution?

$C_2H_6O \rightarrow HC_2H_3O_2$

A. 6 electrons to the right side
B. 4 electrons to the right side
C. 3 electrons to the left side
D. 2 electrons to the left side

64. Which of the statements is true regarding the following redox reaction occurring in an electrolytic cell?

$$3\ C\ (s) + 4\ AlCl_3\ (l) \xrightarrow{\text{Electricity}} 4\ Al\ (l) + 3\ CCl_4\ (g)$$

A. Al^{3+} is produced at the cathode
B. CCl_4 gas is produced at the anode
C. Reduction half-reaction: $C + 4\ Cl^- \rightarrow CCl_4 + 4\ e^-$
D. Oxidation half-reaction: $Al^{3+} + 3\ e^- \rightarrow Al$

65. What is the term for an electrochemical cell that spontaneously produces electrical energy?

I. half-cell II. electrolytic cell III. dry cell

A. I and II only
B. II only
C. III only
D. II and III only

66. The anode of a battery is indicated with a negative (–) sign, because the anode is:

A. where electrons are adsorbed
B. positive
C. negative
D. where electrons are generated

67. Which is the strongest reducing agent for the following half-reaction potentials?

Sn^{4+} (*aq*) + 2 e^- → Sn^{2+} (*aq*) E° = –0.13 V

Ag^+ (*aq*) + e^- → Ag (*s*) E° = +0.81 V

Cr^{3+} (*aq*) + 3 e^- → Cr (*s*) E° = –0.75 V

Fe^{2+} (*aq*) + 2 e^- → Fe (*s*) E° = –0.43 V

A. Cr (*s*)
B. Fe^{2+} (*aq*)
C. Sn^{2+} (*aq*)
D. Ag (*s*)

68. Which is true for the redox reaction occurring in a spontaneous electrochemical cell?

Zn (*s*) + Cd^{2+} (*aq*) → Cd (*s*) + Zn^{2+} (*aq*)

A. Electrons flow from the Cd electrode to the Zn electrode
B. Anions in the salt bridge flow from the Cd half-cell to the Zn half-cell
C. Zn is reduced at the anode
D. Cd^{2+} is oxidized at the cathode

69. Based upon the reduction potential: Zn^{2+} + 2 e^- → Zn (*s*); *E*° = –0.76 V, does a reaction take place when Zinc (*s*) is added to aqueous HCl, under standard conditions?

A. Yes, because the reduction potential for H^+ is negative
B. Yes, because the reduction potential for H^+ is zero
C. No, because the oxidation potential for Cl^- is positive
D. No, because the reduction potential for Cl^- is negative

70. If 1 amp of current passes a cathode for 10 minutes, how much Zn (*s*) forms in the following reaction? (Use the molecular mass of Zn = 65 g/mole)

Zn^{2+} + 2 e^- → Zn (*s*)

A. 0.10 g
B. 10.0 g
C. 0.20 g
D. 2.20 g

71. In the oxidation–reduction reaction Mg (*s*) + Cu^{2+} (*aq*) → Mg^{2+} (*aq*) + Cu (*s*), which atom/ion is reduced, and which atom/ion is oxidized?

A. The Cu^{2+} ion is reduced (gains electrons) to form Cu metal, while Mg metal is oxidized (loses electrons) to form Mg^{2+}

B. Since Mg is transformed from a solid to an aqueous solution and Cu is transformed from an aqueous solution to a solid, no oxidation-reduction reaction occurs

C. The Cu^{2+} ion is oxidized (i.e., gains electrons) from Cu metal, while Mg metal is reduced (i.e., loses electrons) to form Mg^{2+}

D. The Mg^{2+} ion is reduced (i.e., gains electrons) from Cu metal, while Cu^{2+} is oxidized (i.e., loses electrons) to the Mg^{2+}

72. Electrolysis is an example of a(n):

A. acid-base reaction
B. exothermic reaction
C. physical change
D. chemical change

73. What is the term for a process characterized by the loss of electrons?

A. Redox
B. Reduction
C. Electrochemistry
D. Oxidation

74. Which of the following statements is true about a galvanic cell?

A. The standard reduction potential for the anode reaction is always positive
B. The standard reduction potential for the anode reaction is always negative
C. The standard reduction potential for the cathode reaction is always positive
D. E° for the cell is always positive

75. Which of the following statements is true?

A. Galvanic cells were invented by Thomas Edison
B. Galvanic cells generate electrical energy rather than consume it
C. Electrolysis cells generate alternating current when their terminals are reversed
D. Electrolysis was discovered by Lewis Latimer

76. Which statement is correct for an electrolysis cell that has two electrodes?

A. Reduction occurs at the anode, which is positively charged
B. Reduction occurs at the anode, which is negatively charged
C. Reduction occurs at the cathode, which is positively charged
D. Reduction occurs at the cathode, which is negatively charged

77. How long would it take to deposit 4.00 grams of Cu from a $CuSO_4$ solution if a current of 2.5 A is applied? (Use the molecular mass of Cu = 63.55 g/mol and the conversion of 1 mol e^- = 9.65×10^4 C and 1 C = A·s)

A. 1.02 hours
B. 1.36 hours
C. 4.46 hours
D. 6.38 hours

78. Which battery system is completely rechargeable?

A. Alkaline batteries
B. Fuel cells
C. Lead-acid storage batteries
D. Dry-cell batteries

79. In which type of cell does the following reaction occur when electrons are forced into a system by applying an external voltage?

$$Fe^{2+} + 2\ e^- \rightarrow Fe\ (s) \qquad E° = -0.44\ V$$

A. Electrolytic cell
B. Battery
C. Electrochemical cell
D. Galvanic cell

80. What is the term for a reaction that represents separate oxidation or reduction processes?

A. Reduction reaction
B. Redox reaction
C. Oxidation reaction
D. Half-reaction

Notes for active learning

Notes for active learning

Answer Keys and Detailed Explanations: Diagnostic Tests

Diagnostic Test 1 – Answer Key and Detailed Explanations

1	**B**	Electronic Structure & Periodic Table	**31**	**A**	Thermochemistry
2	**A**	Bonding	**32**	**B**	Kinetics Equilibrium
3	**D**	Phases & Phase Equilibria	**33**	**D**	Solution Chemistry
4	**D**	Stoichiometry	**34**	**C**	Acids & Bases
5	**A**	Thermochemistry	**35**	**D**	Electronic Structure & Periodic Table
6	**C**	Kinetics Equilibrium	**36**	**B**	Bonding
7	**D**	Solution Chemistry	**37**	**A**	Phases & Phase Equilibria
8	**D**	Acids & Bases	**38**	**C**	Stoichiometry
9	**D**	Electrochemistry	**39**	**C**	Thermochemistry
10	**B**	Electronic Structure & Periodic Table	**40**	**C**	Kinetics Equilibrium
11	**C**	Bonding	**41**	**B**	Solution Chemistry
12	**B**	Phases & Phase Equilibria	**42**	**D**	Acids & Bases
13	**B**	Stoichiometry	**43**	**A**	Electrochemistry
14	**D**	Thermochemistry	**44**	**D**	Electronic Structure & Periodic Table
15	**C**	Kinetics Equilibrium	**45**	**B**	Bonding
16	**A**	Solution Chemistry	**46**	**D**	Phases & Phase Equilibria
17	**D**	Acids & Bases	**47**	**A**	Stoichiometry
18	**B**	Electronic Structure & Periodic Table	**48**	**C**	Thermochemistry
19	**A**	Bonding	**49**	**A**	Kinetics Equilibrium
20	**C**	Phases & Phase Equilibria	**50**	**A**	Solution Chemistry
21	**B**	Stoichiometry	**51**	**B**	Acids & Bases
22	**A**	Thermochemistry	**52**	**B**	Electronic Structure & Periodic Table
23	**C**	Kinetics Equilibrium	**53**	**B**	Bonding
24	**D**	Solution Chemistry	**54**	**C**	Phases & Phase Equilibria
25	**B**	Acids & Bases	**55**	**D**	Stoichiometry
26	**A**	Electrochemistry	**56**	**B**	Thermochemistry
27	**D**	Electronic Structure & Periodic Table	**57**	**A**	Kinetics Equilibrium
28	**B**	Bonding	**58**	**C**	Solution Chemistry
29	**C**	Phases & Phase Equilibria	**59**	**D**	Acids & Bases
30	**A**	Stoichiometry	**60**	**D**	Electrochemistry

1. B is correct.

Cobalt's atomic number = 27

Atomic weight = 60

Number of protons = number of electrons = atomic number = 27

Atomic weight = atomic number + number of neutrons

Number of neutrons = atomic weight – atomic number

Number of neutrons: 60 – 27 = 33

2. A is correct.

Induced dipoles occur within nonpolar molecules. They are induced (created temporarily) by the difference in electronegativity of atoms. Because they are induced and not permanent dipoles, the attraction is very weak compared to other forms of attraction.

3. D is correct.

Temperature is the measure of the average kinetic energy of the molecules. The greater is the kinetic energy of the gas; the higher is the temperature.

An increase in kinetic energy results in a more rapid motion of the molecules, and this rapid motion results in the breaking of intermolecular bonds. The breaking of bonds allows molecules to escape from the solution, thus affecting the solubility of gas molecules.

4. D is correct.

Assume 100 g of a compound.

C: 49.3 g × (1 mole / 12 g) = 4.1 moles

O: 43.8 g × (1 mole / 16 g) = 2.7 moles

H: 6.9 g × (1 mole / 1 g) = 6.9 moles

Therefore, the compound is $C_{4.1}O_{2.7}H_{6.9}$

Divide each element by 2.7: $C_{1.5}O_1H_{2.5}$

Multiply by 2 to obtain whole numbers: $C_3O_2H_5$

The compound weighs 73 g/mole (3 carbons × 2 oxygens × 5 hydrogens), which is 3 times less than the molecular weight of the unknown compound (219 g/mol).

Therefore, multiply by 3 to obtain the molecular formula: $C_9O_6H_{15}$

5. A is correct.

ΔH_{fusion} = mass × heat of fusion

ΔH_{fusion} = 17.8 × 205 J/g

ΔH_{fusion} = 3,649 J

6. C is correct.

In exothermic reactions, heat released is a product of the forward reaction.

If the temperature of the system is raised, heat and the other product of the forward reaction (e.g., methanol) combine to make the reaction go in the reverse direction.

The perturbation of adding heat to this reaction causes the system, originally at equilibrium, to adjust to the perturbation, by consuming heat and shifting the equilibrium to the left.

7. D is correct.

First, determine the number of moles:

Moles of $CuSO_4$ = mass of $CuSO_4$ / molecular mass of $CuSO_4$

Moles of $CuSO_4$ = 0.88 g / [(63.55 g/mol) + (32.06 g/mol) + (4 × 16 g/mol)]

Moles of $CuSO_4$ = 0.88 g / (159.61 g/mol)

Moles of $CuSO_4$ = 5.5×10^{-3} moles

Convert volume to liters:

24 mL × 0.001 L/mL = 0.024 L

Divide moles by volume to calculate molarity:

Molarity = moles/volume

Molarity = 5.5×10^{-3} moles / 0.024 L

Molarity = 0.23 M

8. D is correct.

Those compounds will undergo a double-exchange reaction – anions and cations switching pairs. From there, it's about recognizing the ions that are insoluble.

$NiBr_2$ and $AgNO_3$ produce AgBr, which is insoluble in water.

9. D is correct.

Electrochemical reactions are also equilibrium reactions.

$Ni\ (s)\ |\ Ni^{2+}\ (1\ M)\ ||\ H^{+}\ (1\ M)\ |\ H_2\ (2\ atm)\ |\ Pt\ (s)$

The above cell notation translates into this equilibrium reaction:

$Ni\ (s) + H^{+}\ (1\ M) \leftrightarrow Ni^{2+}\ (1\ M) + H_2\ (2\ atm)$

Cell voltage indicates the tendency of a reaction to shift towards the product side. To decrease the cell voltage, the equilibrium shifts towards the reactants.

The pressure of H_2 is proportional to the amount of H_2 present. If the pressure is increased, the equilibrium shifts to the left, which decreases the cell voltage.

10. B is correct.

The nucleus, the dense central core of an atom, contains most of the atom's mass. It consists of protons and neutrons, each having a mass of approximately 1 amu.

A cloud of orbiting electrons surrounds the nucleus, and each electron has a mass of approximately 1/1,837 amu. The electrons contribute to only a tiny fraction of the mass of an atom.

11. C is correct.

Phosphorous (like nitrogen) has 5 valence electrons. Three substituents on a central atom that has an unbonded pair of electrons (e.g., nitrogen in ammonia) indicate a trigonal pyramidal geometry.

12. B is correct.

Boyle's Law:

$(P_1V_1) = (P_2V_2)$

Solve for the final pressure:

$P_2 = (P_1V_1) / V_2$

$P_2 = [(260\ torr) \times (10.00\ L\ O_2)] / (16.0\ L\ O_2)$

$P_2 = 163\ torr$

Notice that pressure units are left in torr.

13. B is correct.

Calculate the moles of each gas in the mixture:

Moles of CH_4:

moles CH_4 = mass of CH_4 / molecular mass of CH_4

moles CH_4 = 32.0 g / [(12 g/mol) + (4 × 1 g/mol)]

moles CH_4 = 2 mol

Moles of O_2:

moles O_2 = mass of O_2 / molecular mass of O_2

moles O_2 = 24.0 g / (2 × 16 g/mol)

moles O_2 = 0.75 mol

Moles of SO_2:

moles SO_2 = mass of SO_2 / molecular mass of SO_2

moles SO_2 = 32.0 g / [32 g/mol + (2 × 16 g/mol)]

moles SO_2 = 0.50 mol

Moles of CO_2:

moles CO_2 = mass of CO_2 / molecular mass of CO_2

moles CO_2 = 44.0 g / [(12 g/mol) + (2 × 16 g/mol)]

moles CO_2 = 1.0 mol

Total moles:

2 mol CH_4 + 0.75 mol O_2 + 0.50 mol SO_2 + 1.0 mol CO_2 = 4.25 mol

14. D is correct.

Heat = mass × specific heat × change in temperature:

$q = m \times c \times \Delta T$

Rearrange to solve for mass:

$m = q / (c \times \Delta T)$

m = 9.22 J / [0.90 J/g·°C × (28.5 °C – 21.2 °C)]

m = 9.22 J / [0.90 J/g·°C × (7.3 °C)]

m = 1.4 g

15. C is correct.

General formula for the equilibrium constant:

aA + bB ↔ cC + dD

$K_{eq} = ([C]^c \times [D]^d) / ([A]^a \times [B]^b)$

For equilibrium constant calculation, only include species in aqueous or gas phases (all in this problem).

$K_{eq} = [CH_4] \times [H_2S]^2 / [CS_2] \times [H_2]^4$

16. A is correct.

A saturated solution is a solution in which the maximum amount of solvent has been dissolved. Any more solute added will precipitate as crystals on the bottom of the beaker.

17. D is correct.

Basic solutions have the hydroxide [$^-$OH] > 10^{-7}, which means the acid [H_3O^+] < 10^{-7}

18. B is correct.

Metalloids are semimetallic elements (between metals and nonmetals). The metalloids are boron (B), silicon (Si), germanium (Ge), arsenic (As), antimony (Sb) and tellurium (Te). Some literature reports polonium (Po) and astatine (At) as metalloids.

They have properties between metals and nonmetals. They typically have a metallic appearance but are only fair conductors of electricity (as opposed to metals which are excellent conductors), which makes them useable in the semiconductor industry.

Metalloids tend to be brittle, and chemically they behave more like nonmetals.

19. A is correct.

Hydrogens, bonded directly to F, O or N, participate in hydrogen bonds. The hydrogen is partially positive (i.e., delta plus) due to the bond to these electronegative atoms.

The lone pair of electrons on the F, O or N interacts with the partial positive (∂+) hydrogen to form a hydrogen bond.

20. C is correct.

As the volume of the container is decreased, the molecules approach each other more closely, which leads to an increase in the number and strength of intermolecular attractions among the molecules.

As these attractions increase, the number and intensity of collisions between the molecules and the wall decrease leading to a decrease in the pressure exerted on the walls of the container by the gas.

21. B is correct.

Balanced equation (double replacement):

$$CH_4\,(g) + 2\,O_2\,(g) \xrightarrow{\text{Spark}} CO_2\,(g) + 2\,H_2O\,(g)$$

22. A is correct.

In a chemical reaction, bonds within reactants are broken and new bonds form to yield products.

For bond dissociation:

$$\Delta H_{reaction} = \text{(sum of bond energy in reactants)} - \text{(sum of bond energy in products)}$$

This is the opposite of ΔH_f problems, where:

$\Delta H_{reaction}$ = [sum of ΔH_f product – sum of ΔH_f reactant]

For $N_2 + O_2 \rightarrow 2NO$:

$\Delta H_{reaction}$ = sum of bond energy in reactants – sum of bond energy in products

$\Delta H_{reaction} = [(N{\equiv}N) + (O{=}O)] - 2(N{-}O)$

$\Delta H_{reaction} = (226\text{ kcal} + 199\text{ kcal}) - (2 \times 145\text{ kcal})$

$\Delta H_{reaction} = 135\text{ kcal}$

23. C is correct.

Increasing the concentration of reactants (or increasing pressure) increases the rate of a reaction because there is an increased probability that any two reactant molecules collide with sufficient energy to overcome the energy of activation and form products.

24. D is correct.

Solutions can be gases, liquids or solids. All three statements are possible types of solutions.

An example of a gas-in-liquid solution (statement III) is a hydrochloride/HCl.

In its pure form, HCl is a gas and is passed through water to create an aqueous solution of HCl.

25. B is correct.

Methyl red has a pK_a of 5.1 and is a pH indicator dye that changes color in acidic solutions: turns red in pH under 4.4, orange in pH 4.4-6.2 and yellow in pH over 6.2.

I: phenolphthalein is used as an indicator for acid-base titrations. It is a weak acid, which can dissociate protons (H^+ ions) in solution. The phenolphthalein molecule is colorless, and the phenolphthalein ion is pink. It turns colorless in acidic solutions and pink in basic solutions. With basic conditions, the phenolphthalein (neutral) $\rightleftharpoons$ ions (pink) equilibrium shifts to the right, leading to more ionization as H^+ ions are removed.

III: bromothymol blue is a pH indicator that is often used for solutions with neutral pH near 7 (e.g., managing the pH of pools and fish tanks). Bromothymol blue acts as a weak acid in solution that can be protonated or deprotonated. It appears yellow when protonated (lower pH), blue when deprotonated (higher pH) and bluish green in neutral solution.

26. A is correct.

Determine the oxidation number of each species.

The oxidized substance has an increased oxidation number.

In $Ni(OH)_2$, the oxidation number of Ni = +2

In NiO_2, the oxidation number of Ni = +4

27. D is correct.

Electron shells represent the orbit that electrons follow around an atom's nucleus.

Each shell is composed of one or more subshells, which are named using lowercase letters (*s*, *p*, *d*, *f*). The first shell, n = 1, has one subshell (1*s*), the second shell, n = 2, has two subshells (2s and 2*p*), the third shell, n = 3, has three subshells (3*s*, 3*p* and 3*d*), etc.

28. B is correct.

Covalent bonds involve the sharing of electrons. When the atoms involved in a covalent bond are of the same, or about the same, electronegativity (difference of 0 to 0.4 Pauling units), the electrons are shared equally, resulting in a nonpolar covalent bond.

When the atoms are of different electronegativity (difference of 0.4 to 1.7 Pauling units), the electrons are shared unequally, resulting in a polar covalent bond.

When the difference in electronegativity is greater than 1.7 Pauling units, the compound forms ionic bonds. Ionic bonds involve the transfer of an electron from the electropositive element (along the left-hand column/group) to the electronegative element (along the right-hand column/groups) on the periodic table.

The electropositive atom loses an electron(s) to become a cation (with a complete octet), and the electronegative atom gains an electron(s) to become an anion (with a complete octet).

Covalent bonds form between two non-metal atoms. Find the compound that doesn't have any metal atoms, which is HF. The other elements form ionic bonds due to the gain/loss of electrons to obtain a complete octet. Metals lose electrons, while nonmetals gain electrons.

29. C is correct.

Solids have a definite shape and volume. A solid granite does not change its shape or its volume regardless of the container in which it is placed.

Molecules in a solid are very tightly packed due to the strong intermolecular attractions, which prevents the molecules from moving around.

30. A is correct.

Calculate the molar mass of phosphorous trichloride:

Molar mass of PCl_3 = atomic mass of P + (3 × atomic mass of Cl)

Molar mass of PCl_3 = 30.97 g/mole + (3 × 35.45 g/mole)

Molar mass of PCl_3 = 137.32 g/mole

Then, use this information to calculate the number of moles:

Moles of PCl_3 produced = mass of PCl_3 / molar mass of PCl_3

Moles of PCl_3 = 412 g / 137.32 g/mole

Moles of PCl_3 = 3 moles

The reaction equation is:

$P_4 + 6\ Cl_2 \rightarrow PCl_3$

Always confirm that a reaction equation is balanced before doing mole comparisons. Presence of coefficient(s) does not always indicate a fully-balanced equation; this reaction is one of those.

To balance the equation, assign 4 as the coefficient of PCl_3:

$P_4 + 6\ Cl_2 \rightarrow 4\ PCl_3$

With a balanced equation, moles of P_4 is determined by comparison of the coefficients:

Moles P_4 = (coefficient P_4 / coefficient PCl_3) × moles PCl_3

Moles P_4 = (1 / 4) × 3 moles

Moles P_4 = 0.75 moles

From the balanced reaction, a ½ mole of P_4 (*s*) is needed.

Mass of P_4 = (moles of P_4) × (molar mass of P_4)

Mass of P_4 = 0.75 moles × (4 × 30.97 g/mole)

Mass of P_4 = 93 g

31. A is correct.

Balanced decomposition reaction:

$2\ LiHCO_3\ (s) \rightarrow Li_2CO_3 + H_2O + CO_2$

Decomposition to form water and carbon dioxide is a reaction characteristic of bicarbonate (HCO_3^-).

32. B is correct.

The main function of catalysts is lowering the reaction's activation energy (energy barrier), thus increasing its rate (*k*).

Although they do decrease the amount of energy required to reach the rate-limiting transition state, they do *not* decrease the relative energy of the products and reactants.

Therefore, a catalyst does not affect ΔG.

33. D is correct.

The addition of a solute always elevates the boiling point temperature of a liquid.

Since NaCl dissociates into 2 ions in solution, i = 2:

$\Delta T_b = k_b mi$

ΔT_b = (0.512 °C/m)·(3 m)·(2)

ΔT_b = 3.07 °C

Thus, ΔT_b^1 = (100 °C) + 3.07 °C = 103.07 °C

34. C is correct.

Balanced reaction: 2 $HC_2H_3O_2\,(aq)$ + $Ca(OH)_2\,(aq)$ → $Ca(C_2H_3O_2)_2$ + 2 H_2O

With a neutralization reaction, the cations and anions on both reactants switch pairs, resulting in salt and water.

35. D is correct.

Look at the periodic table and learn the names of major sub-sections (e.g., metals, nonmetals, transition metals, lanthanides, and actinides).

36. B is correct.

Elements are classified based on valence electrons (i.e., electrons in their outer shell).

All elements in the same group (i.e., a vertical column on the periodic table) have the same number of valence electrons, while all elements in the same period (i.e., horizontal rows) have the same number of electron shells.

37. A is correct.

Start by applying the ideal-gas equation: PV = nRT

Convert temperature from °C to Kelvin:

$$(23.5\ ^\circ C + 273) = 296.5\ K$$

Set the initial and final P/V/T conditions equal:

$$(P_1V_1 / T_1) = (P_2V_2 / T_2)$$

Solve for the volume at STP (STP conditions: 0 °C or 273 K and 1 atm or 760 torr):

$$V_2 = (P_1V_1T_2) / (P_2T_1)$$

$$V_2 = [(273\ K) \times (578\ torr) \times (1.8\ L)] / [(760\ torr) \times (296.5\ K)]$$

$$V_2 = 1.26\ L$$

38. C is correct.

Balanced reaction:

$$5\ N_2O_4\,(l) + 4\ N_2H_3(CH_3)\ (l) \rightarrow 12\ H_2O\ (g) + 9\ N_2\,(g) + 4\ CO_2\,(g)$$

If 26 moles of water are produced, calculate how many moles of N_2O_4 are required.

The problem did not specify the amount of the other reagent of $N_2H_3(CH_3)$, so that reagent cannot be used in the calculation.

Use the coefficients to calculate moles:

$$N_2O_4 = (5 / 12) \times 26\ moles = 10.83\ moles$$

Because the amount of N_2O_4 used is less than the starting amount, N_2O_4 is not the limiting reagent.

Therefore, the limiting reagent is the other reagent $N_2H_3(CH_3)$.

39. C is correct.

Bond dissociation energy:

ΔH_{rxn} = sum of energy for bonds broken – sum of energy for bonds formed

$\Delta H_{rxn} = [2(495 \text{ kJ/mole}) + 4(410 \text{ kJ/mole})] – [2(720 \text{ kJ/mole}) + 4(460 \text{ kJ/mole})]$

$= [990 \text{ kJ/mole} + 1{,}640 \text{ kJ/mole}] – [1{,}440 \text{ kJ/mole} + 1{,}840 \text{ kJ/mole}]$

$\Delta H_{rxn} = –650 \text{ kJ/mole}$

A negative ΔH_{rxn} indicates that the reaction is exothermic. The reaction releases energy with the products more stable than the reactants.

40. C is correct.

Equilibrium systems always work to *minimize* the impact of changes to the conditions of the reaction.

If the pressure is increased, the volume of the system decreases.

To minimize the impact of a perturbation of pressure, a reaction shifts towards the side with fewer molecules (or the side with the smaller sum of coefficients).

41. B is correct.

The solvent is the component with the highest % volume composition.

42. D is correct.

The ionic product constant for water K_w is equal to 1×10^{-14}:

$K_w = [H_3O^+] \times [^-OH]$

$1 \times 10^{-14} = [H_3O^+] \times [^-OH]$

Rearrange to solve for $[OH^-]$:

$[^-OH] = [1 \times 10^{-14} \text{ M}] / [H_3O^+]$

$[^-OH] = [1 \times 10^{-14} \text{ M}] / [4.5 \times 10^{-3} \text{ M}]$

$[^-OH] = 2.22 \times 10^{-12} \text{ M}$

43. A is correct.

Half-reaction:

$Cl_2O_7 \rightarrow HClO$

Balancing half-reaction in acidic conditions:

Step 1: Balance all atoms except for H and O

$Cl_2O_7 \rightarrow 2\ HClO$

Step 2: To balance oxygen, add H_2O to the side with fewer oxygen atoms

$Cl_2O_7 \rightarrow 2\ HClO + 5\ H_2O$

Step 3: To balance hydrogen, add H^+ to the opposing side of H_2O added in the previous step

$Cl_2O_7 + 12\ H \rightarrow 2\ HClO + 5\ H_2O$

Step 4: Balance charges by adding electrons to the side with a greater/more positive total charge

Total charge on the left side: 12(+1) = +12

Total charge on the right side: 0

Add 12 electrons to the left side:

$Cl_2O_7 + 12\ H^+ + 12\ e^- \rightarrow 2\ HClO + 5\ H_2O$

44. D is correct.

In the ground state, the 2*s* subshell should be filled and then followed by the 2*p* subshell for the configuration of a neutral atom. $1s^22s^23s^1$ represents the excited state for boron.

45. B is correct.

Electronegativity increase towards the right along a period and up a group on the periodic table.

Fluorine (Pauling unit of 4.0) is the most electronegative element while Fr and Cs (similar Pauling units of 0.7 and 0.79, respectively) are the least electronegative elements on the periodic table.

46. D is correct.

Charles's law (i.e., the law of volumes) describes how a gas behaves when it is heated.

It states that volume is directly proportional to temperature.

$V_1 / T_1 = V_2 / T_2$

or

$V_1T_2 = V_2T_1$

47. A is correct.

48. C is correct.

To determine enthalpy:

ΔH = [reactants: energy required to break bonds] – [products: energy of bonds formed]

$\Delta H = [(H–H) + (Cl–Cl)] – [2 \times (H–Cl)]$

$\Delta H = = (436\ kJ + 243\ kJ) – (2 \times 431\ kJ)$

$\Delta H = –183\ kJ$

The negative number indicates that the reaction is exothermic and the energy released is greater than 75 kJ.

49. A is correct.

K_{eq}, or equilibrium constant, approximates the ratio of products and reactants in a reaction.

$K_{eq} > 1$: mostly products are present

$K_{eq} < 1$: mostly reactants are present

50. A is correct.

Start by calculating the moles of NaCl required:

moles NaCl = molarity × volume

moles NaCl = 4 M × (100 mL × 0.001 L/mL)

moles NaCl = 0.4 mol

Then, use this information to calculate the mass:

mass NaCl = moles NaCl × molar mass of NaCl

mass NaCl = 0.4 mol × (23.00 g/mol + 35.45 g/mol)

mass NaCl = 23.4 g

51. B is correct.

Only the extent of dissociation of the dissolved acid or base determines its classification as weak or strong. Strong acids and bases completely dissociate into their constituent ions, while weak acids and bases only partially dissociate.

52. B is correct.

Average atomic mass = sum of (relative abundance × atomic mass)

Average atomic mass = (22% × 16.0 amu) + (44% × 17.0 amu) + (34% × 18.0 amu)

Average atomic mass = (3.52 amu) + (7.48 amu) + (6.12 amu)

Average atomic mass = 17.1 amu

53. B is correct.

Ice contains 4 hydrogen bonds. Liquid water contains between 1 (at temperatures close to boiling) and 3 (at temperatures close to freezing) hydrogen bonds.

Steam has 0 hydrogen bonds between the gaseous water molecules.

54. C is correct.

Hydrocarbons are nonpolar molecules, which means that the dominant intermolecular force is London dispersion. This force gets stronger as the number of atoms in each molecule increases. Stronger force results in a higher boiling point.

Branching decreases the surface area of contact and decreases the boiling point.

55. D is correct.

The answer choices differ only in the coefficients. Analyze each and find the balanced equation, where each atom is present in equal amounts on both sides of the reaction.

56. B is correct.

Calculation of standard free-energy (ΔG) uses the same method as the heat of formation (ΔH_f) problems:

$$\Delta G = (\text{sum of } \Delta G \text{ products}) - (\text{sum of } \Delta G \text{ reactants})$$

From the Gibbs free energy expression:

$$\Delta G^\circ = \Delta H^\circ - T\Delta S^\circ$$

In this problem, the temperature is not provided.

However, the presence of degree symbols (ΔG°, ΔH° and ΔS°) indicates standard values (measured at 25 °C or 298.15 K).

Check the units of ΔH° and ΔS° before calculating; one may be in kilojoules and the other in joules.

$$H_2 + I_2 \rightarrow 2\ HI$$

H_2: $\Delta G^\circ = \Delta H^\circ - T\Delta S^\circ$

$$\Delta G^\circ = 0 - (298.15\ K \times 130.6\ J/mol\ K)$$

$$\Delta G^\circ = -38.9\ kJ/mol$$

I_2: $\Delta G^\circ = \Delta H^\circ - T\Delta S^\circ$

$$\Delta G^\circ = 0 - (298.15\ K \times 116.12\ J/mol\ K)$$

$$\Delta G^\circ = -34.6\ kJ/mol$$

HI: $\Delta G^\circ = \Delta H^\circ - T\Delta S^\circ$

$$\Delta G^\circ = 26{,}000\ J/mol - (298.15\ K \times 206\ J/mol\ K)$$

$$\Delta G^\circ = -35.4\ kJ$$

ΔG° for total reaction:

$$\Delta G^\circ_{reaction} = \Delta G^\circ \text{ products} - \Delta G^\circ \text{ reactants}$$

$$\Delta G^\circ_{reaction} = (2 \times \Delta G^\circ\ HI) - (\Delta G^\circ\ H_2 + \Delta G^\circ\ I_2)$$

$$\Delta G^\circ_{reaction} = (2 \times -35.4\ kJ) - (-34.6\ kJ - 38.9\ kJ)$$

$$\Delta G^\circ_{reaction} = 2.7\ kJ$$

57. A is correct.

Chemical equilibrium refers to a dynamic process whereby the *rate* at which a reactant molecule is being transformed into a product. This is the same *rate* for a product molecule to be transformed into a reactant.

58. C is correct.

In coagulation, heating or adding an electrolyte to a colloid causes the particles to bind together. This increases the colloid particles' mass, and they precipitate due to gravity.

59. D is correct.

The Brønsted-Lowry acid-base theory focuses on the ability to accept and donate protons (H^+).

A Brønsted-Lowry acid is a term for a substance that donates a proton (H^+) in an acid-base reaction.

A Brønsted-Lowry base is a term for a substance that accepts a proton.

HCN dissociates a proton (H^+), and this proton is abstracted by HS to form H_2S.

60. D is correct.

$$Na \rightarrow Na^+$$

Balance the charges by adding an electron:

$$Na \rightarrow Na^+ + e^-$$

According to the balanced reaction, Na lost 1 electron to form the Na^+ ion.

Notes for active learning

Diagnostic Test 2 – Answer Key and Detailed Explanations

1	**D**	Electronic Structure & Periodic Table	**31**	**A**	Thermochemistry
2	**C**	Bonding	**32**	**A**	Kinetics Equilibrium
3	**A**	Phases & Phase Equilibria	**33**	**B**	Solution Chemistry
4	**B**	Stoichiometry	**34**	**D**	Acids & Bases
5	**D**	Thermochemistry	**35**	**D**	Electronic Structure & Periodic Table
6	**B**	Kinetics Equilibrium	**36**	**D**	Bonding
7	**C**	Solution Chemistry	**37**	**C**	Phases & Phase Equilibria
8	**A**	Acids & Bases	**38**	**A**	Stoichiometry
9	**C**	Electrochemistry	**39**	**B**	Thermochemistry
10	**B**	Electronic Structure & Periodic Table	**40**	**A**	Kinetics Equilibrium
11	**B**	Bonding	**41**	**D**	Solution Chemistry
12	**C**	Phases & Phase Equilibria	**42**	**C**	Acids & Bases
13	**B**	Stoichiometry	**43**	**B**	Electrochemistry
14	**B**	Thermochemistry	**44**	**D**	Electronic Structure & Periodic Table
15	**B**	Kinetics Equilibrium	**45**	**D**	Bonding
16	**A**	Solution Chemistry	**46**	**C**	Phases & Phase Equilibria
17	**A**	Acids & Bases	**47**	**A**	Stoichiometry
18	**D**	Electronic Structure & Periodic Table	**48**	**D**	Thermochemistry
19	**C**	Bonding	**49**	**C**	Kinetics Equilibrium
20	**A**	Phases & Phase Equilibria	**50**	**C**	Solution Chemistry
21	**D**	Stoichiometry	**51**	**B**	Acids & Bases
22	**A**	Thermochemistry	**52**	**C**	Electronic Structure & Periodic Table
23	**C**	Kinetics Equilibrium	**53**	**A**	Bonding
24	**D**	Solution Chemistry	**54**	**D**	Phases & Phase Equilibria
25	**B**	Acids & Bases	**55**	**B**	Stoichiometry
26	**A**	Electrochemistry	**56**	**A**	Thermochemistry
27	**A**	Electronic Structure & Periodic Table	**57**	**D**	Kinetics Equilibrium
28	**C**	Bonding	**58**	**B**	Solution Chemistry
29	**B**	Phases & Phase Equilibria	**59**	**C**	Acids & Bases
30	**D**	Stoichiometry	**60**	**B**	Electrochemistry

1. D is correct.

Metals are excellent conductors of both heat and electricity because the molecules in metal are very closely packed together (i.e., have a high density).

Metals are generally in a solid state at room temperature.

Metals are very malleable, meaning that they can be pressed or hammered into different shapes without breaking.

2. C is correct.

The total number of valence electrons in a molecule = sum of valence electrons in each atom.

Check the periodic table to determine valence electrons of each element:

S = 6 electrons; O = 6 electrons; F = 7 electrons

SOF_2: 6 e^- + 6 e^- + (2 × 7 e^-) = 26 electrons

3. A is correct.

Deposition is the thermodynamic process where gas transitions directly into a solid, without first becoming a liquid.

An example of deposition occurs in sub-freezing air when water vapor changes directly into ice without becoming a liquid.

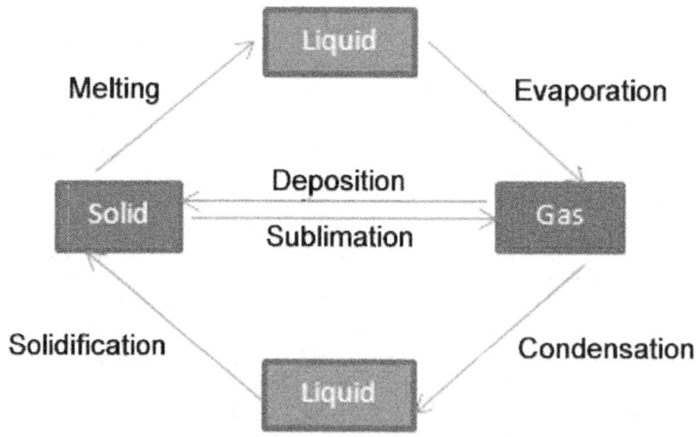

Interconversion of states of matter

4. B is correct.

A spectator ion is an ion that exists as both a reactant and a product in a chemical equation.

5. D is correct.

Heat = mass × specific heat × change in temperature.

$$q = m \times c \times \Delta T$$

Rearrange the equation to solve for mass:

$$m = q / (c \times \Delta T)$$

$$m = 488\ J / [0.130\ J/g{\cdot}°C \times (34.4\ °C - 21.8\ °C)]$$

$$m = 488\ J / (0.130\ J/g{\cdot}°C \times 12.6\ °C)$$

$$m = 298\ g$$

6. B is correct.

This is a propagation reaction because the free radical (i.e., unpaired electron) has been transferred.

Initiation: a neutral molecule undergoes homolytic cleavage to generate radicals.

Propagation: a radical species exists on both sides of the reaction.

Termination: radicals combine to generate a neutral molecule.

7. C is correct.

Most ionic compounds, or compounds with a metal cation and nonmetal anion, produce ions when dissolved in water.

Here are some types of ionic compounds that dissolve easily in water:

- most group IA compounds (e.g., sodium, potassium, etc.)
- nitrate (NO_3) compounds

However, organic compounds (mainly consisting of C, H, O) generally do not produce ions when dissolved in water.

Organic compound formaldehyde (CH_2O) is an aldehyde and does not dissociate into ions.

The resulting anion would have a high pK_a because it would be too unstable.

8. A is correct.

Balanced reaction:

$$H_2SO_4\ (aq) + 2\ NaOH\ (aq) \rightarrow Na_2SO_4\ (aq) + 2\ H_2O\ (l)$$

In a neutralization reaction, cations and anions from acid and base exchange partners, which results in the formation of salt and water.

9. C is correct.

In standard notation for electrolysis reactions, the oxidation reaction is written first, and the reduction reaction second.

Zinc is undergoing oxidation, so it is on the left side. Therefore, Ag is on the right side.

It is important to note the format of the reactions provided. Both reactions provided are in the form of oxidation reactions. Therefore, for the reduction of (Ag), reverse the reaction from Ag (*s*)|Ag^+ (*aq*) to Ag^+ (*aq*)|Ag (*s*).

10. B is correct.

The n = 3 shell contains subshells 3*s*, 3*p* and 3*d*.

An *s* shell can contain 2 electrons, a *p* shell can contain up to 6 electrons, and a *d* shell can contain up to 10 electrons. 2 + 6 + 10 = 18, so the maximum number of electrons in the n = 3 shell is 18.

11. B is correct.

Carbon is the central atom and is bonded to three substituents (i.e., two chlorine atoms and a double bond to oxygen).

The hybridization of carbon is sp^2 with a bond angle of 120°. Atoms that are sp^2 hybridized have a trigonal planar geometry.

12. C is correct.

Gases do *not* have definite volumes and shapes; rather they take up the volume and shape of the container that they are in. This is different from liquids (which have a definite volume but take up the shape of the container that they are in) and solids (which have both a definite volume and a definite shape).

13. B is correct.

An electrolyte is a substance that produces an electrically conducting solution when dissolved in a polar solvent (e.g., water).

The dissolved electrolyte separates into positively-charged cations and negatively-charged anions.

Strong electrolytes dissociate completely (or nearly completely) because the resulting ions are stable in the solution.

14. B is correct.

Enthalpy change is defined by the following equation: $\Delta H = H_f - H_i$

If the standard enthalpy of the products is less than the standard enthalpy of the reactants, the standard enthalpy of reaction is negative, and the reaction is exothermic (i.e., releases heat).

If the standard enthalpy of the products is more than the standard enthalpy of the reactants, the standard enthalpy of reaction is positive, and the reaction is endothermic (i.e., absorbs heat).

In contrast to exothermic reactions where heat is a product, endothermic reactions absorb heat (as a reactant) from the surroundings.

15. B is correct.

Increasing the temperature of a reaction increases the reaction rate.

An increase in the rate constant equals a faster reaction rate.

16. A is correct.

Find concentrations of ions of $Mg(NO_3)_2$ and K_2SO_4 upon complete dissociation:

$Mg(NO_3)_2$	$\rightarrow$	Mg^{2+}	+	$2\ NO_3^-$
0.04 mole		0.04 mole		0.08 mole

1 liter $Mg(NO_3)_2 \times$ (0.04 mole/liter) = 0.04 moles $Mg(NO_3)_2$

1 liter $K_2SO_4 \times$ (0.08 mole/liter) = 0.24 moles K_2SO_4

Calculate the concentration of Mg^{2+} in the mixture:

$[Mg^{2+}]$ = (0.04 mole / 4 liters total volume) = 0.01 M

Calculate the concentration SO_4^{2+}:

$[SO_4^{2-}]$ = (0.24 mole / 4 liters total volume) = 0.06 M

The ion product is Q:

$Q = [Mg^{2+}]\cdot[SO_4^{2-}]$

$Q = [0.01\ M]\cdot[0.06\ M]$

$Q = [1 \times 10^{-2}\ M]\cdot[6 \times 10^{-2}\ M]$

$Q = 6 \times 10^{-4}\ M$

Because $Q > K_{sp}$ ($6 \times 10^{-4} > 4 \times 10^{-5}$), more solid forms to decrease the ion product to the K_{sp} value.

Therefore, precipitation occurs.

17. A is correct.

Strong bases are unstable anions that become more stable by abstracting a proton (Arrhenius or Brønsted-Lowry definitions) or by sharing electrons (Lewis definition).

The hydroxides of Group I and II metals are considered strong bases. Examples include LiOH (lithium hydroxide), NaOH (sodium hydroxide), KOH (potassium hydroxide), RbOH (rubidium hydroxide), CsOH (cesium hydroxide), $Ca(OH)_2$, calcium hydroxide, $Sr(OH)_2$ strontium hydroxide and $Ba(OH)_2$ barium hydroxide.

18. D is correct.

Isotopes are variants of a particular element, which differ in the number of neutrons. All isotopes of the element have the same number of protons and occupy the same position on the periodic table.

The number of protons within the atom's nucleus is the atomic number (Z) and is equal to the number of electrons in the neutral (non-ionized) atom.

Each atomic number identifies a specific element, but not the isotope; an atom of a given element may have a wide range in its number of neutrons.

The number of both protons and neutrons (i.e., nucleons) in the nucleus is the atom's mass number (A), and each isotope of an element has a different mass number.

19. C is correct.

HCN is polar due to the electronegativity of the nitrogen atom.

Nitrogen is partial negative, while hydrogen is partial positive.

20. A is correct.

Gay-Lussac's Law:

$$(P_1 / T_1) = (P_2 / T_2)$$

or

$$(P_1T_2) = (P_2T_1)$$

According to Gay-Lussac Law:

$$P / T = \text{constant}$$

If the pressure of a gas sample is doubled, the temperature is doubled as well.

21. D is correct.

Balanced equation (synthesis):

$$4\ P\ (s) + 5\ O_2\ (g) \rightarrow 2\ P_2O_5\ (s)$$

22. A is correct.

The total enthalpy (i.e., internal energy) of a system cannot be measured directly; the *enthalpy change* of a system is measured instead.

Enthalpy change is defined by the following equation: $\Delta H = H_f - H_i$

If the standard enthalpy of the products is less than the standard enthalpy of the reactants, the standard enthalpy of reaction is negative, and the reaction is exothermic (i.e., releases heat).

If the standard enthalpy of the products is greater than the standard enthalpy of the reactants, the standard enthalpy of reaction is positive, and the reaction is endothermic (i.e., absorbs heat).

23. C is correct.

If the temperature increases, the reaction shifts toward the endothermic side. Because the reaction is endothermic, the reaction shifts towards the product.

For exothermic reactions, when the temperature is increased, it shifts towards the reactants.

24. D is correct.

Calculate the concentration of each option.

Because all solutions have the same solute and solvent, concentration expressed in g/mL can be used to compare concentration.

A: 2.8 g / 2 mL = 1.4 g/mL

B: 2.8 g / 5 mL = 0.56 g/mL

C: 25 g / 50 mL = 0.5 g/mL

D: 40 g / 160 mL = 0.25 g/mL

25. B is correct.

$pH = -\log[H^+]$

$pH = -\log(1.4 \times 10^{-3})$

$pH = 2.86$

26. A is correct.

Half-reaction:

$NO_3^- \rightarrow NH_4^+$

Balancing half-reaction in acidic conditions:

Step 1: Balance all atoms except for H and O

$NO_3^- \rightarrow NH_4^+$ (N is already balanced)

Step 2: To balance oxygen, add H_2O to the side with fewer oxygen atoms

$NO_3^- \rightarrow NH_4^+ + 3\ H_2O$

Step 3: To balance hydrogen, add H^+ to the opposing side of H_2O added in the previous step

$NO_3^- + 10\ H^+ \rightarrow NH_4^+ + 3\ H_2O$

Step 4: Balance charges by adding electrons to the side with a greater/more positive total charge

Total charge on the left side: 10(+1) – 1 = +9

Total charge on the right side: +1

Add 8 electrons to the left side: $NO_3^- + 10\ H^+ + 8\ e^- \rightarrow NH_4^+ + 3\ H_2O$

27. A is correct.

Metalloids are semimetallic elements (i.e. between metals and nonmetals). The metalloids are boron (B), silicon (Si), germanium (Ge), arsenic (As), antimony (Sb) and tellurium (Te). Some literature reports polonium (Po) and astatine (At) as metalloids.

They have properties between metals and nonmetals. They typically have a metallic appearance but are only fair conductors of electricity (as opposed to metals which are excellent conductors), which makes them useable in the semiconductor industry.

Metalloids tend to be brittle, and chemically they behave more like nonmetals.

28. C is correct.

Hydrogen bonds (H is bonded directly to fluorine, oxygen or a nitrogen atom) are the strongest intermolecular forces (between molecules), followed by dipole-dipole and then van der Waals forces (i.e., London dispersion).

Hydrogens, bonded directly to N, F and O, participate in hydrogen bonds. The hydrogen is partially positive (i.e., delta plus) due to the bond to these (F, O, N) electronegative atoms. The lone pair of electrons on the F, O or N interacts with the partially positive hydrogen to form a hydrogen bond.

Polar molecules have stronger dipole-dipole force between molecules because the partially positive end of one molecule bonds with the partially negative end of another molecule.

Nonpolar molecules have induced dipole-induced dipole intermolecular force, which is weak compared to dipole-dipole force. The strength of induced dipole-induced dipole interaction correlates with the number of electrons and protons of an atom.

29. B is correct.

Vapor pressure is inversely proportional to the strength of the intermolecular force.

With stronger intermolecular forces, the molecules are more likely to stick together in the liquid form, and less of them participate in the liquid-vapor equilibrium.

30. D is correct.

Assume that H and O are in their most common oxidation states, which is +1 and –2, respectively.

Because the sum of oxidation numbers in a neutral molecule is zero:

$$0 = \text{oxidation state of H} + \text{oxidation state of Cl} + (4 \times \text{oxidation state of O})$$

$$0 = 1 + \text{oxidation state of Cl} + 4(-2)$$

$$\text{oxidation state of Cl} = +7$$

31. A is correct.

The overall reaction is the sum of the two individual reactions.

Enthalpy change is defined by the following equation: $\Delta H = H_f - H_i$

$$\Delta H_{rxn}: (-806 \text{ kJ/mlol}) + (-86 \text{ kJ/mol}) = -892 \text{ kJ/mol}$$

32. A is correct.

Most nitrate $(NO_3)_2$ ions are soluble in aqueous solutions.

Therefore, the addition of $Pb(NO_3)_2$ increases the concentration of Pb^{2+} (product), which shifts the equilibrium towards reactants.

33. B is correct.

Molarity = moles of solute / liter of solution

To obtain molarity: start with the chemical formula. Using the formula, the molecular weight of the molecule can be determined using the periodic table. Once the molecular weight is determined, the mass of the solute is required to calculate the number of moles. Finally, divide moles by the volume of solution to obtain molarity.

34. D is correct.

Acids are known to have a sour taste (e.g., lemon juice) because the sour taste receptors on the tongue detect the dissolved hydrogen (H^+) ions.

A pH greater than 7 and feels slippery are all qualities of bases, not acids. An acid is a chemical substance with a pH less than 7 which produces H^+ ions in water. An acid can be neutralized by a base (i.e., a substance with a pH above 7) to form a salt.

Litmus paper is red under acidic conditions and blue under basic conditions. However, acids are not known to have a slippery feel; this is a characteristic of bases. Bases feel slippery because they dissolve the fatty acids and oils from skin and therefore reduce the friction between the skin cells.

35. D is correct.

The atomic number (Z) refers to the number of protons and defines the element.

36. D is correct.

With little or no difference in electronegativity between the atoms (i.e., Pauling units < 0.4), the bond is a nonpolar covalent bond, whereby the electrons are shared equally between the two bonding atoms.

Polar covalent bonded compounds are covalently bonded atoms that participate in unequal sharing of electrons due to large electronegativity differences between the atoms (i.e., Pauling units of 0.4 to 1.7).

When the difference in electronegativity is greater than 1.7 Pauling units, the compounds form ionic bonds. Ionic bonds involve the transfer of an electron from the electropositive element (along the left-hand column/group) to the electronegative element (along the right-hand column/groups) on the periodic table.

In both covalent and polar covalent bonds, the electrons are shared so that each atom acquires a noble gas configuration.

Ionic bonds (in dry conditions) are stronger than covalent bonds. In aqueous conditions, ionic bonds are weak, and the compound spontaneously dissociates into ions (e.g., table salt in a glass of water).

The electropositive atom loses an electron(s) to become a cation (with a complete octet), and the electronegative atom gains an electron(s) to become an anion (with a complete octet).

37. C is correct.

The kinetic theory assumes random motion of molecules, elastic collisions, no significant volume occupied by molecules and little attraction between molecules.

At the same temperature, the average kinetic energy of molecules of different gases is the same, regardless of differences in mass.

38. A is correct.

A single-replacement reaction is a chemical reaction in which one element is substituted for another element in a compound, making a new compound and an element.

A double-replacement reaction occurs when parts of two ionic compounds are exchanged to make two new compounds.

$BaCl_2 + H_2SO_4 \rightarrow BaSO_4 + 2\ HCl$ is a double-replacement reaction with two products ($BaSO_4$ and 2 HCl).

$F_2 + 2\ NaCl \rightarrow Cl_2 + 2\ NaF$ and $Fe + CuSO_4 \rightarrow Cu + FeSO_4$ are correctly classified as single-replacement reactions because they each produce a compound (2NaF and $FeSO_4$, respectively) and an element (Cl_2 and Cu).

$2\ NO_2 + H_2O_2 \rightarrow 2\ HNO_3$ is correctly classified as a synthesis reaction because two species are combining to form a more complex chemical compound as the product.

39. B is correct.

Lattice energy is the amount of energy released when two gaseous ions combine to create a molecule.

The combined energy of both ions is lower than the sum of the ions' initial energies. The resulting molecule has lower potential energy and is more stable.

40. A is correct.

Reagents that do not affect the rate are 0th order.

41. D is correct.

The *like dissolves like* rule applies when a solvent is miscible with a solute that has similar properties.

A polar solute is miscible with a polar solvent. An ionic compound (separates into polar electrolytes) is miscible in a polar solvent.

42. C is correct.

If a solution is a good conductor, it has a high concentration of ions. Therefore, it is highly ionized.

Conductivity does not have any correlation with reactivity.

43. B is correct.

The starting materials are NaCl and H_2O. There are 4 ions in the solution: Na^+, H^+, Cl^- and ^-OH.

Analyze the cations and anions separately.

Comparing Na^+ and H^+: according to the electrochemical series, H^+ has a higher reduction potential than Na^+. H is reduced (i.e., gains electrons) to form H_2 while Na^+ remains unchanged.

Comparing Cl^- and ^-OH: according to the electrochemical series, Cl^- is more electronegative than ^-OH, so Cl^- is oxidized (i.e., loss of electrons) to form Cl_2, while ^-OH remains unchanged.

44. D is correct.

Electron configuration: $1s^22s^22p^63s^2$

The fastest way to identify an element using the periodic table is by using the element's atomic number.

The atomic number is equal to the number of protons (and equals the same number for electrons in a neutral atom).

In this problem, the total number of electrons can be determined by adding up all the electrons in the provided electron configuration: $2 + 2 + 6 + 2 = 12$.

Locate element #12 in the periodic table: Mg.

45. D is correct.

The valence shell is the outermost shell of electrons around an atom.

Atoms with a complete valence shell (i.e., containing the maximum number of electrons), such as noble gases, are the most non-reactive elements.

Atoms with only one electron in their valence shells (alkali metals) or those just missing one electron from a complete valence shell (halogens) are the most reactive elements.

46. C is correct.

Start by applying the ideal gas equation:

$$PV = nRT$$

Next, convert the given temperature units from Celsius to Kelvin:

$$T_1 = 24\ °C + 273 = 297\ K$$

$$T_2 = 64\ °C + 273 = 337\ K$$

Then, convert the given volume units from mL to L:

$$V_1 = 350\ mL = 0.350\ L$$

$$V_2 = 300\ mL = 0.300\ L$$

Set the initial and final P, V and T conditions equal:

$(P_1V_1 / T_1) = (P_2V_2 / T_2)$

Solve for the final pressure:

$P_2 = (P_1V_1T_2) / (V_2T_1)$

$P_2 = [(1.2 \text{ atm}) \times (0.350 \text{ L}) \times (337 \text{ K})] / [(0.300 \text{ L}) \times (297 \text{ K})]$

Notice that the question does not ask to solve for the final pressure.

47. A is correct.

Balance the overall reaction:

$6 \text{ HCl } (aq) + 2 \text{ Fe } (s) \rightarrow 2 \text{ FeCl}_3 (aq) + 3 \text{ H}_2 (g)$

Identify the component that undergoes oxidation (increasing oxidation number).

For this problem, it would be Fe because it went from 0 in elemental iron to Fe^{3+} in $FeCl_3$.

Write the reactions as a half reaction with its coefficients: $2 \text{ Fe} \rightarrow 2 \text{ Fe}^{3+}$

Add electrons to balance the charges: $2 \text{ Fe} \rightarrow 2 \text{ Fe}^{3+} + 6 \text{ e}^-$

48. D is correct.

Thermal radiation is the emission of electromagnetic waves (i.e., the energy carried via photons) from all matter that has a temperature (i.e., kinetic energy) greater than absolute zero.

A mirror-like surface reflects radiation, keeping heat inside the thermos.

49. C is correct.

Temperature is the only change that would affect the constant. Changing pressure, volume or concentration shift the equilibrium, but the equilibrium constant would not change.

A catalyst increases the rate of the reaction (k) but has no effect on the stability of the products and therefore does not change the equilibrium constant (K_{eq}).

50. C is correct.

For dilution problems:

concentration (C) × volume (V) of all the solutions before mixing = concentration × the volume of the final solution.

$(CV)_{\text{Soln 1}} + (CV)_{\text{Soln 2}} = (CV)_{\text{Final soln}}$

$(0.03 \text{ M})\cdot(20 \text{ mL}) + (0.06 \text{ M})\cdot(15 \text{ mL}) = C(35 \text{ mL})$

$[(0.03 \text{ M})\cdot(20 \text{ mL}) + (0.06 \text{ M})\cdot(15 \text{ mL})] / 35 \text{ mL} = C$

$(0.6 \text{ M}\cdot\text{mL}) + (0.9 \text{ M}\cdot\text{mL}) / 35 \text{ mL} = C$

$C = 0.043 \text{ M}$

51. B is correct.

For K_a of an acid: a higher value indicates a stronger acid (i.e., more stable anion).

HF (hydrofluoric acid) has the highest K_a value and is, therefore, the strongest acid.

The order of acid strength (from strongest to weakest) is HF (hydrofluoric acid) HNO_2 (nitrous acid), H_2CO_3 (carbonic acid), HClO (hypochlorous acid) and then HCN (hydrocyanic acid).

52. C is correct.

Charged plates are a pair of plates with a positive charge on one plate and negative on the other. Particles are shot at high speed in between those plates. If a particle is deflected, it travels in a curve towards one of the plates.

A particle is deflected if it has a positive or negative charge and travels towards a plate with the opposite charge.

A hydrogen atom has an equal number of protons and electrons, which means that the overall charge on the atom is zero. Thus, it is not affected by positive or negative charges on the plates.

Alpha particles, protons, and cathode rays are charged particles and are deflected by charged plates.

53. A is correct.

Not all elements can form double or triple covalent bonds because there must be at least two vacancies in an atom's valence electron shell for the double bond formation, and at least three vacancies for triple bond formation.

The elements of Group VIIA have seven valence electrons with one vacancy. Therefore, all covalent bonds formed by these elements are single covalent bonds.

54. D is correct.

The molecules of an ideal gas exert no attractive forces. Therefore, a real gas behaves most nearly like an ideal gas when it is at high temperature and low pressure because under these conditions, the molecules are far apart from each other and exert little or no attractive forces on each other.

55. B is correct.

The mass number is an approximation of the atomic weight of the element as amu (or g/mole)

14 grams = 1 mole

28 grams = 2 moles

56. A is correct.

In this problem, heat flows from Au to Ag.

q released by Au = q captured by Ag

Let Au be sample 1 and Ag be sample 2.

$$(m_1 \times c_1 \times \Delta T_1) = (m_2 \times c_2 \times \Delta T_2)$$

Solving for the unknown (mass of Ag or m_2):

$$m_2 = (m_1 \times c_1 \times \Delta T_1) / (c_2 \times \Delta T_2)$$

$$m_2 = [26 \text{ g} \times 0.130 \text{ J/g·°C} \times (97.2 \text{ °C} - 29.4 \text{ °C})] / [0.240 \text{ J/g·°C} \times (34 \text{ °C} - 21.4 \text{ °C})]$$

$$m_2 = (26 \text{ g} \times 0.130 \text{ J/g·°C} \times 67.8 \text{ °C}) / (0.240 \text{ J/g·°C} \times 12.6 \text{ °C})$$

$$m_2 = 75.8 \text{ g}$$

57. D is correct.

All aspects of collision affect the rate of chemical reactions.

Another important factor in reaction rates is temperature.

58. B is correct.

%v/v of alcohol = (volume of alcohol / volume of wine) × 100%

%v/v of alcohol = (25 mL / 225 mL) × 100%

%v/v of alcohol = 11.1%

59. C is correct.

When given a choice of multiple explanations, choose the most descriptive one.

However, always check the statement because sometimes the most descriptive option is not accurate.

In this case, the longest option is also accurate, so that would be the most specific description.

60. B is correct.

Oxidation of Mg (*s*):	$Mg\ (s) \rightleftarrows Mg^{2+} + 2\ e^-$	E° = 2.35 V
Reduction of Pb (*s*):	$Pb^{2+} + 2\ e^- \rightleftarrows Pb\ (s)$	E° = –0.13 V
Net Reaction:	$Mg\ (s) + Pb^{2+} \rightleftarrows Pb\ (s) + Mg^{2+}$	E° = 2.22 V

Note: when a reaction reverses, the sign of E° changes.

Notes for active learning

Notes for active learning

Diagnostic Test 3 – Answer Keys and Detailed Explanations

1	**A**	Electronic Structure & Periodic Table	**31**	**D**	Thermochemistry
2	**A**	Bonding	**32**	**C**	Kinetics Equilibrium
3	**C**	Phases & Phase Equilibria	**33**	**A**	Solution Chemistry
4	**B**	Stoichiometry	**34**	**D**	Acids & Bases
5	**D**	Thermochemistry	**35**	**B**	Electronic Structure & Periodic Table
6	**B**	Kinetics Equilibrium	**36**	**A**	Bonding
7	**D**	Solution Chemistry	**37**	**C**	Phases & Phase Equilibria
8	**A**	Acids & Bases	**38**	**B**	Stoichiometry
9	**B**	Electrochemistry	**39**	**A**	Thermochemistry
10	**C**	Electronic Structure & Periodic Table	**40**	**D**	Kinetics Equilibrium
11	**D**	Bonding	**41**	**B**	Solution Chemistry
12	**D**	Phases & Phase Equilibria	**42**	**B**	Acids & Bases
13	**D**	Stoichiometry	**43**	**B**	Electrochemistry
14	**B**	Thermochemistry	**44**	**A**	Electronic Structure & Periodic Table
15	**A**	Kinetics Equilibrium	**45**	**B**	Bonding
16	**C**	Solution Chemistry	**46**	**D**	Phases & Phase Equilibria
17	**A**	Acids & Bases	**47**	**C**	Stoichiometry
18	**D**	Electronic Structure & Periodic Table	**48**	**C**	Thermochemistry
19	**C**	Bonding	**49**	**C**	Kinetics Equilibrium
20	**B**	Phases & Phase Equilibria	**50**	**B**	Solution Chemistry
21	**A**	Stoichiometry	**51**	**C**	Acids & Bases
22	**D**	Thermochemistry	**52**	**D**	Electronic Structure & Periodic Table
23	**C**	Kinetics Equilibrium	**53**	**B**	Bonding
24	**A**	Solution Chemistry	**54**	**B**	Phases & Phase Equilibria
25	**D**	Acids & Bases	**55**	**D**	Stoichiometry
26	**B**	Electrochemistry	**56**	**B**	Thermochemistry
27	**A**	Electronic Structure & Periodic Table	**57**	**C**	Kinetics Equilibrium
28	**A**	Bonding	**58**	**B**	Solution Chemistry
29	**C**	Phases & Phase Equilibria	**59**	**A**	Acids & Bases
30	**D**	Stoichiometry	**60**	**D**	Electrochemistry

1. A is correct.

Neutrons are the neutral particles located inside the nucleus of an atom.

1 amu is 1/12 the mass of ^{12}C atoms.

1 amu is approximately the mass of 1 proton.

Neutrons have approximately the same mass as a proton.

2. A is correct.

Covalent bonds are chemical bonds that involve the sharing of electron pairs between atoms. These electrons can be shared equally (when atoms have the same electronegativity) or unequally (when atoms have different electronegativity).

With little or no difference in electronegativity between the atoms (Pauling units < 0.4), it is a nonpolar covalent bond, whereby the electrons are shared equally between the two bonding atoms.

Polar covalent bonded atoms are covalently bonded compounds that involve unequal sharing of electrons due to large electronegativity differences between the atoms (Pauling units of 0.4 to 1.7).

However, the correct answer must be "covalent" because the other choices do not involve the sharing of electrons.

Ionic bonds involve the transfer of electrons within the molecule.

Dipole, London and van der Waals are weak intermolecular (i.e., between molecules) forces.

3. C is correct.

Vapor pressure is proportional to temperature.

Vapor pressure is the pressure exerted by a vapor in thermodynamic equilibrium with its condensed phases (i.e., solid or liquid) at a given temperature in a closed system.

The equilibrium vapor pressure is an indicator of a liquid's evaporation rate.

4. B is correct.

Definition: mole = weight / molecular mass

Usually, the moles are calculated for each answer choice.

Since all answer choices have the same weight, compare the molecular masses instead.

The molecule with the largest molecular mass has the lowest number of moles.

From the periodic table: CH_4 = 16.04 g/mol   CO = 28.01 g/mol

AlH_3 = 30.01 g/mol N_2 = 28.01 g/mol

5. D is correct.

6. B is correct.

Endergonic reactions are nonspontaneous with the energy of the products higher than the energy of the reactants. ΔG is greater than 0.

Exergonic reactions are spontaneous with the energy of the reactants higher than the energy of the products. ΔG is less than 0.

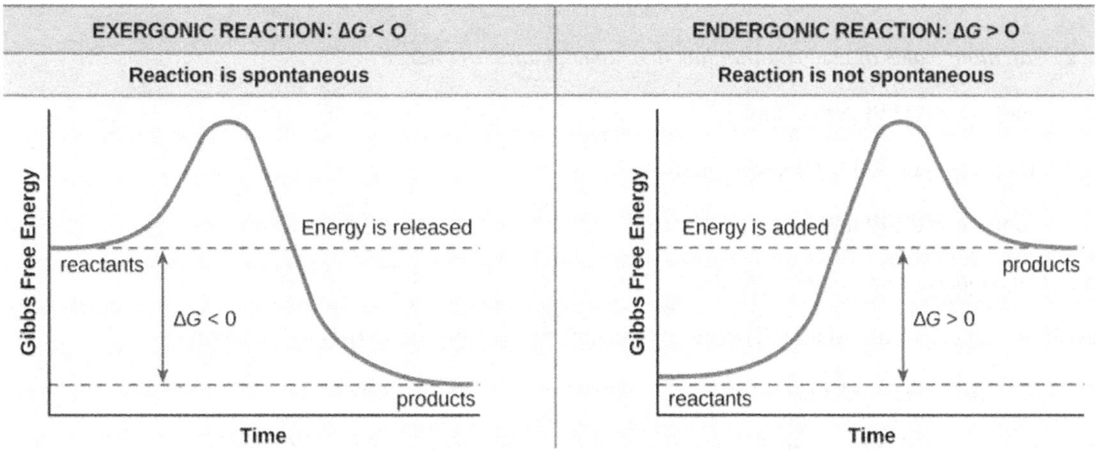

7. D is correct.

A colloid is a solution that has particles larger than those of the solution, but not large enough to precipitate due to gravity.

Colloidal particles are too small to be extracted by simple filtration.

8. A is correct.

Phenolphthalein is used as an indicator for acid-base titrations. It is a weak acid, which can dissociate protons (H^+ ions) in solution. The phenolphthalein molecule is colorless, and the phenolphthalein ion is pink. It turns colorless in acidic solutions and pink in basic solutions.

With basic conditions, the phenolphthalein (neutral) $\rightleftharpoons$ ions (pink) equilibrium shifts to the right, leading to more ionization as H^+ ions are removed.

Bromothymol blue is a pH indicator that is often used for solutions with neutral pH near 7 (e.g., managing the pH of pools and fish tanks). Bromothymol blue acts as a weak acid in a solution that can be protonated or deprotonated.

Bromothymol blue appears yellow when protonated (lower pH), blue when deprotonated (higher pH) and bluish green in neutral solution.

Methyl red has a pK_a of 5.1 and is a pH indicator dye that changes color in acidic solutions: turns red in pH under 4.4, orange in pH 4.4-6.2 and yellow in pH over 6.2.

9. B is correct.

Calculating the mass of metal deposited in the cathode.

Step 1: Calculate total charge using current and time

Q = current × time

Q = 4.8 A × (50.0 minutes) × (60 s / 1 minute)

Q = 14,400 A·s = 14,400 C

Step 2: Calculate moles of electron that has that same amount of charge

moles e^- = Q / 96,500 C/mol

moles e^- = 14,400 C / 96,500 C/mol

moles e^- = 0.149 mol

Step 3: Calculate moles of metal deposit

The solution contains chromium (III) sulfate. Half-reaction of chromium (III) ion reduction:

Cr^{3+} (*aq*) + $3e^-$ → Cr (*s*)

Use the moles of an electron from the previous calculation to calculate moles of Cr:

moles Cr = (coefficient Cr / coefficient electron) × moles electron

moles Cr = (1 / 3) × 0.149 mol

moles Cr = 0.0497 mol

Step 4: Calculate the mass of metal deposit

mass Cr = moles Cr × molecular mass of Cr

mass Cr = 0.0497 mol × 52.00 g/mol

mass Cr = 2.58 g

10. C is correct.

Atoms in excited states have electrons in lower energy orbitals promoted to higher energy orbitals.

$1s^22s^22p^63s^23p^64s^1$ is the electron configuration of potassium in the ground state.

$1s^22s^22p^63s^23p^6$ has only 18 electrons, not the 19 electrons of potassium.

$1s^22s^22p^63s^23p^64s^2$ has 20 electrons, not the 19 electrons of potassium.

$1s^22s^22p^63s^23p^7$ has 19 electrons, but $3p^7$ exceeds the limit of 6 electrons for the *p* orbital.

11. D is correct.

The charge on Ca is +2, while the charge on SO_4 is −2.

The proper formula is $CaSO_4$.

12. D is correct.

Boyle's Law:

$$(P_1V_1) = (P_2V_2)$$

or

$P \times V$ = constant: pressure and volume are inversely proportional.

If the volume of a gas increases, its pressure decreases proportionally.

13. D is correct.

An electrolyte is a substance that produces an electrically conducting solution when dissolved in a polar solvent (e.g., water).

The dissolved electrolyte separates into positively-charged cations and negatively-charged anions.

Strong electrolytes dissociate completely (or nearly completely) because the resulting ions are stable in the solution.

14. B is correct.

From the equation for Gibbs free energy:

$$\Delta G = \Delta H - T\Delta S$$

With no change in entropy, $T\Delta S = 0$.

Therefore, $\Delta G = \Delta H$

The reaction is spontaneous if ΔG is negative.

The reaction is exothermic (releases heat as a product) if ΔH is negative.

Endothermic reactions are nonspontaneous with heat as a reactant. The products are less stable than the reactants and ΔG is positive.

Exothermic reactions are spontaneous with heat as a product. The products are more stable than the reactants and ΔG is negative.

Endergonic refers to a positive ΔG, while exergonic refers to a negative ΔG.

	$\Delta H < 0$	$\Delta H > 0$
$\Delta S > 0$	Spontaneous at all T ($\Delta G < 0$)	Spontaneous at high T ($T\Delta S$ is large)
$\Delta S < 0$	Spontaneous at low T ($T\Delta S$ is small)	Nonspontaneous at all T ($\Delta G > 0$)

15. A is correct.

Increasing the concentration of reactants increases the frequency of collision and therefore the relative rate of the reaction.

16. C is correct.

Spectator ions appear on both sides of the ionic equation.

Ionic equation of the reaction:

$2\ K^+(aq) + SO_4^{2-}(aq) + Ba^{2+}(aq) + 2\ NO_3^-(aq) \rightarrow BaSO_4(s) + 2\ K^+(aq) + 2\ NO_3^-(aq)$

17. A is correct.

Neutralization reactions form water and salt (e.g., $Cr_2(SO_4)_3$).

18. D is correct.

Nonmetals tend to be highly volatile (i.e., easily vaporized), have low density and are good insulators of heat and electricity. They tend to have high ionization energy and electronegativity and share (or gain) an electron when bonding with other elements.

Seventeen elements are generally classified as nonmetals. Most are gases (hydrogen, helium, nitrogen, oxygen, fluorine, neon, chlorine, argon, krypton, xenon and radon). One nonmetal is a liquid (bromine), and a few are solids (carbon, phosphorus, sulfur, selenium, and iodine).

19. C is correct.

Electronegativity is a chemical property to describe an atom's tendency to attract electrons to itself. The most common use of electronegativity pertains to polarity along the sigma (single) bond.

The greater the difference in electronegativity between two atoms, the more polar of a bond these atoms form, whereby the atom with the higher electronegativity is the partial (delta) negative end of the dipole.

In the periodic table, electronegativity increases from left to right.

Metals are located towards the left side and nonmetals towards the right.

Therefore, nonmetals should have higher electronegativity values than metals.

20. B is correct.

The sodium is a cation (positive ion), while water has a dipole due to the differences in electronegativity between the electronegative oxygen and the bonded hydrogens.

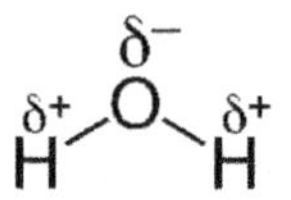

Dipole in a water molecule

The delta minus on the electronegative oxygen is attracted to the positive sodium ion.

21. A is correct.

The mass of an electron is less than 1/10th of one percent of a proton (or neutron) mass.

Neutrons and protons have approximately the same mass.

The element with the greatest sum for protons and neutrons has the greatest mass:

Mass number: 35 protons + 35 neutrons = 70

22. D is correct.

Heat = mass × specific heat × change in temperature:

$$q = m \times c \times \Delta T$$

Rearrange to solve for ΔT:

$$\Delta T = q / (m \times c)$$

$$\Delta T = 340\ J / (30\ g \times 4.184\ J/g{\cdot}°C)$$

$$\Delta T = 2.7\ °C$$

The problem indicates that heat is removed from H_2O, so temperature decreases:

$$\text{Final } T = \text{initial } T - \Delta T$$

$$\text{Final } T = 19.8\ °C - 2.7\ °C$$

$$\text{Final } T = 17.1\ °C$$

23. C is correct.

Catalysts provide an alternative pathway for the reaction to proceed to product formation. It lowers the energy of activation (i.e., relative energy between reactants and transition state) and therefore speeds the rate of the reaction.

Catalysts do not affect the Gibbs free energy (ΔG: stability of products vs. reactants) or the enthalpy (ΔH: bond breaking in reactants or bond making in products).

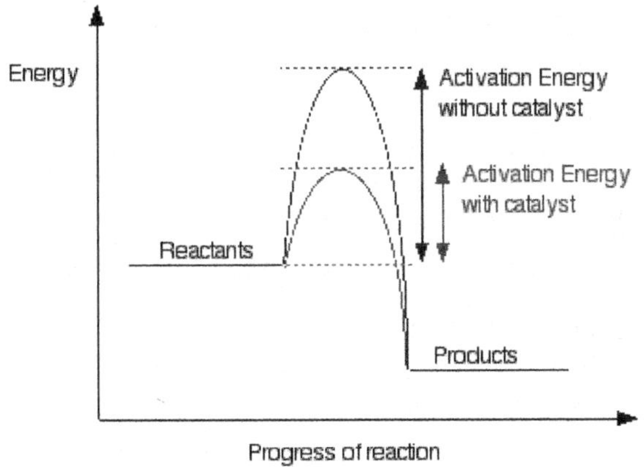

24. A is correct.

Mass volume percent:

(mass of solute / volume of solution) × 100%

(14.6 g / 260 mL) × 100% = 5.62%

25. D is correct.

In the Arrhenius theory, acids are defined as substances that dissociate in aqueous solutions to donate H^+ (hydrogen ions).

In the Arrhenius theory, bases are defined as substances that dissociate in aqueous solutions to donate ^-OH (hydroxide ions).

A strong acid dissociates completely (or nearly so) because the resulting anion is stable (i.e., low pK_a for the acid), while a weak acid does not dissociate completely because the resulting anion is unstable (i.e., high pK_a for the acid).

26. B is correct.

A galvanic (i.e., voltaic or electrochemical) cell involves a spontaneous redox reaction ($\Delta G < 0$) that uses the energy released to generate electricity.

An electrolytic cell utilizes electrical energy from an external source to drive a nonspontaneous redox reaction to occur ($\Delta G > 0$).

Oxidation always occurs at the anode, while reduction always occurs at the cathode.

In reaction II, each I^- anion loses 1 e^- to produce neutral I_2 during oxidation.

27. A is correct.

The three coordinates that come from Schrodinger's wave equations are the principal (*n*), angular (*l*) and magnetic (*m*) quantum numbers. These quantum numbers describe the size, shape, and orientation in the space of the orbitals on an atom, respectively.

The principal quantum number (*n*) describes the size of the orbital and the energy of an electron and the most probable distance of the electron from the nucleus. The angular momentum quantum number (*l*) describes the shape of the orbital (i.e., subshells).

The magnetic quantum number (*m*) determines the number of orbitals and their orientation within a subshell. Consequently, its value depends on the orbital angular momentum quantum number (*l*). Given a certain *l*, *m* is an interval ranging from $-l$ to $+l$ (i.e., it can be zero, a negative integer, or a positive integer).

The *s* is the spin quantum number (e.g., +½ or −½).

28. A is correct.

The valence shell is the outermost shell (i.e., highest principal quantum number *n*) of an atom.

Valence electrons are those of the outermost electron shell that can participate in a chemical bond.

The number of valence electrons for an element can be determined by its group (i.e., vertical column) on the periodic table. Except the transition metals (i.e., groups 3-12), the group number identifies how many valence electrons are associated with a particular element: all elements of the same group have the same number of valence electrons.

Electron dot structure (also known as the Lewis dot structure) is a visual representation of the valence electron configuration around an atom.

It is drawn by placing dots that represent valence electrons around a chemical symbol.

29. C is correct.

The ideal gas law:

$$PV = nRT$$

where P is pressure, V is volume, n is the number of molecules, R is the ideal gas constant and T is the temperature of the gas.

The volume and temperature are directly proportional; if one increases, the other increases as well.

30. D is correct.

$ClCH_2$ total mass:

$$[Cl = 35.5 \text{ g/mol} + C = 12 \text{ g/mol} + (4 \times H = 4 \text{ g/mol})] \approx 50 \text{ g/mol}$$

Two hydrogens have a % by mass of about 2 / 50, which is 4%.

ClC_2H_5 total mass:

$$(35.5 \text{ g/mol}) + (2 \times 12 \text{ g/mol}) + (5 \times 1 \text{ g/mol}) = 65 \text{ g/mol}$$

Five hydrogens have a % by mass of about 5 / 65, which is greater than 4%.

Cl_2CH_2 total mass:

$$(2 \times 35.5 \text{ g/mol}) + 12 \text{ g/mol} + (2 \times 1 \text{ g/mol}) = 85 \text{ g/mol}$$

Two hydrogens have a % by mass of 2 / 85 for the H, which is less than 4%.

$ClCH_3$ total mass:

$$(35.5 \text{ g/mol}) + 12 \text{ g/mol} + 3 \times 1 \text{ g/mol}) = 50 \text{ g/mol}$$

Three hydrogens have a % mass of 3 / 50 for the H, which is about 6%.

31. D is correct.

The total enthalpy (i.e., internal energy) of a system cannot be measured directly; the enthalpy change of a system is measured instead.

Enthalpy change is defined by the following equation:

$$\Delta H = H_f - H_i$$

If the standard enthalpy of the products is less than the standard enthalpy of the reactants, the standard enthalpy of reaction is negative, and the reaction is exothermic (i.e., releases heat).

If the standard enthalpy of the products is more than the standard enthalpy of the reactants, the standard enthalpy of reaction is positive, and the reaction is endothermic (i.e., absorbs heat).

If the bonds formed are stronger than bonds broken, that would be an exothermic reaction because some energy is released to the surroundings.

32. C is correct.

Chemical equilibrium refers to a dynamic process whereby the *rate* at which a reactant molecule is being transformed into a product is the same as the *rate* for a product molecule to be transformed into a reactant.

33. A is correct.

The addition of dissolved F^- (by adding NaF) decreases the solubility of CaF_2.

The common ion effect is responsible for the reduction in the solubility of an ionic precipitate when a soluble compound containing one of the ions of the precipitate is added to the solution in equilibrium with the precipitate.

If the concentration of any of the ions is increased, some of the ions in excess combine with the oppositely charged ions and are effectively removed from the solution (consistent with the Le Châtelier's principle). Then, some of the salt is precipitated until the ion product is equal to the solubility product.

34. D is correct.

The production of H^+ ions in water is characteristic of acid, not a base.

A base is a chemical substance with a pH greater than 7 and produces hydroxide (OH^-) ions in water.

A base is neutralized by an acid (i.e., a substance with a pH below 7) to form a salt.

Bases are known to have a slippery or soapy feel, because they dissolve the fatty acids and oils from the skin, cutting down the amount of friction.

They also have a bitter taste; for example, coffee is bitter because it contains caffeine, which is a base.

35. B is correct.

The alkaline earth metals (group IIA) include beryllium (Be), magnesium (Mg), calcium (Ca), strontium (Sr), barium (Ba) and radium (Ra).

Representative elements, from the periodic table, are groups IA & IIA (on the left) and groups IIIA – VIIIA (on the right).

The transition elements have characteristics that are not found in other elements, which results from the partially filled *d* shell. These include the formation of compounds whose color is due to *d–d* electronic transitions, the formation of compounds in many oxidation states, due to the relatively low reactivity of unpaired *d* electrons.

The alkali metals (group IA) include lithium (Li), potassium (K), sodium (Na), rubidium (Rb), cesium (Cs) and francium (Fr).

Alkali metals lose one electron to become +1 cations, and the resulting ion has a complete octet of valence electrons. The alkali metals have low electronegativity and react violently with water (e.g., the violent reaction of sodium metal with water).

On the periodic table, the lanthanides and the actinides are shown as two additional rows below the main body of the table.

36. A is correct.

The Pauli exclusion principle states that two identical electrons cannot have the same quantum numbers. It is not possible for two electrons to have the same values of the four quantum numbers: the principal quantum number (n), the angular momentum quantum number (ℓ), the magnetic quantum number (m_ℓ), and the spin quantum number (m_s).

For two electrons residing in the same orbital, n, ℓ, and m_ℓ are the same, m_s must be different, and the electrons have opposite half-integer spins, $+\frac{1}{2}$ and $-\frac{1}{2}$.

37. C is correct.

Hydrogens, bonded directly to F, O or N, participate in hydrogen bonds. The hydrogen is partially positive (i.e., $\partial+$) due to the bond to these electronegative atoms. The lone pair of electrons on the F, O or N interacts with the partial positive ($\partial+$) hydrogen to form a hydrogen bond.

38. B is correct.

Balanced equation (double replacement):

$$2\ C_2H_6\ (g) + 7\ O_2\ (g) \xrightarrow{\text{Spark}} 4\ CO_2\ (g) + 6\ H_2O\ (g)$$

39. A is correct.

If V, R, and T are constants, then the ideal gas law, PV = nRT implies that P is proportional to n.

Therefore, increasing n causes an increase in P.

40. D is correct.

The reaction is exothermic, which means that heat is a product. Lowering the temperature causes more products to form.

Summary of temperature effects:

> Increasing the temperature of a system in dynamic equilibrium favors the endothermic reaction because the system counteracts the change by absorbing the extra heat.
>
> Decreasing the temperature of a system in dynamic equilibrium favors the exothermic reaction because the system counteracts the change by producing more heat.

41. B is correct.

% (m/m) means that both solute and solvent are measured by their mass.

If the solution contains 14% (m/m), it means that the rest of it is water, or:

$$100\% - 14\% = 86\%$$

Use this percentage to determine the mass:

$$\text{mass of water} = 86\% \times 160 \text{ grams}$$

$$\text{mass of water} = 138 \text{ g}$$

42. B is correct.

First, determine the reaction product by writing down the full equation.

When an acid reacts with a base, it forms a salt and water.

Exchange the cations and anions from acid and base to form the new salt:

$$NaOH + H_2SO_4 \rightarrow Na_2SO_4 + H_2O$$

Always check that a reaction is balanced before starting calculations. That equation was not balanced, so balance it first:

$$2\ NaOH + H_2SO_4 \rightarrow Na_2SO_4 + 2\ H_2O$$

The residue mentioned in the problem must be Na_2SO_4 residue. Calculate moles of Na_2SO_4:

$$\text{moles } Na_2SO_4 = \text{mass } Na_2SO_4 / \text{molecular mass } Na_2SO_4$$

$$\text{moles } Na_2SO_4 = (840 \text{ mg} \times 0.001 \text{ g/mg}) / [(2 \times 22.99 \text{ g/mol}) + 32.06 \text{ g/mol} + (4 \times 16 \text{ g/mol})]$$

$$\text{moles } Na_2SO_4 = (840 \text{ mg} \times 0.001 \text{ g/mg}) / (45.98 \text{ g/mol}) + 32.06 \text{ g/mol} + (64 \text{ g/mol})]$$

$$\text{moles } Na_2SO_4 = (840 \text{ mg} \times 0.001 \text{ g/mg}) / (142.04 \text{ g/mol})$$

$$\text{moles } Na_2SO_4 = 5.9 \times 10^{-3} \text{ mole}$$

Because coefficient Na_2SO_4 = coefficient H_2SO_4, moles Na_2SO_4 = moles $H_2SO_4 = 6.06 \times 10^{-3}$ mole.

Use this information to determine the molarity:

$$\text{Molarity of } H_2SO_4 = \text{moles of } H_2SO_4 / \text{volume of } H_2SO_4$$

$$\text{Molarity of } H_2SO_4 = 5.9 \times 10^{-3} \text{ mole} / (40 \text{ mL} \times 0.001 \text{ L / mL})$$

$$\text{Molarity of } H_2SO_4 = 0.148 \text{ M}$$

43. B is correct.

Half-reaction: $C_2H_6O \rightarrow HC_2H_3O_2$

Balancing half-reaction in acidic conditions:

Step 1: Balance all atoms except for H and O

$C_2H_6O \rightarrow HC_2H_3O_2$ (C is already balanced)

Step 2: To balance oxygen, add H_2O to the side with fewer oxygen atoms

$C_2H_6O + H_2O \rightarrow HC_2H_3O_2$

Step 3: To balance hydrogen, add H^+ to the opposing side of H_2O added in the previous step

$C_2H_6O + H_2O \rightarrow HC_2H_3O_2 + 4\ H^+$

Step 4: Balance charges by adding electrons to the side with a greater/more positive total charge

Total charge on the left side: 0

Total charge on the right side: $4(+1) = +4$

Add 4 electrons to the right side:

$C_2H_6O + H_2O \rightarrow HC_2H_3O_2 + 4\ H^+ + 4\ e^-$

44. A is correct.

Isotopes are variants of a particular element, which differ in the number of neutrons. All isotopes of the element have the same number of protons and occupy the same position on the periodic table.

The number of protons within the atom's nucleus is the atomic number (Z) and is equal to the number of electrons in the neutral (non-ionized) atom.

Each atomic number identifies a specific element, but not the isotope; an atom of a given element may have a wide range in its number of neutrons.

The number of both protons and neutrons (i.e., nucleons) in the nucleus is the atom's mass number (A), and each isotope of an element has a different mass number.

The ^1H is the most common hydrogen isotope with an abundance of more than 99.98%. The ^1H is protium, and its nucleus consists of a single proton.

The ^2H isotope is deuterium, while the ^3H tritium.

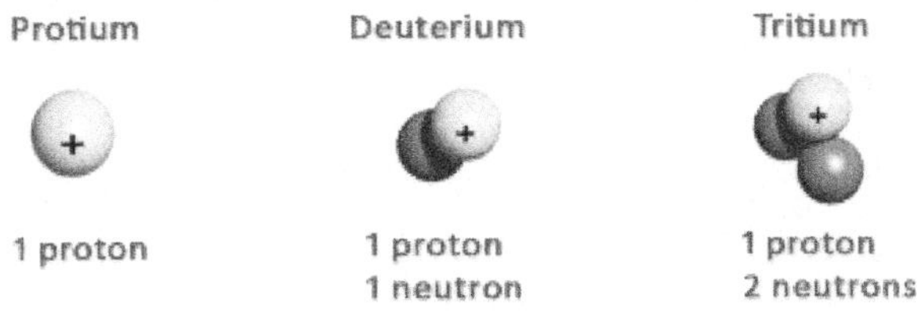

Three isotopes of hydrogen

45. B is correct.

Resonance spreads the double-bond character equally among all three S–O bonds.

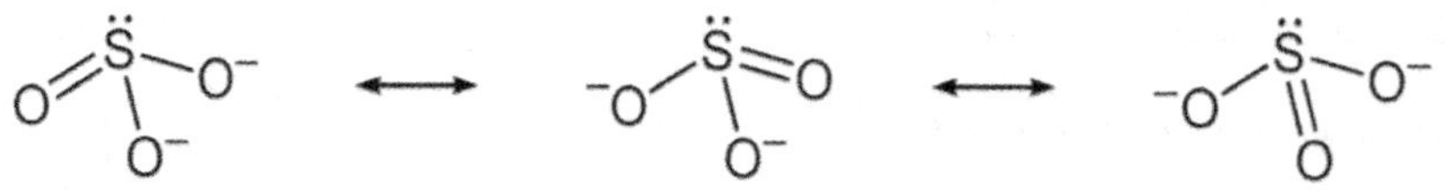

Resonance structures of the sulfite ion

46. D is correct.

The boiling point depends on the intermolecular bonding of the molecules. All of the options are nonpolar halogen molecules. The dominant intermolecular force is dispersion force (or induced dipole).

As the number of electron shells increases, the valence electrons are further from the nucleus, and these molecules would be more prone to the temporary induced dipole.

Therefore, larger molecules will have stronger bonds and higher boiling points.

47. C is correct.

The oxidizing agent is the substance that causes the other reactant(s) to be oxidized.

The oxidizing agent also undergoes reduction.

In this reaction, HCl is the oxidizing agent because it is reduced to H_2 and it oxidizes Co.

48. C is correct.

Convection is defined as the movement of heat through liquids and gases.

The lid prevents fluid (e.g., air) from escaping the vessel.

49. C is correct.

General formula for the equilibrium constant of a reaction:

$$aA + bB \leftrightarrow cC + dD$$

$$K_{eq} = ([C]^c \times [D]^d) / ([A]^a \times [B]^b)$$

For equilibrium constant calculation, only include species in aqueous or gas phases (which, in this case, is all of them):

$$K_{eq} = [NO_2]^2 / [N_2O_4]$$

Substitute in the values for each concentration:

$$K_{eq} = (0.400)^2 / 0.800$$

$$K_{eq} = 0.200$$

50. B is correct.

Most hydroxide salts are only slightly soluble.

In general, hydroxide salts of Group I elements are soluble.

Hydroxide salts of Group II elements (Ca, Sr and Ba) are slightly soluble.

Hydroxide salts of transition metals and Al^{3+} are insoluble (e.g., $Fe(OH)_3$, $Al(OH)_3$, $Co(OH)_2$).

Salts containing Cl^-, Br^-, I^- are generally soluble. Exceptions to this rule are halide salts of Ag^+, Pb^{2+}, and $(Hg_2)^{2+}$. For example, AgCl, $PbBr_2$, and Hg_2Cl_2 are all insoluble.

51. C is correct.

Use this formula for pH:

$$pH = -\log[H^+]$$

$$pH = -\log[0.000001]$$

$$pH = 6$$

52. D is correct.

Electrons cannot be precisely located in space at any point in time, and orbitals describe probability regions for finding the electrons.

The one *s* subshell is spherically symmetrical, the three *p* orbitals have dumbbell shapes, and the five *d* orbitals have four lobes.

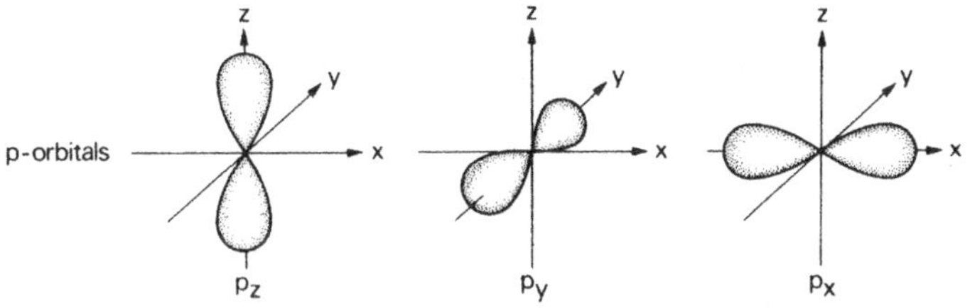

Representation of the three *p* orbitals

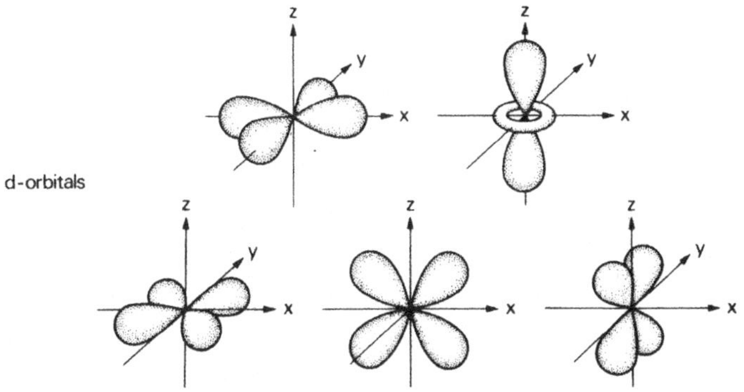

Representation of the five *d* orbitals

53. B is correct.

The trigonal pyramidal molecular geometry results from three bonds (i.e., three substituents) and one lone pair on the central atom in the molecule.

Examples of trigonal pyramidal molecules (shown below) include ammonia (NH_3), xenon trioxide (XeO_3), the chlorate ion (ClO_3^-) and the sulfite ion (SO_3^{2-}).

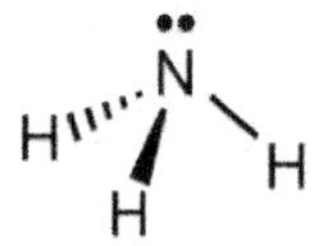

Ammonia

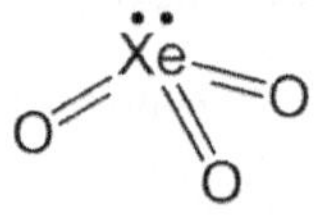

Xenon trioxide

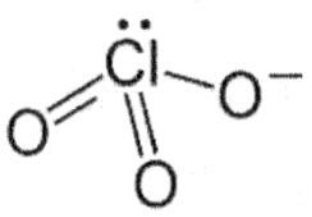

Chlorate ion

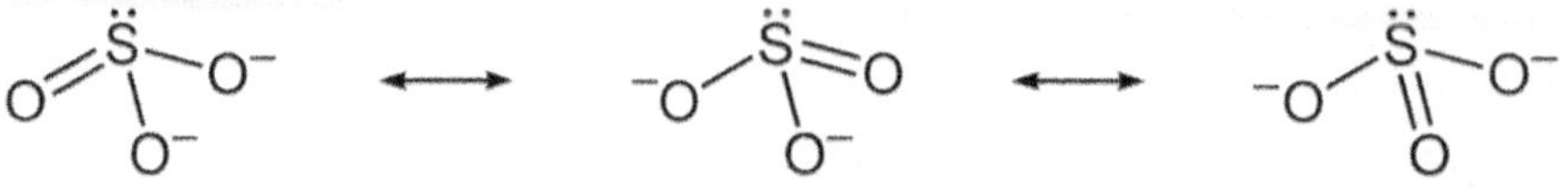

Three resonance structures for the sulfite ion

In ammonia, the trigonal pyramidal molecule undergoes rapid inversion of the lone pair of electrons on the central nitrogen atom.

Molecules with tetrahedral electron pair geometries (and four substituents) have sp^3 hybridization at the central atom and do not undergo inversion.

54. B is correct.

The mixture is 40% CO_2.

Partial pressure of CO_2:

$$0.4 \times 700 \text{ torr} = 280 \text{ torr}$$

55. D is correct.

Start by calculating the molecular mass (MW) of glucose:

(6 × atomic mass of C) + (12 × atomic mass of H) + (6 × atomic mass of O)

(6 × 12.01 g/mole) + (12 × 1.01 g/mole) + (6 × 16.00 g/mole)

MW glucose = 180.18 g/mole

Then multiply the number of moles by the molecular mass:

moles of glucose × MW of glucose

8.50 moles × 180.18 g/mole

Mass of glucose = 1,532 g

56. B is correct.

A spontaneous process releases free energy (usually as heat) and moves to a lower, more thermodynamically stable energy state.

A release of free energy from the system corresponds to a negative change in free energy, but a positive change for the surroundings.

For a reaction at STP, ΔG is the Gibbs free energy:

$$\Delta G = \Delta H - T\Delta S$$

The sign of ΔG depends on the signs of the changes in enthalpy (ΔH) and entropy (ΔS), and the absolute temperature (T, in Kelvin).

ΔG changes from positive to negative (or vice versa) when $T = \Delta H / \Delta S$.

For heterogeneous systems, where all of the species of the reaction are in different phases and can be mechanically separated, the following is true.

When ΔG is negative, a process or chemical proceeds spontaneously in the forward direction.

When ΔG is positive, the process proceeds spontaneously in reverse.

When ΔG is zero, the process is already in equilibrium, with no net change.

Consider the signs of ΔH and ΔS.

When ΔS is positive, and ΔH is negative, a process is always spontaneous.

When ΔS is positive, and ΔH is positive, the relative magnitudes of ΔS and ΔH determine if the reaction is spontaneous. High temperatures make the reaction more favorable.

When ΔS is negative, and ΔH is negative, the relative magnitudes of ΔS and ΔH determine if the reaction is spontaneous. Low temperatures make the reaction more favorable.

When ΔS is negative, and ΔH is positive, a process is not spontaneous at any temperature, but the reverse process is spontaneous.

57. C is correct.

Increasing the concentration of reactants (or increasing pressure) increases the rate of a reaction because there is an increased probability that any two reactant molecules will collide with sufficient energy to overcome the energy of activation and to form products.

58. B is correct.

First, determine the moles of NaCl:

Moles of NaCl = mass of NaCl / molecular mass of NaCl

Moles of NaCl = 30 g / (58.45 g/mol)

Moles of NaCl = 0.513 mole

Then, use this information to calculate the molarity:

Molarity = moles/volume

Molarity = 0.513 mole / 0.675 L

Molarity 0.76 M

59. A is correct.

A salt is a combination of an acid and a base.

The pH is determined by the acid and base that created the salt.

Combination of:

weak acid and strong base: salt is basic

strong acid and weak base: salt is acidic

strong acid and strong base: salt is neutral

weak acid and weak base: could be anything (acidic, basic, or neutral)

In this problem, only NaF is a salt created from a strong base (NaOH) and a weak acid (HF), which means NaF is basic.

Examples of strong acids and bases that commonly appear in chemistry problems:

Strong acids:

HCl (hydrochloric acid) HNO_3 (nitric acid)

H_2SO_4 (sulfuric acid) HBr (hydrobromic acid)

HI (hydroiodic acid) $HClO_3$ (chloric acid)

$HClO_4$ (perchloric acid)

Strong bases:

LiOH (lithium hydroxide) NaOH (sodium hydroxide)

KOH (potassium hydroxide) $Ca(OH)_2$ (calcium hydroxide)

RbOH (rubidium hydroxide) $Sr(OH)_2$, (strontium hydroxide)

CsOH (cesium hydroxide) $Ba(OH)_2$ (barium hydroxide)

60. D is correct.

CCl_4 is an oxidation product because the oxidation number of C increases from 0 to +4.

Oxidations always occur at the anode.

Notes for active learning

Notes for active learning

Answer Keys and Detailed Explanations: Topical Practice Questions

Topical Practice Questions – Answer Keys

Electronic and Atomic Structure; Periodic Table

1: B	11: B	21: C	31: A	41: B	51: C	61: D	71: D
2: C	12: D	22: D	32: D	42: A	52: D	62: B	72: B
3: A	13: A	23: A	33: B	43: D	53: D	63: C	73: A
4: B	14: C	24: C	34: C	44: A	54: A	64: B	74: D
5: C	15: A	25: D	35: A	45: D	55: D	65: A	75: B
6: B	16: C	26: D	36: C	46: A	56: D	66: B	76: D
7: C	17: D	27: C	37: B	47: C	57: A	67: D	77: C
8: D	18: C	28: D	38: B	48: D	58: A	68: A	78: A
9: B	19: C	29: A	39: C	49: B	59: A	69: B	79: B
10: A	20: B	30: B	40: D	50: A	60: C	70: C	80: C

Chemical Bonding

1: D	11: B	21: D	31: B	41: C	51: D	61: A	71: C
2: A	12: D	22: C	32: C	42: A	52: D	62: D	72: B
3: C	13: C	23: B	33: B	43: C	53: D	63: A	73: C
4: D	14: A	24: D	34: A	44: D	54: C	64: B	74: C
5: B	15: D	25: D	35: D	45: C	55: D	65: C	75: B
6: B	16: C	26: B	36: D	46: A	56: A	66: A	76: C
7: A	17: C	27: A	37: B	47: B	57: C	67: B	77: A
8: A	18: A	28: C	38: A	48: A	58: C	68: D	78: B
9: D	19: D	29: A	39: B	49: D	59: B	69: B	79: D
10: C	20: C	30: D	40: A	50: B	60: C	70: B	80: D

Phases and Phase Equilibria

1: A	11: B	21: D	31: B	41: C	51: B	61: D	71: D
2: C	12: C	22: A	32: D	42: D	52: C	62: A	72: B
3: C	13: A	23: C	33: C	43: B	53: A	63: B	73: A
4: A	14: D	24: D	34: B	44: D	54: D	64: C	74: C
5: A	15: C	25: D	35: C	45: D	55: C	65: D	75: A
6: B	16: A	26: A	36: D	46: A	56: B	66: B	76: B
7: B	17: C	27: D	37: A	47: C	57: B	67: A	77: C
8: B	18: C	28: B	38: C	48: C	58: B	68: C	78: D
9: D	19: C	29: C	39: B	49: C	59: A	69: D	79: D
10: D	20: B	30: A	40: C	50: B	60: C	70: A	80: D

Stoichiometry

1: C	11: C	21: D	31: A	41: B	51: D	61: B	71: A
2: C	12: D	22: A	32: C	42: D	52: A	62: D	72: D
3: C	13: C	23: B	33: B	43: D	53: C	63: D	73: B
4: D	14: D	24: C	34: A	44: A	54: B	64: B	74: C
5: C	15: D	25: C	35: A	45: D	55: D	65: C	75: D
6: B	16: A	26: D	36: C	46: B	56: B	66: D	76: A
7: A	17: B	27: A	37: D	47: C	57: C	67: B	77: D
8: C	18: C	28: B	38: A	48: A	58: D	68: C	78: C
9: B	19: D	29: D	39: B	49: C	59: D	69: C	79: D
10: C	20: D	30: C	40: A	50: B	60: D	70: B	80: B

Thermochemistry

1: A	11: C	21: A	31: B	41: A	51: C	61: C	71: D
2: B	12: A	22: C	32: C	42: C	52: C	62: C	72: C
3: D	13: B	23: A	33: D	43: B	53: C	63: C	73: D
4: C	14: D	24: B	34: B	44: D	54: C	64: B	74: A
5: A	15: C	25: C	35: D	45: A	55: D	65: A	75: C
6: D	16: B	26: C	36: A	46: D	56: A	66: B	76: D
7: D	17: A	27: A	37: B	47: C	57: C	67: D	77: B
8: A	18: B	28: D	38: D	48: C	58: D	68: C	78: D
9: C	19: C	29: A	39: D	49: B	59: B	69: A	79: A
10: D	20: C	30: D	40: D	50: B	60: A	70: B	80: A

Kinetics and Equilibrium

1: A	11: D	21: B	31: C	41: A	51: C	61: A	71: B
2: B	12: B	22: A	32: B	42: D	52: B	62: B	72: B
3: A	13: D	23: D	33: A	43: B	53: B	63: D	73: B
4: A	14: B	24: B	34: A	44: D	54: C	64: A	74: D
5: C	15: B	25: C	35: B	45: B	55: A	65: D	75: D
6: D	16: D	26: D	36: D	46: C	56: D	66: C	76: B
7: B	17: B	27: B	37: A	47: B	57: C	67: D	77: D
8: A	18: D	28: A	38: A	48: D	58: B	68: D	78: C
9: D	19: A	29: D	39: D	49: C	59: D	69: B	79: B
10: D	20: C	30: A	40: C	50: D	60: C	70: A	80: D

Solution Chemistry

1: B	11: C	21: B	31: D	41: D	51: B	61: D	71: D
2: C	12: A	22: A	32: B	42: B	52: C	62: A	72: A
3: A	13: C	23: C	33: B	43: B	53: B	63: A	73: B
4: A	14: C	24: B	34: B	44: A	54: A	64: A	74: A
5: D	15: B	25: A	35: D	45: D	55: A	65: C	75: A
6: A	16: D	26: C	36: A	46: B	56: B	66: D	76: C
7: B	17: B	27: A	37: A	47: A	57: B	67: C	77: C
8: D	18: A	28: D	38: C	48: C	58: B	68: B	78: C
9: A	19: A	29: D	39: D	49: D	59: D	69: D	79: B
10: C	20: D	30: A	40: B	50: C	60: C	70: C	80: D

Acids and Bases

1: A	11: B	21: C	31: A	41: A	51: C	61: A	71: A
2: B	12: D	22: A	32: C	42: A	52: A	62: C	72: D
3: A	13: C	23: D	33: A	43: D	53: A	63: D	73: C
4: C	14: B	24: D	34: D	44: B	54: D	64: A	74: D
5: C	15: C	25: B	35: A	45: C	55: A	65: A	75: B
6: C	16: A	26: B	36: C	46: D	56: C	66: D	76: A
7: D	17: D	27: B	37: B	47: A	57: A	67: A	77: C
8: A	18: C	28: D	38: B	48: D	58: C	68: A	78: A
9: B	19: D	29: C	39: B	49: B	59: B	69: B	79: B
10: B	20: C	30: C	40: D	50: B	60: C	70: D	80: C

Electrochemistry

1: D	11: A	21: A	31: A	41: C	51: A	61: D	71: A
2: C	12: C	22: D	32: B	42: A	52: C	62: C	72: D
3: A	13: A	23: A	33: B	43: C	53: B	63: B	73: D
4: B	14: B	24: A	34: B	44: B	54: C	64: B	74: D
5: C	15: C	25: B	35: D	45: B	55: D	65: C	75: B
6: A	16: D	26: C	36: A	46: B	56: C	66: D	76: D
7: C	17: B	27: D	37: A	47: A	57: D	67: A	77: B
8: D	18: C	28: A	38: B	48: B	58: A	68: B	78: C
9: D	19: C	29: B	39: D	49: D	59: A	69: B	79: A
10: B	20: D	30: C	40: B	50: B	60: B	70: C	80: D

Notes for active learning

Electronic and Atomic Structures; Periodic Table – Detailed Explanations

===

Practice Set 1: Questions 1–20

===

1. B is correct.

Ionization energy (IE) is the amount of energy required to remove the most loosely bound electron of an isolated gaseous atom to form a cation. This is an endothermic process.

Ionization energy is expressed as:

$$X + \text{energy} \rightarrow X^+ + e^-$$

where X is an atom (or molecule) capable of being ionized (i.e., having an electron removed), X^+ is that atom or molecule after an electron is removed, and e^- is the removed electron.

The principal quantum number (n) describes the size of the orbital and the energy of an electron, and the most probable distance from the nucleus. It refers to the size of the orbital and the energy level of an electron.

The elements with larger shell sizes (n is large) listed at the bottom of the periodic table have low ionization energies. This is due to the shielding (by the inner shell electrons) from the positive charge of the nucleus.

The greater the distance between the electrons and the nucleus, the less energy is needed to remove the outer valence electrons.

2. C is correct.

Accepting electrons to form anions is a characteristic of non-metals to obtain the electron configuration of the noble gases (i.e., complete octet).

Except for helium (which has a complete octet with 2 electrons, $1s^2$), the noble gases have complete octets with ns^2 and np^6 orbitals.

Donating electrons to form cations (e.g., Ca^{2+}, Fr^+, Na^+) is a characteristic of metals to obtain the electron configuration of the noble gases (i.e., complete octet).

3. A is correct.

Metalloids are semimetallic elements (i.e., between metals and nonmetals). The metalloids are boron (B), silicon (Si), germanium (Ge), arsenic (As), antimony (Sb), and tellurium (Te). Some literature reports polonium (Po) and astatine (At) as metalloids.

Seventeen elements are generally classified as nonmetals. Eleven are gases: hydrogen (H), helium (He), nitrogen (N), oxygen (O), fluorine (F), neon (Ne), chlorine (Cl), argon (Ar), krypton (Kr), xenon (Xe) and radon (Rn). One nonmetal is a liquid – bromine (Br) – and five are solids: carbon (C), phosphorus (P), sulfur (S), selenium (Se), and iodine (I).

4. B is correct.

An element is a pure chemical substance consisting of a single type of atom, distinguished by its atomic number (Z) (i.e., the number of protons it contains). One hundred eighteen elements have been identified, of which the first 94 occur naturally on Earth, with the remaining 24 being synthetic elements.

The properties of the elements on the periodic table repeat at regular intervals, creating "groups" or "families" of elements. Each column on the periodic table is a group, and elements within each group have similar physical and chemical characteristics due to the orbital location of their outermost electron. These groups only exist because the elements of the periodic table are listed by increasing atomic numbers.

5. C is correct.

For $n = 2$ shells, it has 2 orbitals: *s, p*

Each orbital can hold two electrons.

The maximum number of electrons in each shell:

The *s* subshell has 1 spherical orbital and can accommodate 2 electrons.

The *p* subshell has 3 dumbbell-shaped orbitals and can accommodate 6 electrons.

Maximum number of electrons in $n = 2$ shell is:

2 (for *s*) + 6 (for *p*) = 8 electrons

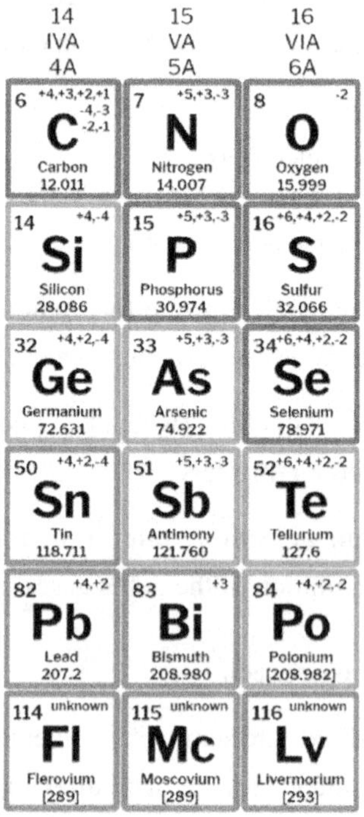

6. B is correct.

Groups IVA, VA and VIA each contain at least one metal and one nonmetal.

Group IVA has three metals (tin, lead, and flerovium) and one nonmetal (carbon).

Group VA has two metals (bismuth and moscovium) and two nonmetals (nitrogen and phosphorous).

Group VIA has one metal (livermorium) and three nonmetals (oxygen, sulfur, and selenium).

All three groups are part of the *p*-block of the periodic table.

7. C is correct.

Most elements on the periodic table (over 100 elements) are metals. Currently, there are 84 metal elements on the Periodic Table.

Seventeen elements are generally classified as nonmetals. Eleven are gases: hydrogen (H), helium (He), nitrogen (N), oxygen (O), fluorine (F), neon (Ne), chlorine (Cl), argon (Ar), krypton (Kr), xenon (Xe) and radon (Rn). One nonmetal is a liquid – bromine (Br) – and five are solids: carbon (C), phosphorus (P), sulfur (S), selenium (Se), and iodine (I).

Therefore, with the ratio of 84:17, there are about five times more metals than nonmetals.

8. D is correct.

English chemist John Dalton is known for his Atomic Theory, which states that *elements are made of small particles called atoms, which cannot be created or destroyed.*

9. B is correct.

Isotopes are variants of a particular element, which differ in the number of neutrons. Isotopes of the element have the same number of protons and occupy the same position on the periodic table.

The number of protons within the atom's nucleus is the atomic number (Z) equal to the number of electrons in the neutral (non-ionized) atom. Each atomic number identifies a specific element, but not the isotope; an atom of a given element may have a wide range in its number of neutrons.

The number of protons and neutrons (i.e., nucleons) in the nucleus is the atom's mass number (A), and each isotope of an element has a different mass number.

The atomic mass unit (amu) was designed using ^{12}C isotope as the reference.

1 amu = 1/12 mass of a ^{12}C atom.

Masses of other elements are measured against this standard.

If the mass of an atom 55.91 amu, the atom's mass is 55.91 × (1/12 mass of ^{12}C).

10. A is correct.

The three coordinates that come from Schrodinger's wave equations are the principal (n), angular (l), and magnetic (m) quantum numbers. These quantum numbers describe the size, shape, and orientation in the space of the orbitals on an atom.

The principal quantum number (n) describes the size of the orbital, the energy of an electron, and the most probable distance of the electron from the nucleus.

The principal quantum number (n) refers to the size of the orbital and the energy level of an electron.

The angular momentum quantum number (l) describes the shape of the orbital within the subshells.

The magnetic quantum number (m) determines the number of orbitals and their orientation within a subshell. Consequently, its value depends on the orbital angular momentum quantum number (l). Given a specific l, m is an interval ranging from $-l$ to $+l$ (i.e., it can be zero, a negative integer, or a positive integer).

The s is the spin quantum number (i.e.., $+½$ or $-½$).

11. B is correct.

Electron shells represent the orbit that electrons allow around an atom's nucleus. Each shell comprises one or more subshells, named using lowercase letters (*s*, *p*, *d*, *f*).

The first shell has one subshell (1*s*), the second shell has two subshells (2*s*, 2*p*), the third shell has three subshells (3*s*, 3*p*, 3*d*), etc.

An *s* subshell holds 2 electrons, and each subsequent subshell in the series can hold 4 more (*p* holds 6, *d* holds 10, *f* holds 14).

The shell number (i.e., principal quantum number) before the *s* (i.e., 4 in this example) does not affect how many electrons can occupy the subshell.

Subshell name	Subshell max electrons	Shell max electrons
1s	2	**2**
2s	2	2 + 6 = **8**
2p	6	
3s	2	2 + 6 + 10 = **18**
3p	6	
3d	10	
4s	2	2 + 6 + + 10 + 14 = **32**
4p	6	
4d	10	
4f	14	

12. D is correct.

The characteristic line spectra for atoms result from photons being emitted when excited electrons drop to lower energy levels.

13. A is correct.

In general, the size of neutral atoms increases down a group (i.e., increasing shell size) and decreases from left to right across the periodic table.

Negative ions (anions) are *much larger* than their neutral elements, while positive ions (cations) are *smaller*.

All examples are isoelectronic because of the same number of electrons.

Atomic numbers:

Br = 35; K = 19; Ar = 18; Ca = 20

The general trend for the atomic radius is to decrease from left to right and increase from top to bottom in the periodic table.

When the ion gains or loses an electron to create a new charged ion, its radius changes slightly, but the general radius trend still applies.

The ions K^+, Ca^{2+}, and Ar have identical numbers of electrons.

However, Br is located below Cl (larger principal quantum number *n*), and its atomic number is almost twice the others. This indicates that Br has more electrons, and its radius must be significantly larger than the other atoms.

14. C is correct.

The ground state configuration of sulfur is $[Ne]3s^2 3p^4$.

According to Hund's rule, the p orbitals are filled separately and pair the electrons by +½ or –½ spin.

The first three p electrons fill separate orbitals, and then the fourth electron pairs with two remaining unpaired electrons.

15. A is correct.

There are two ways to obtain the proper answer to this problem:

1. Using the atomic number.

Calculate the atomic number by adding the electrons:

$2 + 2 + 6 + 2 + 6 + 1 = 19$

Find element number 19 in the periodic table.

Check the group where it is located to see other elements that belong to the same group.

Element number 19 is potassium (K), so the element that belongs to the same group (IA) is lithium (Li).

2. Using subshells.

Identify the outermost subshell and use it to identify its group in the periodic table:

In this problem, the outermost subshell is $4s^1$.

Relationship between outermost subshell and group:

s^1 = Group IA

s^2 = Group IIA

p^1 = Group IIIA

p^2 = Group IVA

…

p^6 = Group VIII A

d = transition element

f = lanthanide/actinide element

16. C is correct.

The number of valence electrons for an element can be determined by its group (i.e., vertical column) on the periodic table. Except for the transition metals (i.e., groups 3-12), the group number identifies how many valence electrons are associated with a particular element: elements of the same group have the same number of valence electrons.

Atoms are most stable when they contain 8 electrons (i.e., complete octet) in the valence shell.

17. D is correct.

The principal quantum number (*n*) describes the size of the orbital and the energy of an electron, and the most probable distance from the nucleus. It refers to the size of the orbital and the energy level of an electron.

The elements with larger shell sizes (*n* is large) listed at the bottom of the periodic table have low ionization energies. This is due to the shielding (by the inner shell electrons) from the positive charge of the nucleus.

The greater the distance between the electrons and the nucleus, the less energy is needed to remove the outer valence electrons.

Ionization energy decreases with increasing shell size (i.e., *n* value) and generally increases to the right across a period (i.e., row) in the periodic table.

Moving down a column corresponds to increasing shell size with electrons further from the nucleus and decreasing nuclear attraction.

18. C is correct.

The *f* subshell has 7 orbitals.

Each orbital can hold two electrons.

The capacity of an *f* subshell is 7 orbitals × 2 electrons/orbital = 14 electrons.

19. C is correct.

The term "electron affinity" does not use the word energy as a reference; it is one of the measurable energies like ionization energy.

20. B is correct.

The attraction of the nucleus on the outermost electrons determines the ionization energy, which increases towards the right and increases up on the periodic table.

Practice Set 2: Questions 21–40

21. C is correct.

The mass number (A) is the sum of protons and neutrons in an atom.

The mass number approximates the atomic weight of the element as amu (grams per mole).

The problem only specifies the atomic mass (A) of Cl: 35 amu.

The atomic number (Z) is not given, but the information is available in the periodic table (atomic number = 17).

number of neutrons = atomic weight – atomic number

number of neutrons = 35 – 17

number of neutrons = 18

22. D is correct.

An element is a pure chemical substance that consists of a single type of atom, distinguished by its atomic number (Z) (i.e., the number of protons that it contains). One hundred eighteen elements have been identified, of which the first 94 occur naturally on Earth, with the remaining 24 being synthetic elements.

The properties of the elements on the periodic table repeat at regular intervals, creating "groups" or "families" of elements. Each column on the periodic table is a group, and elements within each group have similar physical and chemical characteristics due to the orbital location of their outermost electron. These groups only exist because the elements of the periodic table are listed by increasing atomic numbers.

23. A is correct.

Metals are elements that form positive ions by losing electrons during chemical reactions. Thus, metals are electropositive elements.

Metals are characterized by bright luster, hardness, ability to resonate sound and are excellent conductors of heat and electricity.

Metals, except mercury, are solids under normal conditions. Potassium has the lowest melting point of the solid metals at 146 °F.

24. C is correct.

Electron affinity is defined as the energy liberated when an electron is added to a gaseous neutral atom converting it to an anion.

25. D is correct.

Metalloids are semimetallic elements (i.e., between metals and nonmetals). The metalloids are boron (B), silicon (Si), germanium (Ge), arsenic (As), antimony (Sb), and tellurium (Te). Some literature reports polonium (Po) and astatine (At) as metalloids.

They have properties between metals and nonmetals. They typically have a metallic appearance but are only fair conductors of electricity (as opposed to metals that are excellent conductors), making them useable in the semiconductor industry.

Metalloids tend to be brittle, and chemically they behave more like nonmetals.

26. D is correct.

Halogens (Group VIIA) include fluorine (F), chlorine (Cl), bromine (Br), iodine (I), and astatine (At).

Halogens gain one electron to become a –1 anion, and the resulting ion has a complete octet of valence electrons.

27. C is correct.

Dalton's Atomic Theory, developed in the early 1800s, states that atoms of a given element are identical in mass and properties.

The masses of atoms of a particular element may be not identical, although atoms of an element must have the same number of protons; they can have different numbers of neutrons (i.e., isotopes).

28. D is correct.

Elements are defined by the number of protons (i.e., atomic number).

The isotopes are neutral atoms: # electrons = # protons.

Isotopes are variants of a particular element that differ in the number of neutrons. Isotopes of the element have the same number of protons and occupy the same position on the periodic table.

The subscript on the left denotes the number of protons.

The superscript on the left denotes the number of protons and neutrons.

The superscript on the right denotes the charge of an ion.

The number of protons within the atom's nucleus is the atomic number (Z) equal to the number of electrons in the neutral (non-ionized) atom.

Each atomic number (Z) identifies a specific element, but not the isotope; an atom of a given element may have a wide range in its number of neutrons.

The number of protons and neutrons (i.e., nucleons) in the nucleus is the atom's mass number (A), and each isotope of an element has a different mass number.

29. A is correct.

A cathode-ray particle is a different name for an electron.

Those particles (i.e., electrons) are attracted to the positively charged cathode, implying that they are negatively charged.

30. B is correct.

The three coordinates that come from Schrodinger's wave equations are the principal (n), angular (l), and magnetic (m) quantum numbers. These quantum numbers describe the size, shape, and orientation in the space of the orbitals on an atom.

The principal quantum number (n) describes the size of the orbital, the energy of an electron, and the most probable distance of the electron from the nucleus. It refers to the size of the orbital and the energy level of an electron.

The angular momentum quantum number (l) describes the shape of the orbital of the subshells.

Carbon has an atomic number of 6 and an electron configuration of $1s^2$, $2s^2$, $2p^2$.

Therefore, electrons are in the second shell of $n = 2$, and two subshells are in the outermost shell of $l = 1$.

The values of l are 0 and 1 whereby, only the largest value of l ($l = 1$) is reported.

31. A is correct.

The alkaline earth metals (group IIA), in the ground state, have a filled s subshell with 2 electrons.

32. D is correct.

The 3rd shell consists of s, p and d subshells.

Each orbital can hold two electrons.

The s subshell has 1 spherical orbital and can accommodate 2 electrons

The p subshell has 3 dumbbell-shaped orbitals and can accommodate 6 electrons

The d subshell has 5 lobe-shaped orbitals and can accommodate 10 electrons

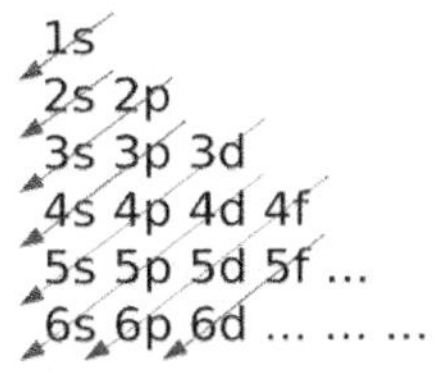

Order of filling orbitals

The $n = 3$ shell can accommodate 18 electrons.

The element with the electron configuration terminating in $3p^4$ is sulfur (i.e., a total of 16 electrons).

33. B is correct.

Ions typically form with the same electron configuration as noble gases. Except for helium (which has a complete octet with 2 electrons, $1s^2$), the noble gases have complete octets with ns^2 and np^6 orbitals.

Halogens (Group VIIA) include fluorine (F), chlorine (Cl), bromine (Br), iodine (I), and astatine (At).

Halogens gain one electron to become –1 anion, and the resulting ion has a complete octet of valence electrons.

34. C is correct.

A continuous spectrum refers to a broad, uninterrupted spectrum of radiant energy.

The visible spectrum refers to the light that humans can see and includes the colors of the rainbow.

The ultraviolet spectrum refers to electromagnetic radiation with a wavelength shorter than visible light but longer than X-rays.

The radiant energy spectrum refers to electromagnetic (EM) waves of all wavelengths. However, the frequency bands in an EM signal may be sharply defined with interruptions or broad.

35. A is correct.

In general, the size of neutral atoms increases down a group (i.e., increasing shell size) and decreases from left to right across the periodic table.

Positive ions (cations) are *much smaller* than the neutral element (greater effective nuclear charge). In comparison, negative ions (anions) are *much larger* (due to the smaller effective nuclear charge and repulsion of valence electrons).

36. C is correct.

In the ground state, the 3*p* orbitals fill before the 3*d* orbitals.

The lowest energy orbital fills before an orbital of a higher energy level.

Aufbau principle to determine the order of energy levels in subshells:

From the table on the right, the orbitals increase in energy from:

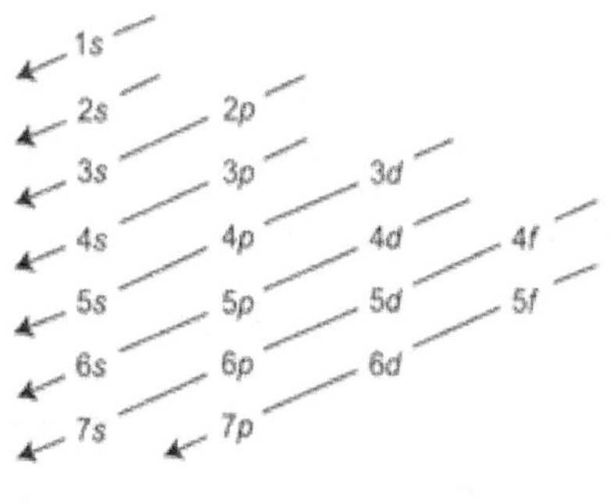

$$1s < 2s < 2p < 3s < 3p < 4s < 3d < 4p < 5s < 4d < 5p < 6s < 4f < 5d < 6p < 7s < 5f < 6d < 7p$$

37. B is correct.

Ionization energy is the energy needed to remove an electron from a neutral atom of the element in the gas phase.

The principal quantum number (n) describes the size of the orbital, the energy of an electron, and the most probable distance from the nucleus. It refers to the size of the orbital and the energy level of an electron.

The elements with larger shell sizes (n is large) listed at the bottom of the periodic table have low ionization energies. This is due to the shielding (by the inner shell electrons) from the positive charge of the nucleus. The greater the distance between the electrons and the nucleus, the less energy is needed to remove the outer valence electrons.

Ionization energy decreases with increasing shell size (i.e., n value) and generally increases to the right across a period (i.e., row) in the periodic table.

Neon (Ne) has an atomic number of 10 and a shell size of $n = 2$.

Rubidium (Rb) has an atomic number of 37 and a shell size of $n = 5$.

Potassium (K) has an atomic number of 19 and a shell size of $n = 4$.

Calcium (Ca) has an atomic number of 20 and a shell size is $n = 4$.

38. B is correct.

Seventeen elements are generally classified as nonmetals. Eleven are gases: hydrogen (H), helium (He), nitrogen (N), oxygen (O), fluorine (F), neon (Ne), chlorine (Cl), argon (Ar), krypton (Kr), xenon (Xe) and radon (Rn). One nonmetal is a liquid – bromine (Br) – and five are solids: carbon (C), phosphorus (P), sulfur (S), selenium (Se), and iodine (I).

Alkali metals (group IA) include lithium (Li), potassium (K), sodium (Na), rubidium (Rb), cesium (Cs) and francium (Fr).

Alkali metals lose one electron to become +1 cations, and the resulting ion has a complete octet of valence electrons.

Alkaline earth metals (group IIA) include beryllium (Be), magnesium (Mg), calcium (Ca), strontium (Sr), barium (Ba), and radium (Ra). Alkaline earth metals lose two electrons to become +2 cations, and the resulting ion has a complete octet of valence electrons.

39. C is correct.

The atom has 47 protons, 47 electrons, and 60 neutrons.

Because the periodic table is arranged by atomic number (Z), the fastest way to identify an element is to determine its atomic number. The atomic number equals the number of protons or electrons, which means that this atom's atomic number is 47. Use this information to locate element #47 in the table, which is Ag (silver).

Check the atomic mass (A), equal to the atomic number + number of neutrons.

For this atom, the mass is $60 + 47 = 107$.

The mass of Ag on the periodic table is listed as 107.87, which is the average mass of Ag isotopes.

Usually, isotopes of an element have similar masses (within 1-3 amu).

40. D is correct.

The three coordinates that come from Schrodinger's wave equations are the principal (n), angular (l), and magnetic (m) quantum numbers.

These quantum numbers describe the size, shape, and orientation of orbitals in an atom.

The principal quantum number (n) describes the size of the orbital, the energy of an electron, and the most probable distance of the electron from the nucleus.

==

Practice Set 3: Questions 41–60

==

41. B is correct.

The atomic number (Z) is the sum of protons in an atom that determines an element's chemical properties and location on the periodic table.

The mass number (A) is the sum of protons and neutrons in an atom.

The mass number approximates the atomic weight of the element as amu (grams per mole).

Atomic mass – atomic number = number of neutrons

9 – 4 = 5 neutrons

42. A is correct.

Electrons are the negatively charged particles (charge –1) located in the electron cloud, orbiting around the nucleus of the atom. Electrons are extremely tiny particles, much smaller than protons and neutrons, and they have a mass of about 5×10^{-4} amu.

43. D is correct.

A compound consists of two or more different atoms which associate via chemical bonds.

Calcium chloride ($CaCl_2$) is an ionic compound of calcium and chloride.

Dichloromethane has the molecular formula of CH_2Cl_2

Dichlorocalcium exists as a hydrate with the molecular formula of $CaCl_2{\cdot}(H_2O)_2$.

Carbon chloride (i.e., carbon tetrachloride) has the molecular formula of CCl_4.

44. A is correct.

Congeners are chemical substances related by origin, structure, or function. Regarding the periodic table, congeners are the elements of the same group which share similar properties (e.g., copper, silver, and gold are congeners of Group 11).

Stereoisomers, diastereomers, and epimers are terms commonly used in organic chemistry.

Stereoisomers are chiral molecules (attached to 4 different substituents and are non-superimposable mirror images. They have the same molecular formula and the same sequence of bonded atoms but are oriented differently in 3-D space (e.g., *R* / *S* enantiomers).

Diastereomers are chiral molecules that are not mirror images. The most common form is a chiral molecule with more than 1 chiral center. Additionally, *cis* / *trans* (*Z* / *E*) geometric isomers are diastereomers.

Epimers are diastereomers differing in absolute configuration at only one chiral center.

Anomers are a type of stereoisomer used in carbohydrate chemistry to describe the orientation of the glycosidic bond of adjacent saccharides (e.g., α and β linkage of sugars). A refers to the hydroxyl group – of the anomeric carbon – pointing downward while β points upward.

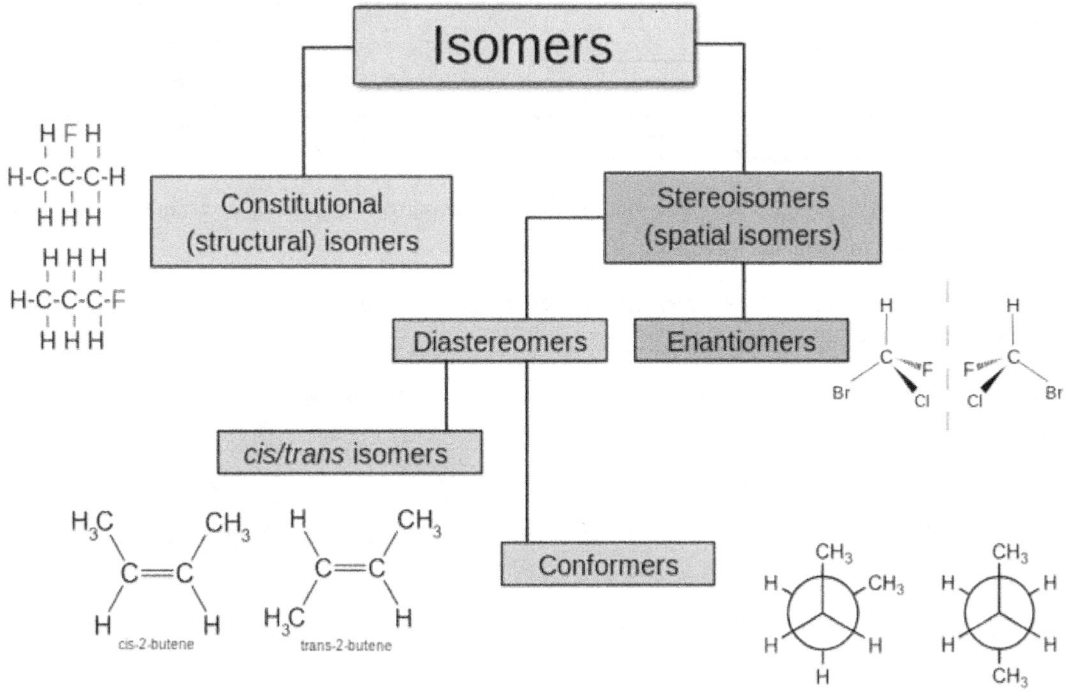

Summary of isomers

45. D is correct.

A group (or family) is a vertical column, and elements within each group share similar properties.

A period is a horizontal row in the periodic table of elements.

46. A is correct.

Metals, except mercury, are solids under normal conditions.

Potassium has the lowest melting point of the solid metals at 146 °F.

The relatively low melting temperature for potassium is due to its fourth shell ($n = 4$), which means its valence electrons are further from the nucleus; therefore, there is less attraction between its electrons and protons.

47. C is correct.

Alkaline earth metals (group IIA) include beryllium (Be), magnesium (Mg), calcium (Ca), strontium (Sr), barium (Ba), and radium (Ra). Alkaline earth metals lose two electrons to become +2 cations, and the resulting ion has a complete octet of valence electrons.

Transition metals (or transition elements) are defined as a partially-filled *d* or *f* subshell in a common oxidative state.

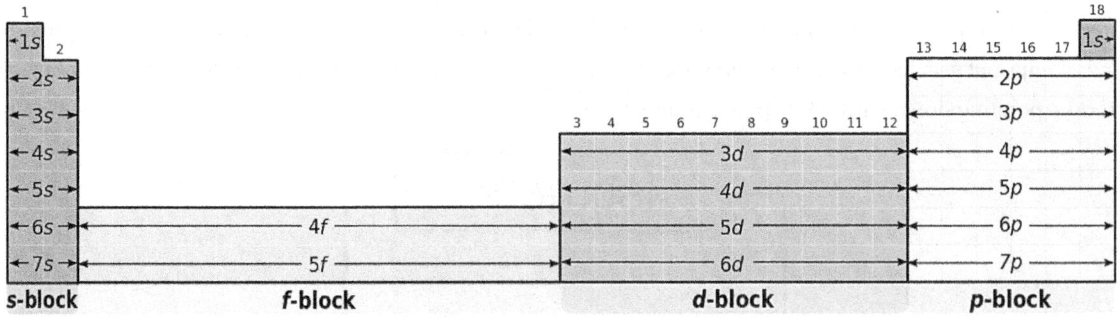

Transition metals occur in groups (vertical columns) 3–12 of the period table. They occur in periods (horizontal rows) 4–7. This group of elements includes silver, iron, and copper.

The *f*-block lanthanides (i.e., rare earth metals) and actinides (i.e., radioactive elements) are considered transition metals and are *inner transition metals*.

Noble gases (group VIIIA) include helium (He), neon (Ne), argon (Ar), krypton (Kr), xenon (Xe), radon (Rn), and oganesson (Og).

Alkali metals (group IA) include lithium (Li), potassium (K), sodium (Na), rubidium (Rb), cesium (Cs) and francium (Fr). Alkali metals lose one electron to become +1 cations, and the resulting ion has a complete octet of valence electrons.

Halogens (Group VIIA) include fluorine (F), chlorine (Cl), bromine (Br), iodine (I), and astatine (At). Halogens gain one electron to become a –1 anion, and the resulting ion has a complete octet of valence electrons.

48. D is correct.

The metalloids have some properties of metals and some properties of nonmetals.

Metalloids are semimetallic elements (i.e., between metals and nonmetals). The metalloids are boron (B), silicon (Si), germanium (Ge), arsenic (As), antimony (Sb), and tellurium (Te). Some literature reports polonium (Po) and astatine (At) as metalloids.

They have properties between metals and nonmetals. They typically have a metallic appearance but are only fair conductors of electricity (as opposed to metals that are excellent conductors), making them useable in the semiconductor industry.

Metalloids tend to be brittle, and chemically they behave more like nonmetals.

However, the elements in the IIIB group are transition metals, not metalloids.

49. B is correct.

Halogens (Group VIIA) include fluorine (F), chlorine (Cl), bromine (Br), iodine (I), and astatine (At).

Halogens gain one electron to become a –1 anion, and the resulting ion has a complete octet of valence electrons.

50. A is correct.

Isotopes are variants of a particular element that differ in the number of neutrons. Isotopes of the element have the same number of protons and occupy the same position on the periodic table.

The number of protons within the atom's nucleus is the atomic number (Z) equal to the number of electrons in the neutral (non-ionized) atom. Each atomic number identifies a specific element, but not the isotope; an atom of a given element may have a wide range in its number of neutrons.

The number of protons and neutrons (i.e., nucleons) in the nucleus is the atom's mass number (A), and each isotope of an element has a different mass number.

From the periodic table, the atomic mass of a natural Si sample is 28.1, which is less than the mass of ^{29}Si or ^{30}Si. Therefore, ^{28}Si is the most abundant isotope.

51. C is correct.

The initial explanation was that the ray was present in the gas, and the cathode activated it.

Even when gas was not present, the ray was observed, so the conclusion was that the ray must have been coming from the cathode itself.

52. D is correct.

The choices correctly describe the spin quantum number (s).

The three coordinates that come from Schrodinger's wave equations are the principal (n), angular (l), and magnetic (m) quantum numbers. These quantum numbers describe the size, shape, and orientation in the space of the orbitals of an atom.

The principal quantum number (n) describes the size of the orbital, energy of an electron, and the most probable distance of the electron from the nucleus. It refers to the size of the orbital and the energy level of an electron.

The angular momentum quantum number (l) describes the shape of the orbitals of the subshells.

The magnetic quantum number (m) determines the number of orbitals and their orientation within a subshell. Consequently, its value depends on the orbital angular momentum quantum number (l).

Given a specific l, m is an interval ranging from $-l$ to $+l$ (i.e., zero, a negative integer, or a positive integer).

The s is the spin quantum number (e.g., $+½$ or $-½$).

53. D is correct.

The three coordinates that come from Schrodinger's wave equations are the principal (n), angular (l), and magnetic (m) quantum numbers. These quantum numbers describe the size, shape, and orientation in the space of the orbitals on an atom.

The principal quantum number (n) describes the size of the orbital, the energy of an electron, and the most probable distance of the electron from the nucleus. It refers to the size of the orbital and the energy level of an electron.

The angular momentum quantum number (l) describes the shape of the orbital of the subshells.

The magnetic quantum number (m) determines the number of orbitals and their orientation within a subshell. Consequently, its value depends on the orbital angular momentum quantum number (l). Given a specific l, m is an interval ranging from $-l$ to $+l$ (i.e., zero, a negative integer, or a positive integer).

The fourth quantum number is s, which is the spin quantum number (e.g., $+\frac{1}{2}$ or $-\frac{1}{2}$).

Electrons cannot be precisely located in space at a point in time, and orbitals describe probability regions for finding the electrons.

The values needed to locate an electron are n, m and l. The spin can be $+\frac{1}{2}$ or $-\frac{1}{2}$, so four values are needed to describe a single electron.

54. A is correct.

The lowest energy orbital fills before an orbital of a higher energy level.

Aufbau principle to determine the order of energy levels in subshells:

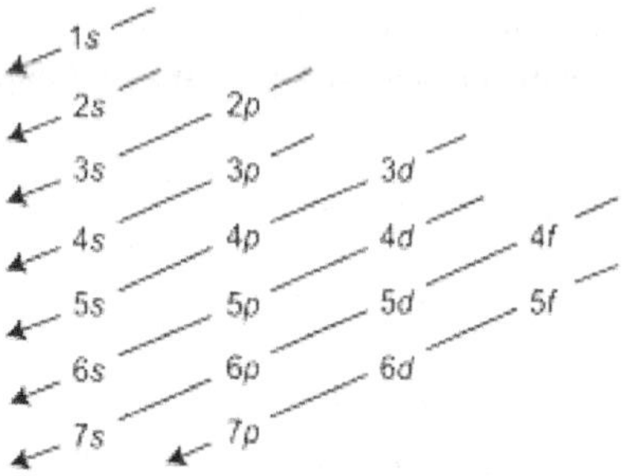

From the table on the right, the orbitals increase in energy from:

$$1s < 2s < 2p < 3s < 3p < 4s < 3d < 4p < 5s < 4d < 5p < 6s < 4f < 5d < 6p < 7s < 5f < 6d < 7p$$

55. D is correct.

In general, the size of neutral atoms increases down a group (i.e., increasing shell size) and decreases from left to right across the periodic table.

Positive ions (cations) are *much smaller* than the neutral element (due to the greater effective nuclear charge). In comparison, negative ions (anions) are *much larger* (due to the smaller effective nuclear charge and repulsion of valence electrons).

56. D is correct.

Boron's atomic number is 5; therefore, it contains 5 electrons.

Use the Aufbau principle to determine the order of filling orbitals.

Remember that each electron shell (principal quantum number n) starts with a new s orbital.

57. A is correct.

Identify an element using the periodic table is its atomic number.

The atomic number is equal to the number of protons or electrons.

The total number of electrons can be found by adding the electrons in the provided electron configuration:

$$2 + 2 + 6 + 2 + 6 + 2 + 10 + 6 + 2 + 10 + 2 = 50.$$

Element #50 in the periodic table is tin (Sn).

58. A is correct.

Ionization energy (IE) is the amount of energy required to remove an isolated gaseous atom's most loosely bound electron to form a cation. This is an endothermic process.

Ionization energy is expressed as:

$$X + \text{energy} \rightarrow X^+ + e^-$$

where X is any atom (or molecule) capable of being ionized (i.e., having an electron removed), X^+ is that atom or molecule after an electron is removed, and e^- is the removed electron.

The principal quantum number (n) describes the size of the orbital, the energy of an electron, and the most probable distance of the electron from the nucleus. It refers to the size of the orbital and the energy level of an electron.

The elements with larger shell sizes (n is large) at the bottom of the periodic table have low ionization energies. This is due to the shielding (by inner shell electrons) from the positive charge of the nucleus. The greater the distance between the electrons and the nucleus, the less energy is needed to remove the outer valence electrons.

Ionization energy decreases with increasing shell size (i.e., n value) and generally increases to the right across a period (i.e., row) in the periodic table.

Argon (Ar) has an atomic number of 18 and a shell size of $n = 3$.

Strontium (Sr) has an atomic number of 38 and a shell size of $n = 5$.

Bromine (Br) has an atomic number of 35 and a shell size of $n = 4$.

Indium (In) has an atomic number of 49 and a shell size is $n = 5$.

59. A is correct.

Electronegativity is defined as the ability of an atom to attract electrons when it bonds with another atom. The most common use of electronegativity pertains to polarity along the sigma (single) bond.

The trend for increasing electronegativity within the periodic table is up and toward the right. The most electronegative atom is fluorine (F), while the least is francium (Fr).

The greater the difference in electronegativity between two atoms, the more polar a bond these atoms form. The atom with the higher electronegativity is the partial (delta) negative end of the dipole.

60. C is correct.

Seventeen elements are generally classified as nonmetals. Eleven are gases: hydrogen (H), helium (He), nitrogen (N), oxygen (O), fluorine (F), neon (Ne), chlorine (Cl), argon (Ar), krypton (Kr), xenon (Xe) and radon (Rn). One nonmetal is a liquid – bromine (Br) – and five are solids: carbon (C), phosphorus (P), sulfur (S), selenium (Se), and iodine (I).

Metals, except mercury, are solids under normal conditions. Potassium has the lowest melting point of the solid metals at 146 °F.

===

Practice Set 4: Questions 61–80

===

61. D is correct.

When an electron absorbs energy, it moves temporarily to a higher energy level. It then drops back to its initial state (or *ground state*) while emitting excess energy. This emission can be observed as visible spectrum lines.

The protons do not move between energy levels, so they cannot absorb or emit energy.

62. B is correct.

The three coordinates that come from Schrodinger's wave equations are the principal (n), angular (l), and magnetic (m) quantum numbers. These quantum numbers describe the size, shape, and orientation in the space of the orbitals on an atom.

The principal quantum number (n) describes the size of the orbital, the energy of an electron, and the most probable distance of the electron from the nucleus.

The angular momentum quantum number (l) describes the shape of the orbital of the subshells.

The magnetic quantum number (m) determines the number of orbitals and their orientation within a subshell. Consequently, its value depends on the orbital angular momentum quantum number (l). Given a specific l, m is an interval ranging from $-l$ to $+l$ (i.e., it can be zero, a negative integer, or a positive integer).

l must be less than n, while m_l must be less than or equal to l.

63. C is correct.

The mass number (A) is the number of nucleons (i.e., protons and neutrons) in an atom.

The atomic number (Z) is the number of protons in an atom.

The number of neutrons in an atom can be calculated by subtracting the atomic number (Z) from the mass number (A).

Mass number – atomic number = number of neutrons

64. B is correct.

A compound consists of two or more different types of atoms that associate via chemical bonds.

An element is a pure chemical substance consisting of a single atom, defined by its atomic number (Z), the number of protons.

One hundred eighteen elements have been identified, of which the first 94 occur naturally on Earth.

65. A is correct.

Alkali metals (group IA) include lithium (Li), potassium (K), sodium (Na), rubidium (Rb), cesium (Cs) and francium (Fr).

Alkaline earth metals (group IIA) include beryllium (Be), magnesium (Mg), calcium (Ca), strontium (Sr), barium (Ba), and radium (Ra).

Halogens (Group VIIA) include fluorine (F), chlorine (Cl), bromine (Br), iodine (I), and astatine (At). Halogens gain one electron to become –1 anion, and the resulting ion has a complete octet of valence electrons.

Noble gases (group VIIIA) include helium (He), neon (Ne), argon (Ar), krypton (Kr), xenon (Xe), radon (Rn), and oganesson (Og). Except for helium (which has a complete octet with 2 electrons, $1s^2$), the noble gases have complete octets with ns^2 and np^6 orbitals.

Representative elements on the periodic table are groups IA and IIA (on the left) and groups IIIA – VIIIA (on the right).

Polonium (Po) element 84 is highly radioactive, with no stable isotopes, and classified as a metalloid or metal.

66. B is correct.

Metalloids are semimetallic elements (i.e., between metals and nonmetals). The metalloids are boron (B), silicon (Si), germanium (Ge), arsenic (As), antimony (Sb), and tellurium (Te). Some literature reports polonium (Po) and astatine (At) as metalloids.

They have properties between metals and nonmetals. They typically have a metallic appearance but are only fair conductors of electricity (as opposed to metals that are excellent conductors), making them useable in the semiconductor industry.

Metalloids tend to be brittle, and chemically they behave more like nonmetals.

Alkali metals (group IA) include lithium (Li), potassium (K), sodium (Na), rubidium (Rb), cesium (Cs) and francium (Fr).

Alkaline earth metals (group IIA) include beryllium (Be), magnesium (Mg), calcium (Ca), strontium (Sr), barium (Ba), and radium (Ra).

Seventeen elements are generally classified as nonmetals. Eleven are gases: hydrogen (H), helium (He), nitrogen (N), oxygen (O), fluorine (F), neon (Ne), chlorine (Cl), argon (Ar), krypton (Kr), xenon (Xe) and radon (Rn). One nonmetal is a liquid – bromine (Br) – and five are solids: carbon (C), phosphorus (P), sulfur (S), selenium (Se), and iodine (I).

Nonmetals tend to be highly volatile (i.e., easily vaporized), have low elasticity, and are good insulators of heat and electricity.

Nonmetals tend to have high ionization energy and electronegativity and share (or gain) an electron when bonding with other elements.

Halogens (Group VIIA) include fluorine (F), chlorine (Cl), bromine (Br), iodine (I), and astatine (At). Halogens gain one electron to become a –1 anion, and the resulting ion has a complete octet of valence electrons.

67. D is correct.

In general, the size of neutral atoms increases down a group (i.e., increasing shell size) and decreases from left to right across the periodic table.

Positive ions (cations) are *much smaller* than the neutral element (due to the greater effective nuclear charge). In comparison, negative ions (anions) are *much larger* (due to the smaller effective nuclear charge and repulsion of valence electrons).

Sulfur (S, atomic number = 16) is smaller than aluminum (Al, atomic number = 13) due to the increase in the number of protons (effective nuclear charge) from left to right across a period (i.e., horizontal rows).

Al^{3+} has the same electronic configuration as Ne ($1s^22s^22p^6$) compared to Al ($1s^22s^22p^63s^23p^1$).

68. A is correct.

The valence shell is an atom's outermost shell (i.e., highest principal quantum number n). Valence electrons are those electrons of the outermost electron shell that can participate in a chemical bond.

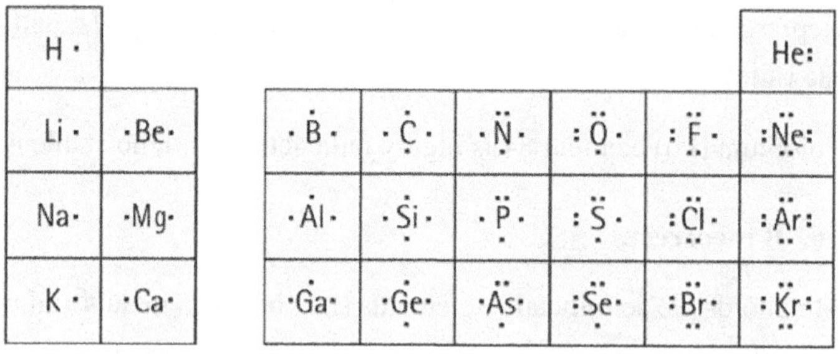

Lewis dot structures showing valence electrons for elements

The number of valence electrons for an element can be determined by its group (i.e., vertical column) on the periodic table. Except for the transition metals (i.e., groups 3-12), the group number identifies how many valence electrons are associated with a particular element: elements of the same group have the same number of valence electrons.

69. B is correct.

The semimetallic elements are arsenic (As), antimony (Sb), bismuth (Bi), and graphite (a crystalline form of carbon). Arsenic and antimony are considered metalloids (along with boron, silicon, germanium, and tellurium), but the terms semimetal and metalloid are not synonymous.

Semimetals, in contrast to metalloids, can be chemical compounds.

70. C is correct.

Noble gases (group VIIIA) include helium (He), neon (Ne), argon (Ar), krypton (Kr), xenon (Xe), radon (Rn), and oganesson (Og).

Except for helium (which has a complete octet with 2 electrons, $1s^2$), the noble gases have complete octets with ns^2 and np^6 orbitals.

The metalloids are boron (B), silicon (Si), germanium (Ge), arsenic (As), antimony (Sb), and tellurium (Te). Some literature reports polonium (Po) and astatine (At) as metalloids.

71. D is correct.

Isotopes are variants of a particular element that differ in the number of neutrons. Isotopes of the element have the same number of protons and occupy the same position on the periodic table.

The experimental results should depend on the mass of the gas molecules.

Deuterium (D or ^2H) is as *heavy hydrogen.* It is one of two stable isotopes of hydrogen. The deuterium nucleus contains one proton and one neutron, compared to H, with 1 proton and 0 neutrons. The mass of deuterium is 2.0141 daltons, compared to 1.0078 daltons for hydrogen.

Based on the mass difference between the isotopes, the density, rate of gas effusion, and atomic vibrations would be different.

72. B is correct.

Elements are defined by the number of protons (i.e., atomic number).

The isotopes are neutral atoms: # electrons = # protons.

Isotopes are variants of a particular element that differ in the number of neutrons. Isotopes of the element have the same number of protons and occupy the same position on the periodic table.

The superscript on the left denotes the number of protons and neutrons.

Since naturally occurring lithium has a mass of 6.9 g/mol and protons and neutrons have a mass of approximately 1 g/mol, ^{7}lithium is the predominant isotope.

73. A is correct.

The charge on 1 electron is negative.

One mole of electrons:

Avogadro's number × e^-.

$(6.02 \times 10^{23}) \times (1.60 \times 10^{-19})$ = 96,485 coulombs (values are rounded).

74. D is correct.

Each orbital can hold two electrons.

The *f* subshell has 7 orbitals and can accommodate 14 electrons.

The *d* subshell has 5 lobed orbitals and can accommodate 10 electrons.

The $n = 3$ shell contains only *s*, *p* and *d* subshells.

75. B is correct.

Ionization energy (IE) is the amount of energy required to remove an isolated gaseous atom's most loosely bound electron to form a cation. This is an endothermic process.

Ionization energy is expressed as:

$$X + \text{energy} \rightarrow X^+ + e^-$$

where X is any atom (or molecule) capable of being ionized (i.e., having an electron removed), X^+ is that atom or molecule after an electron is removed, and e^- is the removed electron.

The principal quantum number (n) describes the size of the orbital, the energy of an electron, and the most probable distance of the electron from the nucleus. It refers to the size of the orbital and the energy level of an electron.

The elements with larger shell sizes (n is large) listed at the bottom of the periodic table have low ionization energies. This is due to the shielding (by the inner shell electrons) from the positive charge of the nucleus. The greater the distance between the electrons and the nucleus, the less energy is needed to remove the outer valence electrons.

Ionization energy decreases with increasing shell size (i.e., n value) and generally increases to the right across a period (i.e., row) in the periodic table.

Chlorine (Cl) has an atomic number of 10 and a shell size of $n = 3$.

Francium (Fr) has an atomic number of 87 and a shell size of $n = 7$.

Gallium (Ga) has an atomic number of 31 and a shell size of $n = 4$.

Iodine (I) has an atomic number of 53 and a shell size is $n = 5$.

76. D is correct.

Electrons are electrostatically (i.e., negative and positive charge) attracted to the nucleus. An atom's electrons generally occupy outer shells only if other electrons have filled the more inner shells. However, there are exceptions to this rule, with some atoms having two or even three incomplete outer shells.

The Aufbau (German for building up) principle is based on the Madelung rule for filling the subshells with the lowest energy levels.

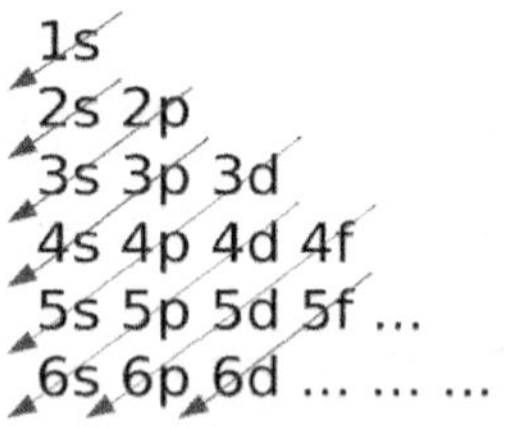

Order of filling electron's orbitals

77. C is correct.

In Bohr's atom model, electrons can jump to higher energy levels, gain energy, or drop to lower energy levels, releasing energy. When electric current flows through an element in the gas phase, glowing light is produced.

A pattern of lines known as the atomic spectra can be seen by directing this light through a prism.

These lines are produced by excited electrons dropping to lower energy levels. Since the energy levels in each element are different, each element has a unique set of lines it produces, which is why the spectrum is called the "atomic fingerprint" of the element.

78. A is correct.

Obtain the atomic number of Mn from the periodic table.

Mn is a transition metal, and it is in Group VIIB/7; its atomic number is 25.

Use the Aufbau principle to fill up the orbitals of Mn: $1s^22s^22p^63s^23p^64s^23d^5$

The transition metals occur in groups 3–12 (vertical columns) of the periodic table. They occur in periods 4–7 (horizontal rows).

Transition metals are defined as elements with a partially-filled *d* or *f* subshell in a common oxidative state. This group of elements includes silver, iron, and copper.

The transition metals are elements whose atom has an incomplete *d* sub-shell, giving rise to cations with an incomplete *d* sub-shell. By this definition, elements in groups 3–11 (or 12 by some literature) are transition metals.

The transition elements have characteristics not found in other elements, resulting from the partially filled *d* shell. These include the formation of compounds whose color is due to *d*–*d* electronic transitions and compounds in many oxidation states due to the relatively low reactivity of unpaired *d* electrons.

The transition elements form many paramagnetic (i.e., attracted to an externally applied magnetic field) compounds due to the presence of unpaired *d* electrons. By exception to their unique traits, a few compounds of main group elements are paramagnetic (e.g., nitric oxide and oxygen).

79. B is correct.

Electronegativity is defined as the ability of an atom to attract electrons when it bonds with another atom. The most common use of electronegativity pertains to polarity along the sigma (single) bond.

The trend for increasing electronegativity within the periodic table is up and toward the right. The most electronegative atom is fluorine (F), while the least is francium (Fr).

80. C is correct.

Alkali metals (group IA) include lithium (Li), potassium (K), sodium (Na), rubidium (Rb), cesium (Cs) and francium (Fr). Alkali metals lose one electron to become +1 cations, and the resulting ion has a complete octet of valence electrons.

Alkaline earth metals (group IIA) include beryllium (Be), magnesium (Mg), calcium (Ca), strontium (Sr), barium (Ba), and radium (Ra). Alkaline earth metals lose two electrons to become +2 cations, and the resulting ion has a complete octet of valence electrons.

Notes for active learning

Chemical Bonding – Detailed Explanations

Practice Set 1: Questions 1–20

---

1. D is correct.

The valence shell is an atom's outermost shell (i.e., highest principal quantum number *n*).

Valence electrons are those electrons of the outermost electron shell that can participate in a chemical bond.

The number of valence electrons for an element can be determined by its group (i.e., vertical column) on the periodic table. Except for the transition metals (i.e., groups 3-12), the group number identifies how many valence electrons are associated with a particular element: elements of the same group have the same number of valence electrons.

2. A is correct.

Three degenerate *p* orbitals exist for an atom with an electron configuration in the second shell or higher. The first shell only has access to *s* orbitals.

The *d* orbitals become available from n = 3 (third shell).

3. C is correct.

The valence shell is an atom's outermost shell (i.e., highest principal quantum number *n*).

Valence electrons are those electrons of the outermost electron shell that can participate in a chemical bond.

The number of valence electrons for an element can be determined by its group (i.e., vertical column) on the periodic table. Except for the transition metals (i.e., groups 3-12), the group number identifies how many valence electrons are associated with a particular element: elements of the same group have the same number of valence electrons.

To find the number of valence electrons in a sulfite ion, SO_3^{2-}, add the valence electrons of each atom:

Sulfur = 6; Oxygen = (6 × 3) = 18

Total = 24

This ion has a net charge of –2, which indicates that it has 2 extra electrons. Therefore, the number of valence electrons would be 24 + 2 = 26 electrons.

4. D is correct.

London dispersion forces result from the momentary flux of valence electrons and are present in all compounds; they are the attractive forces that hold molecules.

They are the weakest of the intermolecular forces, and their strength increases with increasing size (i.e., surface area contact) and polarity of the molecules involved.

5. B is correct.

Each hydroxyl group (alcohol or ~OH) has oxygen with 2 lone pairs and one attached hydrogen.

Therefore, each hydroxyl group can participate in 3 hydrogen bonds:

5 hydroxyl groups × 3 bonds = 15 hydrogen bonds.

The oxygen of the ether group (C–O–C) in the ring has 2 lone pairs for an additional 2 H–bonds.

6. B is correct.

The valence shell is an atom's outermost shell (i.e., highest principal quantum number *n*).

Valence electrons are those electrons of the outermost electron shell that can participate in a chemical bond.

The number of valence electrons for an element can be determined by its group (i.e., vertical column) on the periodic table. Except for the transition metals (i.e., groups 3-12), the group number identifies how many valence electrons are associated with a particular element: elements of the same group have the same number of valence electrons.

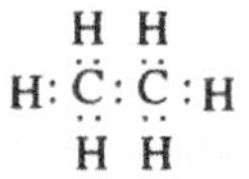

Lewis dot structure for ethane

7. A is correct.

In covalent bonds, the electrons can be shared equally or unequally.

Polar covalent bonded atoms are covalently bonded compounds that involve unequal sharing of electrons due to large electronegativity differences (Pauling units of 0.4 to 1.7) between the atoms.

An example of this is water, where there is a polar covalent bond between oxygen and hydrogen. Water is a polar molecule with the oxygen partial negative while the hydrogens are partial positive.

8. A is correct.

Van der Waals forces involve nonpolar (hydrophobic) molecules, such as hydrocarbons. The van der Waals force is the total of attractive or repulsive forces between molecules, and therefore can be attractive or repulsive. It can include the force between two permanent dipoles, the force between a permanent dipole and a temporary dipole, or the force between two temporary dipoles.

Hydrogens, bonded directly to F, O or N, participate in hydrogen bonds. The hydrogen is partial positive (i.e., delta plus or ∂+) due to the bond to these electronegative atoms. The lone pair of electrons on the F, O or N interacts with the ∂+ hydrogen to form a hydrogen bond.

Hydrogen bonds are a type of dipole-dipole and are the strongest intermolecular forces (i.e., between molecules), followed by other dipole-dipole, dipole-induced dipole, and van der Waals forces (i.e., London dispersion).

9. D is correct.

Representative structures include:

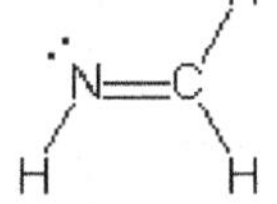

$HNCH_2$: one single and one double bond

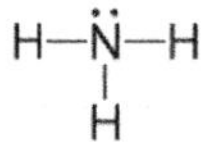

NH_3 : three single bonds

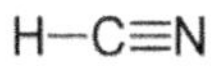

HCN : one triple bond and one single bond

10. C is correct.

The valence shell is an atom's outermost shell (i.e., highest principal quantum number *n*).

Valence electrons are those electrons of the outermost electron shell that can participate in a chemical bond.

The number of valence electrons for an element can be determined by its group (i.e., vertical column) on the periodic table. Except for the transition metals (i.e., groups 3-12), the group number identifies how many valence electrons are associated with a particular element: elements of the same group have the same number of valence electrons.

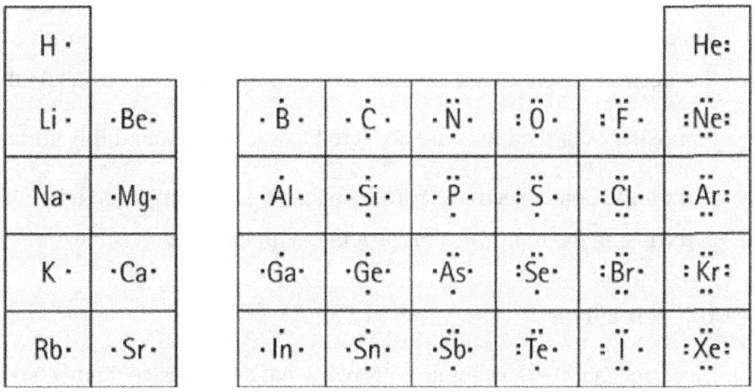

Sample Lewis dot structures for some elements

11. B is correct.

The nitrite ion has the chemical formula NO_2^- with the negative charge distributed between the two oxygen atoms.

Two resonance structures of the nitrite ion

12. D is correct

Nitrogen has 5 valence electrons. In ammonia, nitrogen has 3 bonds to hydrogen, and a lone pair remains on the central nitrogen atom.

The ammonium ion (NH_4^+) has 4 hydrogens, and the lone pair of the nitrogen has been used to bond to the H^+ that has added to ammonia.

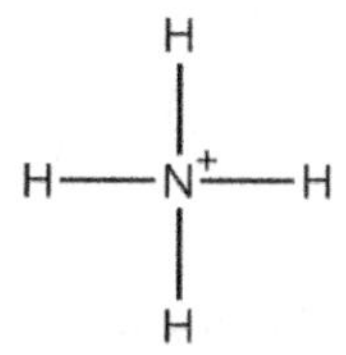

Formal charge shown on the hydronium ion

13. C is correct.

In atoms or molecules that are ions, the number of electrons is not equal to the number of protons, which gives the molecule a charge (either positive or negative).

Ionic bonds form when the difference in electronegativity of atoms in a compound is greater than 1.7 Pauling units.

Ionic bonds involve transferring an electron from the electropositive element (along the left-hand column/group) to the electronegative element (along the right-hand column/groups) on the periodic table.

Oppositely charged ions are attracted to each other, and this attraction results in ionic bonds.

In an ionic bond, electron(s) are transferred from a metal to a nonmetal, giving both molecules a full valence shell and causing the molecules to associate with each other.

14. A is correct.

Electronegativity is a chemical property that describes an atom's tendency to attract electrons to itself.

The most electronegative atom is F, while the least electronegative atom is Fr. The trend for increasing electronegativity within the periodic table is up and toward the right (i.e., fluorine).

15. D is correct.

Hydrogen bonds are the strongest intermolecular forces (i.e., between molecules), followed by dipole-dipole, dipole-induced dipole, and van der Waals forces (i.e., London dispersion).

London dispersion forces are present in compounds; they are the attractive forces that hold molecules. They are the weakest of the intermolecular forces, and their strength increases with the increasing size and polarity of the molecules involved.

16. C is correct.

For asymmetrical molecules (e.g., water is bent), use the atoms' geometry and difference of electronegativity values. For water, H (2.1 Pauling units) is more electropositive than O (3.5 Pauling units).

For sulfur dioxide, S (2.5 Pauling units) is more electropositive than O (3.5 Pauling units). The reported dipole moment for water is 1.8 D compared to 1.6 D for sulfur dioxide.

CO_2 does not have nonbonding electrons on the central carbon atom, and it is symmetrical, so it is non-polar and has zero dipole moment. CCl_4 is symmetrical as a tetrahedron and therefore does not exhibit a net dipole.

17. C is correct.

Cohesion is the property of like molecules sticking together. Hydrogen bonds join water molecules.

Adhesion is the attraction between unlike molecules (e.g., the meniscus observed from water molecules adhering to the graduated cylinder).

Polarity is the differences in electronegativity between bonded molecules. Polarity gives rise to the delta plus (on H) and the delta minus (on O), permitting hydrogen bonds to form between water molecules.

18. A is correct.

With little or no difference in electronegativity between the elements (i.e., Pauling units < 0.4), it is a nonpolar covalent bond, whereby the electrons are shared between the two bonding atoms.

Among the answer choices, the atoms of H, C and O are closest in magnitude for Pauling units for electronegativity.

19. D is correct.

Positively charged nuclei repel each other while each attracts the bonding electrons. These opposing forces reach equilibrium at the bond length.

20. C is correct.

In a carbonate ion (CO_3^{2-}), the carbon atom is bonded to 3 oxygen atoms. Two of those bonds are single covalent bonds, and the oxygen atoms each have an extra (third) lone pair of electrons, which imparts a negative formal charge. The remaining oxygen has a double bond with carbon.

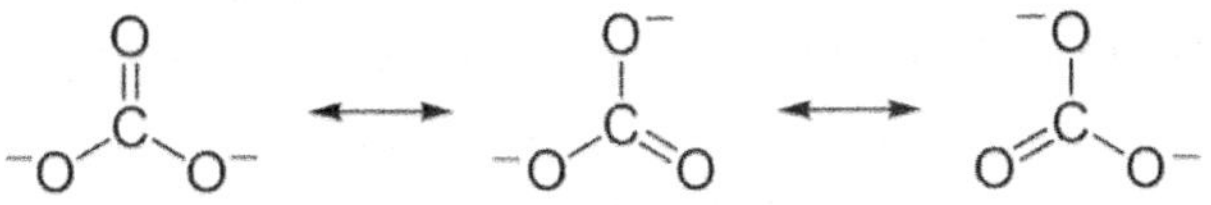

Three resonance structures for the carbonate ion CO_3^{2-}

==

Practice Set 2: Questions 21–40

==

21. D is correct.

A dipole is a separation of full (or partial) positive and negative charges due to differences in the electronegativity of atoms.

22. C is correct.

Ionic bonds involve transferring an electron from the electropositive element (along the left-hand column/group) to the electronegative atom (along the right-hand column/groups) on the periodic table.

Ca is a group II element with 2 electrons in its valence shell.

I is a group VII element with 7 electrons in its valence shell.

Ca becomes Ca^{2+}, and each of the two electrons is joined to I, which becomes I^-.

23. B is correct.

An atom with 4 valence electrons can make a maximum of 4 bonds. 1 double and 1 triple bond equals 5 bonds, which exceeds the maximum allowable bonds.

24. D is correct.

Water molecules stick to each other (i.e., cohesion) due to the collective action of hydrogen bonds between individual water molecules. These hydrogen bonds are constantly breaking and reforming; many molecules are held together by these bonds.

Water also sticks to surfaces (i.e., adhesion) because of water's polarity. The water may form a thin film on a smooth surface (e.g., glass) because the molecular forces between glass and water molecules (adhesive forces) are stronger than the cohesive forces between the water molecules.

25. D is correct.

Hydrogen bonds are the strongest intermolecular forces (i.e., between molecules), followed by dipole-dipole, dipole-induced dipole, and van der Waals forces (i.e., London dispersion).

When ionic compounds are dissolved, each ion is surrounded by more than one water molecule. The combined force ion-dipole interactions of several water molecules are stronger than a single ionic bond.

26. B is correct.

Hydrogens, bonded directly to F, O or N, participate in hydrogen bonds. The hydrogen is partial positive (i.e., delta plus or ∂+) due to the bond to these electronegative atoms. The lone pair of electrons on the F, O or N interacts with the ∂+ hydrogen to form a hydrogen bond.

The two lone pairs of electrons on the oxygen atom can each participate as a hydrogen bond acceptor. The molecule does not have hydrogen bonded directly to an electronegative atom (F, O or N) and cannot be a hydrogen bond donor.

27. A is correct.

Water is a bent molecule with a partial negative charge on the oxygen and a partial positive charge on each hydrogen. Therefore, the water molecule exists as a dipole. The Na^+ is an ion attracted to the partial negative charge on the oxygen in the water molecule.

28. C is correct.

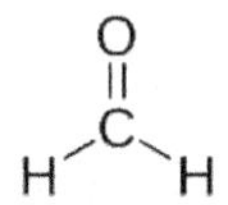

The molecule H_2CO (formaldehyde) is shown below and has one C=O bond and two C–H bonds. The electronegative oxygen pulls electron density away from the carbon atom and creates a net dipole towards the oxygen, resulting in a polar covalent bond. The electronegativity values of carbon and hydrogen are similar and result in a covalent bond (i.e., about equal sharing of bonded electrons).

29. A is correct.

The valence shell is the outermost shell (i.e., highest principal quantum number n) of an atom.

Valence electrons are those electrons of the outermost electron shell that can participate in a chemical bond.

The number of valence electrons for an element can be determined by its group (i.e., vertical column) on the periodic table. Except for the transition metals (i.e., groups 3-12), the group number identifies how many valence electrons are associated with a particular element: elements of the same group have the same number of valence electrons.

30. D is correct.

Group IA elements (e.g., Li, Na, and K) tend to lose 1 electron to achieve a complete octet to be cations with a +1 charge.

Group IIA elements (e.g., Mg and Ca) tend to lose 2 electrons to achieve a complete octet to be cations with a +2 charge.

Group VIIA elements (halogens such as F, Cl, Br, and I) tend to gain 1 electron to achieve a complete octet of anions with a –1 charge.

31. B is correct.

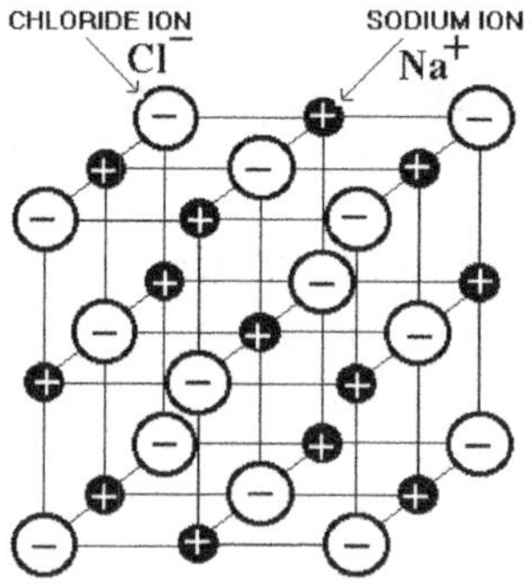

Lattice structure of sodium chloride

32. C is correct.

Each positively charged nuclei attracts the bonding electrons.

33. B is correct.

Electronegativity is a measure of how strongly an element attracts electrons within a bond.

Electronegativity is the relative attraction of the nucleus for bonding electrons. It increases from left to right (i.e., periods) and from bottom to top along a group (like the trend for ionization energy). The most electronegative atom is F, while the least electronegative atom is Fr.

The greater the difference in electronegativity between two atoms in a compound, the more polar of a bond these atoms form. The atom with the higher electronegativity is the partial (delta) negative end of the dipole.

34. A is correct.

Hydrogen bonds are the strongest intermolecular forces (i.e., between molecules), followed by dipole-dipole, dipole-induced dipole, and van der Waals forces (i.e., London dispersion).

Hydrogens, bonded directly to F, O, or N, participate in hydrogen bonds. H–bonding is a polar interaction involving hydrogen forming bonds to the electronegative atoms F, O, or N, which accounts for the high boiling points of water. The hydrogen is partial positive (i.e., delta plus or ∂+) due to the bond to these electronegative atoms. The lone pair of electrons on the F, O, or N interacts with the ∂+ hydrogen to form a hydrogen bond.

Polar molecules have high boiling points because of polar interaction. H_2S is a polar molecule but does not form hydrogen bonds; it forms dipole-dipole interactions.

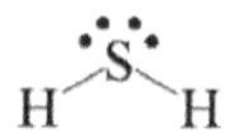

Molecular geometry of H_2S

35. D is correct.

With 4 electron pairs, the starting shape of this element is tetrahedral.

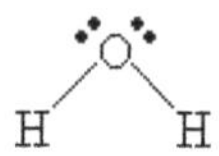

After removing 2 of the 4 groups surrounding the central atom, the central atom has two groups bound (e.g., H_2O). The molecule has the hybridization (and original angles) from a tetrahedral shape, bent rather than linear.

36. D is correct.

Ionic bonds hold salt crystals. Salts are composed of cations and anions and are electrically neutral. When salts are dissolved in solution, they separate into their constituent ions by breaking of noncovalent interactions.

Water: the hydrogen atoms in the molecule are covalently bonded to the oxygen atom.

Hydrogen peroxide: the molecule has one more oxygen atom than a water molecule and is also held together by covalent bonds.

Ester (RCOOR'): undergo hydrolysis and breaks covalent bonds to separate into its constituent carboxylic acid (RCOOH) and alcohol (ROH).

37. B is correct.

An ionic compound consists of a metal ion and a nonmetal ion.

Ionic bonds are formed between elements with an electronegativity difference greater than 1.7 Pauling units (e.g., a metal atom and a non-metal atom).

(K and I) form ionic bonds because K is a metal and I is a nonmetal.

(C and Cl) only have nonmetals, while (Fe and Mg) have only metals.

(Ga and Si) has a metal (Ga) and a metalloid (Si); they *might* form weak ionic bonds.

38. A is correct.

A coordinate bond is a covalent bond (i.e., shared pair of electrons) in which both electrons come from the same atom.

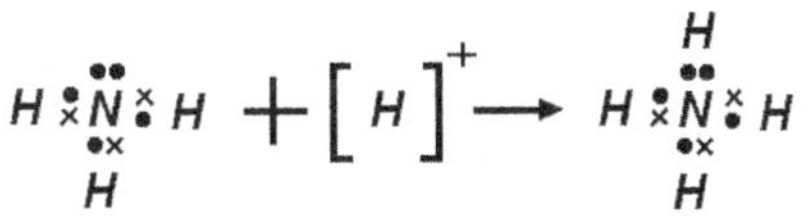

The lone pair of the nitrogen is donated to form the fourth N–H bond

39. B is correct.

Dipole moment depends on the overall shape of the molecule, length of the bond, and whether the electrons are pulled to one side of the bond (or the molecule overall).

For a large dipole moment, one element pulls electrons more strongly (i.e., differences in electronegativity).

Electronegativity is a measure of how strongly an element attracts electrons within a bond.

Electronegativity is the relative attraction of the nucleus for bonding electrons. It increases from left to right (i.e., periods) and from bottom to top along a group (like the trend for ionization energy). The most electronegative atom is F, while the least electronegative atom is Fr.

The greater the difference in electronegativity between two atoms in a compound, the more polar a bond these atoms form. The atom with the higher electronegativity is the partial (delta) negative end of the dipole.

40. A is correct.

The valence shell is the outermost shell (i.e., highest principal quantum number n) of an atom.

The noble gas configuration simply refers to eight electrons in the atom's outermost valence shell, referred to as a complete octet.

Depending on how many electrons it starts with, an atom may have to lose, gain or share an electron to obtain the noble gas configuration.

Practice Set 3: Questions 41–60

41. C is correct.

The bond between the oxygens is nonpolar, while the bonds between the oxygens and hydrogens are polar (due to the differences in electronegativity).

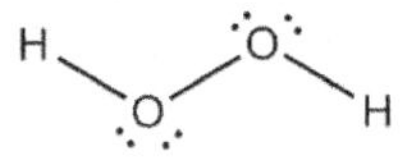

Line bond structure of H_2O_2 with lone pairs shown

42. A is correct.

The octet rule states that atoms of main-group elements tend to combine so that each atom has eight electrons in its valence shell. This occurs because electron arrangements involving eight valence electrons are highly stable, as with the noble gasses.

Selenium (Se) is in group VI and has 6 valence electrons. By gaining two electrons, it has a complete octet (i.e., stable).

43. C is correct.

When one or more new compounds are formed by rearranging atoms is an example of a chemical reaction.

44. D is correct.

HBr experiences dipole-dipole interactions due to the electronegativity difference resulting in a partial negative charge on bromine and a partial positive charge on hydrogen.

45. C is correct.

An ion is an atom (or a molecule) in which the number of electrons is not equal to the number of protons. Therefore, the atom (or molecule) has a net positive or negative electrical charge.

If a neutral atom loses one or more electrons, it has a net positive charge (i.e., cation). If a neutral atom gains one or more electrons, it has a net negative charge (i.e., anion).

Aluminum is a group III atom and has proportionally more protons per electron once the cation forms.

All other elements listed are from group I or II.

46. A is correct.

Dipole-dipole (e.g., $CH_3Cl \ldots CH_3Cl$) attraction occurs between neutral molecules, while ion-dipole interaction involves dipole interactions with charged ions (e.g., $CH_3Cl \ldots {}^{-}OOCCH_3$). Hydrogen bonds are the strongest intermolecular forces (i.e., between molecules), followed by dipole-dipole, dipole-induced dipole, and van der Waals forces (i.e., London dispersion).

47. B is correct.

When more than two H_2O molecules are present (e.g., liquid water), more bonds (between 2 and 4) are possible because the oxygen of one water molecule has two lone pairs of electrons, each of which can form a hydrogen bond with hydrogen on another water molecule.

This bonding can repeat such that every water molecule is H–bonded with up to four other molecules (two through its two lone pairs of O, and two through its two hydrogen atoms).

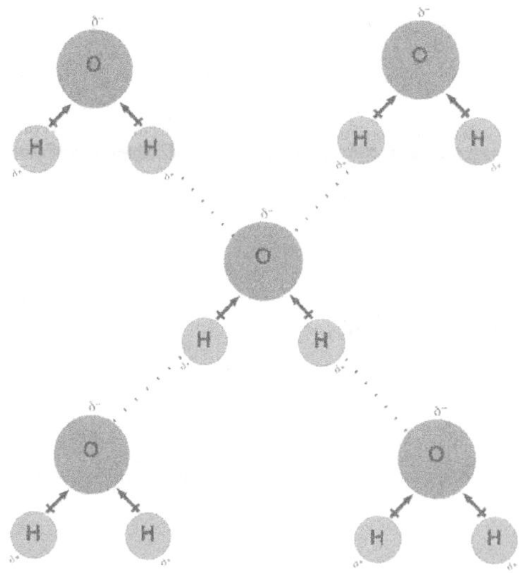

Ice with 4 hydrogen bonds between the water molecules

Hydrogen bonding affects the crystal structure of ice (hexagonal lattice). The density of ice is less than the density of water at the same temperature. Thus, the solid phase of water (ice) floats on the liquid. The creation of the additional H–bonds (up to 4 for ice) forces the individual molecules further, giving them less density, unlike most other substances in the solid phase.

For most substances, the solid form is denser than the liquid phase. Therefore, a block of most solids sinks in the liquid. Regarding pure water, a block of ice (solid phase) floats in liquid water because ice is less dense.

Like other substances, when liquid water is cooled from room temperature, it becomes increasingly dense. However, at approximately 4 °C (39 °F), water reaches its maximum density, and as it is cooled further, it expands and becomes less dense. This phenomenon is *negative thermal expansion* and is attributed to strong intermolecular interactions that are orientation-dependent.

The density of water is about 1 g/cm^3 and depends on the temperature. When frozen, the density of water is decreased by about 9%. This is due to the decrease in intermolecular vibrations, allowing water molecules to form stable hydrogen bonds with other water molecules around. As these hydrogen bonds form, molecules are locking into positions like the hexagonal structure.

Even though hydrogen bonds are shorter in the crystal than in the liquid, this position locking decreases the average coordination number of water molecules as the liquid reaches the solid phase.

48. A is correct.

The phrase "from its elements" in the question stem implies the need to create the formation reaction of the compound from its elements.

Chemical equation:

$$3\ Na^+ + N^{3-} \rightarrow Na_3N$$

Each sodium loses one electron to form Na^+ ions.

Nitrogen gains 3 electrons to form the N^{3-} ion.

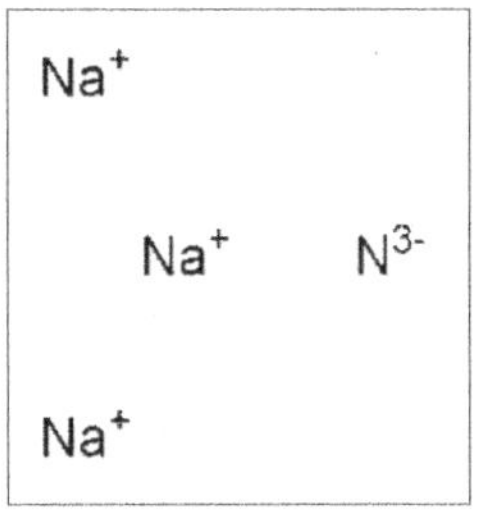

Schematic of an ionic compound

49. D is correct.

The valence shell is the outermost shell (i.e., highest principal quantum number *n*) of an atom.

Valence electrons are those electrons of the outermost electron shell that can participate in a chemical bond.

The number of valence electrons for a neutral element can be determined by its group (i.e., vertical column) on the periodic table. Except for the transition metals (i.e., groups 3-12), the group number identifies how many valence electrons are associated with a particular element: elements of the same group have the same number of valence electrons.

Lewis dot structures show the valence electrons of an individual atom as dots around the symbol of the element. Non-valence electrons (i.e., inner shell electrons) are not represented in Lewis structures.

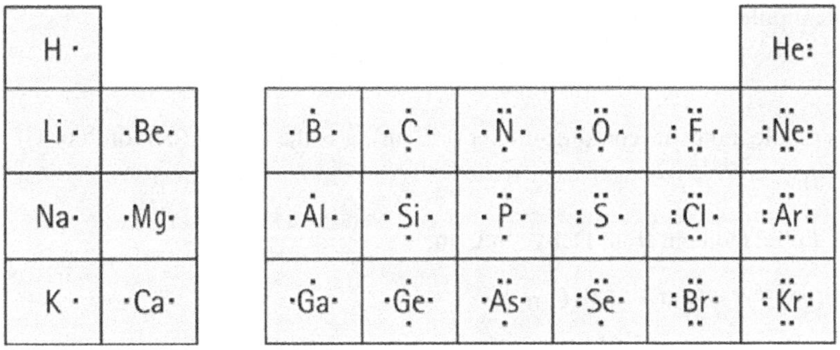

Sample Lewis dot structures for some neutral elements

Mg^{2+} a total of ten electrons but zero valance (outer shell) electrons. The valence shell (i.e., 3s subshell) has zero electrons, and therefore there are no dots shown in the Lewis dot structure: Mg^{2+}

S^{2-} has eight valence electrons (same electronic configuration as Ar).

Ga^+ has two valence electrons (same electronic configuration as Ca).

Ar^+ has seven valence electrons (same electronic configuration as Cl).

50. B is correct.

The octet rule states that atoms of main-group elements tend to combine so that each atom has eight electrons in its valence shell.

51. D is correct.

In a chemical reaction, the bonds being formed are different from the ones broken.

52. D is correct.

Hydrogen sulfide (i.e., sewer gas) is H_2S.

Hydrosulfide (bisulfide) ion is HS^-. Sulfide ion is S^{2-}

53. D is correct.

An ionic compound consists of a metal ion and a nonmetal ion. Ionic bonds are formed between elements with an electronegativity difference greater than 1.7 Pauling units (e.g., a metal atom and a non-metal atom). The charge on bicarbonate ion (HCO_3) is −1, while the charge on Mg is +2.

The proper formula is $Mg(HCO_3)_2$

54. C is correct.

Lattice: crystalline structure – consists of unit cells

Unit cell: the smallest unit of solid crystalline pattern

Covalent unit is not a valid term.

55. D is correct.

Formula to calculate dipole moment:

$$\mu = qr$$

where μ is dipole moment (coulomb-meter or C·m), q is the charge (coulomb or C), and r is the radius (meter or m)

Convert the unit of dipole moment from Debye to C·m:

$$0.16 \times 3.34 \times 10^{-30} = 5.34 \times 10^{-31}\ \text{C·m}$$

Convert the unit of radius to meter:

$$115\ \text{pm} \times 1 \times 10^{-12}\ \text{m/pm} = 115 \times 10^{-12}\ \text{m}$$

Rearrange the dipole moment equation to solve for q:

$$q = \mu / r$$

$$q = (5.34 \times 10^{-31}\ \text{C·m}) / (115 \times 10^{-12}\ \text{m})$$

$$q = 4.64 \times 10^{-21}\ \text{C}$$

Express the charge in terms of electron charge (e).

$$(4.64 \times 10^{-21}\ \text{C}) / (1.602 \times 10^{-19}\ \text{C/e}) = 0.029\ \text{e}$$

In this NO molecule, oxygen is the more electronegative atom. Therefore, the charge experienced by the oxygen atom is negative: –0.029e.

56. A is correct.

The greater the difference in electronegativity between two atoms in a compound, the more polar of a bond these atoms form. The atom with the higher electronegativity is the partial (delta) negative end of the dipole.

57. C is correct.

Hybridization of the central atom and the corresponding molecular geometry:

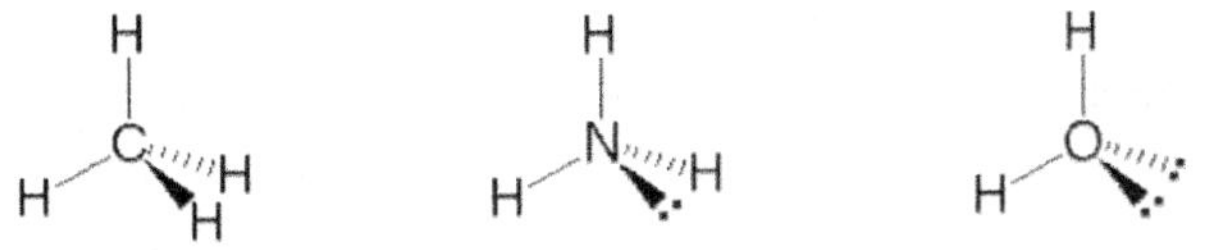

Examples of sp^3 hybridized atoms with 4 substituents (CH_4), 3 substituents (NH_3) and two substituents (H_2O)

Hybridization - Shape

sp – linear

sp^2 – trigonal planar

sp^3 – tetrahedral

sp^3d – trigonal bipyramid

58. C is correct.

Hydrogen bonds are the strongest intermolecular forces (i.e., between molecules), followed by dipole-dipole, dipole-induced dipole, and van der Waals forces (i.e., London dispersion).

Hydrogen is a very electropositive element, and O is a very electronegative element. Therefore, these elements will be attracted to each other, within the same water molecule and between water molecules.

Ionic or covalent bonding can only form within a molecule and not between adjacent molecules.

59. B is correct.

In a compound, the sum of ionic charges must equal zero.

For the charges of K^+ and CO_3^{2-} to balance, there must be 2 K^+ ions for every CO_3^{2-} ion.

The formula of this balanced molecule is K_2CO_3.

60. C is correct.

The substantial difference between the two forms of carbon (i.e., graphite and diamonds) is mainly due to their crystal structure, hexagonal for graphite and cubic for diamond.

The conditions to convert graphite into diamond are high pressure and high temperature. That is why creating synthetic diamonds is time-consuming, energy-intensive, and expensive since carbon is forced to change its bonding structure.

Practice Set 4: Questions 61–80

---

61. A is correct.

Lewis acids are defined as electron-pair acceptors, whereas Lewis bases are electron-pair donors.

Boron has an atomic number of five and has a vacant 2*p* orbital to accept electrons.

62. D is correct.

CH_3SH experiences dipole-dipole interactions due to the electronegativity difference resulting in a partial negative charge on sulfur and a partial positive charge on hydrogen.

CH_3CH_2OH experiences dipole-dipole interactions (due to the electronegativity difference resulting in a partial negative charge on oxygen and a partial positive charge on hydrogen. This example is a type of dipole-dipole known as hydrogen bonding (when hydrogen is bonded directly to fluorine, oxygen, or nitrogen).

63. A is correct.

There is an overlap of sp^2 orbitals and a *p* orbital on the adjacent carbon atoms in carbon-carbon double bonds.

The sp^2 orbitals overlap head-to-head as a *sigma* (σ) bond, whereas the *p* orbitals overlap sideways as a *pi* (π) bond.

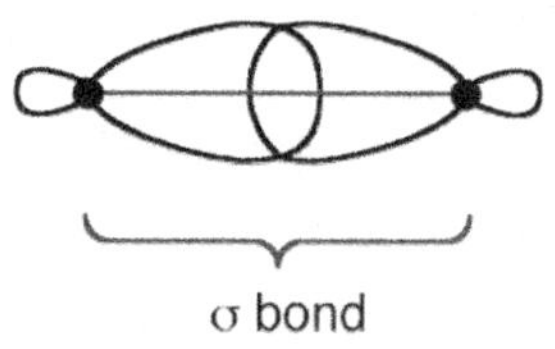

Sigma bond formation showing electron density along the internucleus axis

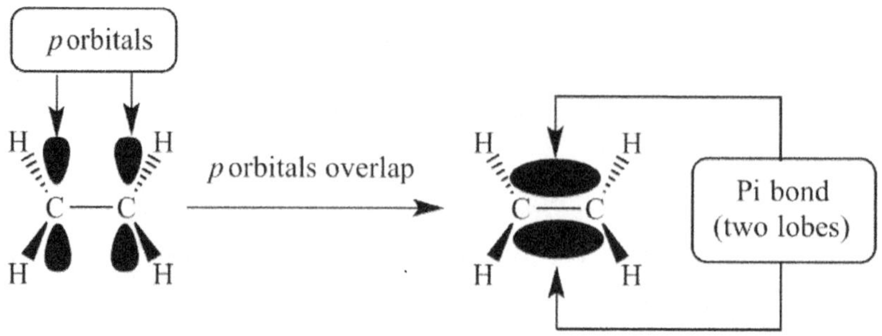

Two pi orbitals showing the pi bond formation during sideways overlap – note the absence of electron density (i.e., node) along the internucleus axis.

Bond lengths and strengths (σ or π) depend on the size and shape of the atomic orbitals and the density to overlap effectively.

The σ bonds are stronger than π bonds because head-to-head orbital overlap involves more shared electron density than sideways overlap. The σ bonds formed from two 2*s* orbitals are shorter than those formed from two 2*p* orbitals or two 3*s* orbitals.

Carbon, oxygen, and nitrogen are in the second period ($n = 2$), while sulfur (S), phosphorus (P), and silicon (Si) are in the third period. Therefore, S, P, and Si use 3*p* orbitals to form π bonds, while C, N, and O use 2*p* orbitals. The 3*p* orbitals are much larger than the 2*p* orbitals, and therefore, there is a reduced probability for an overlap of the 2*p* orbital of C and the 3*p* orbital of S, P, and Si.

B: S, P, and Si can hybridize, but these elements can combine *s* and *p* orbitals and (unlike C, O, and N) have *d* orbitals.

C: S, P, and Si (in their ground state electron configurations) have partially occupied *p* orbitals which form bonds.

D: carbon combines with elements below the second row of the periodic table. For example, carbon forms bonds with higher principal quantum number ($n > 2$) halogens (e.g., F, Cl, Br, and I).

64. B is correct.

The magnesium atom has a +2 charge, and the sulfur atom has a –2 charge.

65. C is correct.

Hydrogen bonds are the strongest intermolecular forces (i.e., between molecules), followed by dipole-dipole, dipole-induced dipole, and van der Waals forces (i.e., London dispersion).

Larger atoms have more electrons; they can exert stronger induced dipole forces than smaller atoms.

Bromine is under chlorine on the periodic table, which means it has more electrons and stronger induced dipole forces.

66. A is correct.

Hydrogen bonds are the strongest intermolecular forces (i.e., between molecules), followed by dipole-dipole, dipole-induced dipole, and van der Waals forces (i.e., London dispersion).

Hydrogens, bonded directly to F, O or N, participate in hydrogen bonds. The hydrogen is partially positive (i.e., delta plus: ∂+) due to the bond to these electronegative atoms. The lone pair of electrons on the F, O or N interacts with the partial positive (∂+) hydrogen to form a hydrogen bond.

When ~NH or ~OH groups are present in a molecule, they form hydrogen bonds with other molecules. Hydrogen bonding is the strongest intermolecular force, and it is the major intermolecular force in this substance.

67. B is correct

Potassium oxide (K_2O) contains metal and nonmetal, which form a salt.

Salts contain ionic bonds and dissociate in aqueous solutions and therefore are strong electrolytes.

A polyatomic (i.e., molecular) ion is a charged chemical species composed of two or more atoms covalently bonded or composed of a metal complex acting as a single unit. An example is the hydroxide ion consisting of one oxygen atom and one hydrogen atom; hydroxide has a charge of −1.

A polyatomic ion does bond with other ions.

A polyatomic ion has various charges.

A polyatomic ion might contain only metals or nonmetals.

A polyatomic ion is not neutral.

oxidation state	−1	+1	+3	+5	+7
anion name	chloride	hypochlorite	chlorite	chlorate	perchlorate
formula	Cl^-	ClO^-	ClO_2^-	ClO_3^-	ClO_4^-

68. D is correct.

The valence shell is an atom's outermost shell (i.e., highest principal quantum number *n*).

Valence electrons are those electrons of the outermost electron shell that can participate in a chemical bond.

69. B is correct.

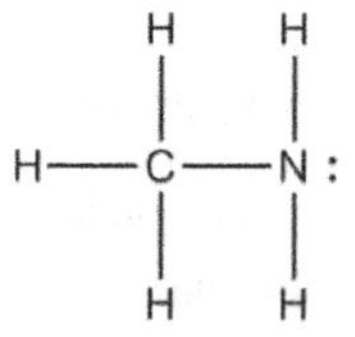

Nitrogen (NH_3) is neutral with three bonds, negative with two bonds ($^-NH_2$), and positive with 4 bonds ($^+NH_4$)

Line bond structure of methylamine with the lone pair on nitrogen shown

70. B is correct.

Sulfide ion is S^{2-}.

Hydrogen sulfide (i.e., sewer gas) is H_2S. Hydrosulfide (bisulfide) ion is HS^-.

Sulfite ion is the conjugate base of bisulfite and has the molecular formula of SO_3^{2-}

Sulfur (S) is a chemical element with an atomic number of 16.

Sulfate ion is a polyatomic anion (i.e., two or more atoms covalently bonded or metal complex) with the molecular formula of SO_4^{2-}

71. C is correct.

Ionic bonds involve electrostatic interactions between oppositely charged atoms.

Electrons from the metallic atom are transferred to the nonmetallic atom, giving both atoms a full valence shell.

72. B is correct.

If the charge of O is −1, the proper formula of its compound with K (+1) should be KO.

73. C is correct.

The lattice energy of a crystalline solid is usually defined as the energy of formation of the crystal from infinitely separated ions (i.e., an exothermic process and hence the negative sign).

Some older textbooks define lattice energy with the opposite sign. The older notation refers to the energy required to convert the crystal into infinitely separated gaseous ions in a vacuum (i.e., an endothermic process and the positive sign).

Lattice energy decreases downward for a group.

Lattice energy increases with charge.

Because NaCl and LiF contain charges of +1 and −1, they would have smaller lattice energy than the higher-charged ions on other options.

Na is lower than Li, and Cl is lower than F in their respective groups so that NaCl would have lower lattice energy than LiF.

The bond between ions of opposite charge is strongest when the ions are small.

The force of attraction between oppositely charged particles is directly proportional to the product of the charges on the two objects (q_1 and q_2) and inversely proportional to the square of the distance between the objects (r^2):

$$F = q_1 \times q_2 / r^2$$

Therefore, the strength of the bond between the ions of opposite charge in an ionic compound depends on the charges on the ions and the distance between the centers of the ions when they pack to form a crystal.

For NaCl, the lattice energy is the energy released by the reaction:

$Na^+ (g) + Cl^- (g) \rightarrow NaCl (s)$, lattice energy = –786 kJ/mol

Other examples:

Al_2O_3, lattice energy = –15,916 kJ/mol

KF, lattice energy = –821 kJ/mol

LiF, lattice energy = –1,036 kJ/mol

74. C is correct.

Formula to calculate dipole moment:

$$\mu = qr$$

where μ is dipole moment (coulomb per meter or C/m), q is charge (coulomb or C), and r is radius (meter or m)

Convert the unit of radius to meters:

$$154 \text{ pm} \times 1 \times 10^{-12} \text{ m/pm} = 154 \times 10^{-12} \text{ m}$$

Convert the charge to coulombs:

$$0.167\ e \times 1.602 \times 10^{-19} \text{ C/e}^- = 2.68 \times 10^{-20} \text{ C}$$

Use these values to calculate the dipole moment:

$$\mu = qr$$

$$\mu = 2.68 \times 10^{-20} \text{ C} \times 154 \times 10^{-12} \text{ m}$$

$$\mu = 4.13 \times 10^{-30} \text{ C·m}$$

Finally, convert the dipole moment to Debye:

$$\mu = 4.13 \times 10^{-30} \text{ C·m} / 3.34 \times 10^{-30} \text{ C·m·D}^{-1}$$

$$\mu = 1.24 \text{ D}$$

75. B is correct.

The greater the difference in electronegativity between two atoms in a compound, the more polar a bond these atoms form. The atom with the higher electronegativity is the partial (delta) negative end of the dipole.

Although each C–Cl bond is very polar, the dipole moments of each of the four bonds in CCl_4 (carbon tetrachloride) cancel because the molecule is a symmetric tetrahedron.

76. C is correct.

The pK_a of the carboxylic acid is about 5 and is deprotonated (exists as an anion) at a pH of 10.

The pK_a of the alcohol is about 15 and is protonated (neutral) at a pH of 10.

77. A is correct.

Hybridization of the central atom and the corresponding electron geometry.

Hybridization	Electron groups	Bonding groups	Lone pairs	Approx. bond angles	Electron geometry	Molecular geometry
sp	2	2	0	180°	Linear	Linear
sp^2	3	3	0	120°	Trigonal planar	Trigonal planar
	3	2	1	<120°	Trigonal planar	Bent
Hybridization	**Electron groups**	**Bonding groups**	**Lone pairs**	**Approximate bond angles**	**Electron geometry**	**Molecular geometry**
sp^3	4	4	0	109.5°	Tetrahedral	Tetrahedral
	4	3	1	<109.5°	Tetrahedral	Trigonal pyramidal
	4	2	2	<<109.5°	Tetrahedral	Bent
sp^3d	5	5	0	120° (equatorial) 90° (axial)	Trigonal bipyramidal	Trigonal bipyramidal
	5	4	1	<120° (equatorial) <90° (axial)	Trigonal bipyramidal	Seesaw
	5	3	2	<90°	Trigonal bipyramidal	T-shaped
	5	2	3	180°	Trigonal bipyramidal	Linear
sp^3d^2	6	6	0	90°	Octahedral	Octahedral
	6	5	1	<90°	Octahedral	Square pyramidal
	6	4	2	90°	Octahedral	Square planar

78. B is correct.

The valence shell is an atom's outermost shell (i.e., highest principal quantum number *n*).

The octet rule states that atoms of main-group elements tend to combine so that each atom has eight electrons in its valence shell. This occurs because electron arrangements involving eight valence electrons are highly stable, as with noble gasses.

79. D is correct.

Solubility depends on the solvent's ability to overcome the intermolecular forces in a solute.

80. D is correct.

Electronegativity is defined as the ability of an atom to attract electrons when it bonds with another atom. The most electronegative atom is F, while the least is Fr. The trend for increasing electronegativity within the periodic table is up and toward the right (i.e., fluorine).

Chlorine has the highest electronegativity of the elements listed.

Electronegativity is a characteristic of non-metals.

Notes for active learning

Notes for active learning

Phases and Phase Equilibria – Detailed Explanations

Practice Set 1: Questions 1–20

1. A is correct.

Solids, liquids, and gases have a vapor pressure that increases from solid to gas.

Vapor pressure is the pressure exerted by a vapor in equilibrium with its condensed phases (i.e., solid or liquid) in a closed system at a given temperature.

Vapor pressure is a colligative property of a substance and depends only on the number of solutes present, not on their identity.

2. C is correct.

Charles' law (i.e., the law of volumes) explains how, at constant pressure, gases behave when the temperature changes:

$$V \alpha T$$

or

$$V / T = \text{constant}$$

or

$$(V_1 / T_1) = (V_2 / T_2)$$

Volume and temperature are proportional.

Doubling the temperature at constant pressure doubles the volume.

3. C is correct.

Colligative properties include lowering of vapor pressure, the elevation of boiling point, depression of freezing point, and increased osmotic pressure.

The addition of solute to a pure solvent lowers the vapor pressure of the solvent; therefore, a higher temperature is required to bring the vapor pressure of the solution in an open container up to the atmospheric pressure. This increases the boiling point.

Because adding solute lowers the vapor pressure, the solution's freezing point decreases (e.g., automobile antifreeze).

4. A is correct.

At a pressure and temperature corresponding to the triple point (point D on the graph) of a substance, all three states (gas, liquid and solid) exist in equilibrium.

The critical point (point E on the graph) is the endpoint of the phase equilibrium curve where the liquid and its vapor become indistinguishable.

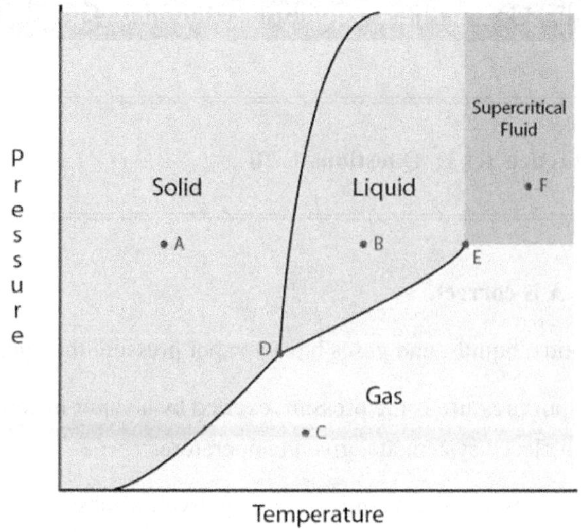

Phase diagram of pressure vs. temperature

5. A is correct.

In the van der Waals equation, *a* is the negative deviation due to attractive forces, and *b* is the positive deviation due to molecular volume.

6. B is correct.

R is the symbol for the ideal gas constant.

R is expressed in (L × atm) / mole × K and the value = 0.0821.

Convert to different units (torr and mL)

$$0.0821[(L \times atm) / (mole \times K)] \times (760\ torr/atm) \times (1{,}000\ mL/L)$$

$$R = 62{,}396\ (torr \times mL) / mole \times K$$

7. B is correct.

Ideal gas law:

$$PV = nRT$$

where P is pressure, V is volume, n is the number of molecules, R is the ideal gas constant, and T is the temperature of the gas.

R and n are constant.

From this equation, if pressure and temperature are halved, there would be no effect on the volume.

8. B is correct.

Kinetic molecular theory of gas molecules states that the average kinetic energy per molecule in a system is proportional to the temperature of the gas.

Since it is given that containers X and Y are at the same temperature and pressure, then molecules of gases must possess the same amount of average kinetic energy.

9. D is correct.

Barometers and manometers are used to measure pressure. They are designed to measure atmospheric pressure, while a manometer can measure lower pressure than atmospheric pressure.

A manometer has both ends of the tube open to the outside (while some may have one end closed), whereas a barometer is a type of closed-end manometer with one end of the glass tube closed and sealed with a vacuum.

The atmospheric pressure is 760 mmHg, so the barometer should be able to accommodate that.

10. D is correct.

Vapor pressure is the pressure exerted by a vapor in equilibrium with its condensed phases (i.e., solid or liquid) in a closed system at a given temperature.

Raoult's law states that the partial vapor pressure of each component of an ideal mixture of liquids equals the vapor pressure of the pure component multiplied by its mole fraction in the mixture.

In exothermic reactions, the vapor pressure deviates negatively from Raoult's law.

Depending on the ratios of the liquids in a solution, the vapor pressure could be lower than either or just lower than X because X is a higher boiling point, thus a lower vapor pressure.

The boiling point increases from adding Y to the mixture because the vapor pressure decreases.

11. B is correct.

$$2\ Na\ (s) + Cl_2\ (g) \rightarrow 2\ NaCl\ (s)$$

In its elemental form, chlorine exists as a gas.

In its elemental form, Na exists as a solid.

12. C is correct.

The ideal gas molecules do not occupy a significant amount of space and exert no intermolecular forces.

In contrast, the molecules of a real gas occupy space and exert (weak attractive) intermolecular forces.

However, an ideal gas and a real gas have pressure, which is created from molecular collisions with the walls of the container.

13. A is correct.

Boyle's law (i.e., pressure-volume law) states that pressure and volume are inversely proportional:

$$(P_1V_1) = (P_2V_2)$$

Solve for the final pressure:

$$P_2 = (P_1V_1) / V_2$$

$$P_2 = [(0.950\ \text{atm}) \times (2.75\ \text{L})] / (0.450\ \text{L})$$

$$P_2 = 5.80\ \text{atm}$$

14. D is correct.

Avogadro's law is an experimental gas law relating the volume of a gas to the amount of substance of gas present; *equal volumes of gases have the same number of molecules at the same temperature and pressure*.

For a given mass of an ideal gas, the volume and amount (i.e., moles) of the gas are directly proportional if the temperature and pressure are constant:

$$V \propto n$$

Where V = volume and n = number of moles of the gas.

Charles' law (i.e., the law of volumes) explains how, at constant pressure, gases behave when the temperature changes:

$$V \propto T$$

or

$$V / T = \text{constant}$$

or

$$(V_1 / T_1) = (V_2 / T_2)$$

Volume and temperature are proportional. Therefore, an increase in one term increases the other.

Gay-Lussac's law (i.e., pressure-temperature law) states that pressure is proportional to temperature:

$$P \propto T$$

or

$$(P_1 / T_1) = (P_2 / T_2)$$

or

$$(P_1T_2) = (P_2T_1)$$

or

$$P / T = \text{constant}$$

If the pressure of a gas increases, the temperature increases.

Boyle's law (i.e., pressure-volume law) states that pressure and volume are inversely proportional.

15. C is correct.

Ideal gas law: PV = nRT

where P is pressure, V is volume, n is the number of molecules, R is the ideal gas constant, and T is the temperature of the gas.

R and n are constant.

If T is constant, the equation becomes PV = constant. They are inversely proportional: if one of the values is reduced, the other increases.

16. A is correct.

Vaporization refers to the change of state from a liquid to a gas. There are two types of vaporization: boiling and evaporation differentiated based on the temperature at which they occur.

Evaporation occurs at a temperature below the boiling point, while boiling occurs at or above the boiling point.

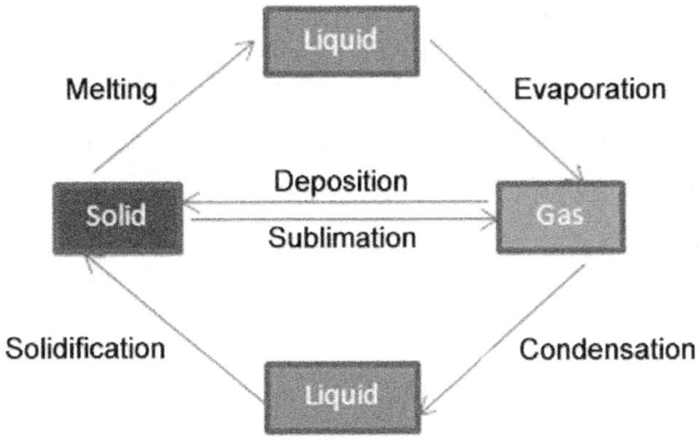

Interconversion of states of matter

17. C is correct.

Ideal gas law:

$$PV = nRT$$

from which simpler gas laws such as Boyle's, Charles' and Avogadro's laws are derived.

The value of n and R are constant.

The common format of the combined gas law:

$$(P_1V_1) / T_1 = (P_2V_2) / T_2$$

Try modifying the equations to recreate the formats of the equation provided by the problem.

$T_2 = T_1 \times P_1 / P_2 \times V_2 / V_1$ should be written as $T_2 = T_1 \times V_1/V_2 \times P_1/P_2$

18. C is correct.

Intermolecular forces act between neighboring molecules. Examples include hydrogen bonding, dipole-dipole, dipole-induced dipole, and van der Waals (i.e., London dispersion) forces.

A dipole-dipole attraction involves asymmetric, polar molecules (based on differences in electronegativity between atoms) that create a dipole moment (i.e., a net vector indicating the force).

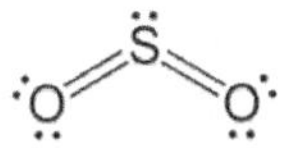

Sulfur dioxide with an indicated bond angle

19. C is correct.

Increasing the pressure of the gas above the liquid puts stress on the equilibrium of the system. Gas molecules start to collide with the liquid surface more often, which increases the rate of gas molecules entering the solution, thus increasing the solubility.

20. B is correct.

Avogadro's law states the correlation between volume and moles (n).

Avogadro's law is an experimental gas law relating the volume of a gas to the amount of substance of gas present. A modern statement of Avogadro's law is:

At the same pressure and temperature, equal volumes of gases have the same number of molecules.

$$V \propto n$$

or

$$V / n = k$$

where V is the volume of the gas, n is the number of moles of the gas, and k is a constant equal to RT/P (where R is the universal gas constant, T is the temperature in Kelvin and P is the pressure).

For comparing the same substance under two sets of conditions, the law is expressed as:

$$V_1 / n_1 = V_2 / n_2$$

Practice Set 2: Questions 21–40

---

21. D is correct.

Colligative properties of solutions depend on the ratio of the number of solute particles to the number of solvent molecules in a solution and not on the type of chemical species present.

Colligative properties include: lowering of vapor pressure, the elevation of boiling point, depression of freezing point, and increased osmotic pressure.

Boiling point (BP) elevation:

$$\Delta BP = iKm$$

where i = the number of particles produced when the solute dissociates, K = boiling elevation constant and m = molality (moles/kg solvent).

In this problem, acid molality is not known.

22. A is correct.

An ideal gas has no intermolecular forces, indicating that its molecules have no attraction to each other. The molecules of a real gas, however, do have intermolecular forces, although these forces are extremely weak.

Therefore, the molecules of a real gas are slightly attracted to one another, although the attraction is nowhere near as strong as the attraction in liquids and solids.

23. C is correct.

Standard temperature and pressure (STP) have a temperature of 273.15 K (0 °C, 32 °F) and a pressure of 10^5 Pa (100 kPa, 750.06 mmHg, 1 bar, 14.504 psi, 0.98692 atm).

The mm of Hg was defined as the pressure generated by a column of mercury one millimeter high. The pressure of mercury depends on temperature and gravity.

This variation in mmHg and torr is a difference in units of about 0.000015%.

In general,

1 torr = 1 mm of Hg = 0.0013158 atm

750.06 mmHg = 0.98692 atm

24. D is correct.

At lower temperatures, the potential energy due to the intermolecular forces are more significant than the kinetic energy; this causes the pressure to be reduced because the gas molecules are attracted to each other.

25. D is correct.

Dalton's law (i.e., the law of partial pressures) states that *the pressure exerted by a mixture of gases equals the sum of the individual gas pressures*.

The pressure due to N_2 and CO_2:

(320 torr + 240 torr) = 560 torr

The partial pressure of O_2 is:

740 torr – 560 torr = 180 torr

180 torr / 740 torr = 24%

26. A is correct.

Ideal gas law:

PV = nRT

where P is pressure, V is volume, n is the number of molecules, R is the ideal gas constant, and T is the temperature of the gas.

If the volume is reduced by ½, the number of moles is reduced by ½.

Pressure is reduced to 90%, so the number of moles is reduced by 90%.

Therefore, the reduction in moles is (½ × 90%) = 45%.

Mass is proportional to the number of moles for a given gas so that mass reduction can be calculated directly:

New mass is 45% of 40 grams = (0.45 × 40 g) = 18 grams

27. D is correct.

Evaporation describes the phase change from liquid to gas.

The mass of the molecules and the attraction of the molecules with their neighbors (to form intermolecular attractions) determine their kinetic energy.

The increase in kinetic energy is required for individual molecules to move from the liquid to the gaseous phase.

28. B is correct.

The molecules of an ideal gas exert no attractive forces. Therefore, a real gas behaves most nearly like an ideal gas at high temperature and low pressure. Under these conditions, the molecules are far apart and exert little or no attractive forces on each other.

29. C is correct.

Hydroxyl (~OH) groups significantly increase the boiling point because they form hydrogen bonds with ~OH groups of neighboring molecules.

Hydrocarbons are nonpolar molecules, which means that the dominant intermolecular force is London dispersion. This force gets stronger as the number of atoms in each molecule increases. The stronger force increases the boiling point.

The branching of the hydrocarbon affects the boiling point. Straight molecules have slightly higher boiling points than branched molecules with the same number of atoms. The reason is that straight molecules can align parallel against each other, and atoms in the molecules are involved in the London dispersion forces.

Another factor is the presence of heteroatoms (i.e., atoms other than carbon and hydrogen). For example, the electronegative oxygen atom between carbon groups or in an ether (C–O–C) slightly increases the boiling point.

30. A is correct.

Kinetic theory explains the macroscopic properties of gases (e.g., temperature, volume, and pressure) by their molecular composition and motion.

The gas pressure is due to the collisions on the walls of a container from molecules moving at different velocities.

Temperature = $\frac{1}{2}mv^2$

31. B is correct.

Methanol (CH_3OH) is alcohol that participates in hydrogen bonding.

Therefore, this gas experiences the strongest intermolecular forces.

32. D is correct.

Density = mass / volume

Gas molecules have a large amount of space between them; therefore, they can be pushed together, and thus gases are very compressible. Because there is such a large amount of space between each molecule in a gas, the extent to which the gas molecules can be pushed together is much greater than the extent to which liquid molecules can be pushed together. Therefore, gases have greater compressibility than liquids.

Gas molecules are further apart than liquid molecules, which is why gases have a smaller density.

33. C is correct.

Vapor pressure is the pressure exerted by a vapor in equilibrium with its condensed phases (i.e., solid or liquid) in a closed system at a given temperature.

Vapor pressure is inversely correlated with the strength of the intermolecular force.

The molecules are more likely to stick together in the liquid form with stronger intermolecular forces, and fewer of them participate in the liquid-vapor equilibrium.

The vapor pressure of a liquid decreases when a nonvolatile substance is dissolved into a liquid.

The decrease in the vapor pressure of a substance is proportional to the number of moles of the solute dissolved in a definite weight of thc solvent. This is *Raoult's law*.

34. B is correct.

Dalton's law (i.e., the law of partial prcssures) states that *the pressure exerted by a mixture of gases equals the sum of the individual gas pressures*. It is an empirical law observed by English chemist John Dalton and related to the ideal gas laws.

35. C is correct.

Solids have a definite shape and volume. For example, a granite block does not change its shape or its volume regardless of the container in which it is placed.

Molecules in a solid are very tightly packed due to the strong intermolecular attractions, which prevent the molecules from moving around.

36. D is correct.

Ideal gas law: PV = nRT

where P is pressure, V is volume, n is the number of molecules, R is the ideal gas constant, and T is the temperature of the gas.

At STP (standard conditions for temperature and pressure), the pressure and temperature for all three flasks are the same. It is known that the volume is the same in each case – 2.0 L.

Therefore, since R is a constant, the number of molecules "n" must be the same for the ideal gas law to hold.

37. A is correct.

Hydrogens, bonded directly to F, O or N, participate in hydrogen bonds. The hydrogen is partially positive (i.e., delta plus: $\partial+$) due to the bond to these electronegative atoms. The lone pair of electrons on the F, O or N interacts with the partial positive ($\partial+$) hydrogen to form a hydrogen bond.

D: the hydrogen on the methyl carbon and that carbon is not attached to N, O, or F. Therefore, that hydrogen cannot form a hydrogen bond, even though there is available oxygen on the other methanol to form a hydrogen bond.

38. C is correct.

Boyle's law (Mariotte's law or the Boyle-Mariotte law) is an experimental gas law that describes how the volume of a gas increases as the pressure decreases (i.e., they are inversely proportional) if the temperature is constant.

Boyle's law (i.e., pressure-volume law) states that pressure and volume are inversely proportional:

$$P_1V_1 = P_2V_2$$

or $$P \times V = \text{constant}$$

If the volume of a gas increases, its pressure decreases proportionally.

Dalton's law (i.e., the law of partial pressures) states that *the pressure exerted by a mixture of gases equals the sum of the individual gas pressures.*

Charles' law (i.e., the law of volumes) explains how, at constant pressure, gases behave when the temperature changes:

$$V \propto T$$

or $$V / T = \text{constant}$$

or $$(V_1 / T_1) = (V_2 / T_2)$$

Volume and temperature are proportional: an increase in one results in an increase in the other.

Gay-Lussac's law (i.e., pressure-temperature law) states that pressure is proportional to temperature:

$$P \propto T$$

or

$$(P_1 / T_1) = (P_2 / T_2)$$

or

$$(P_1T_2) = (P_2T_1)$$

or

$$P / T = \text{constant}$$

If the pressure of a gas increases, the temperature increases.

Avogadro's law is an experimental gas law relating the volume of a gas to the amount of substance of gas present. *Equal volumes of gases at the same temperature and pressure have the same number of molecules.*

39. B is correct.

As the automobile travels the highway, friction is generated between the road and its tires. The heat energy increases the air temperature inside the tires, causing the molecules to have more velocity. These fast-moving molecules collide with the walls of the tire at a higher rate, and thus the pressure is increased.

Gay-Lussac's law (i.e., pressure-temperature law) states that pressure is proportional to temperature:

$$P \propto T$$

or

$$(P_1 / T_1) = (P_2 / T_2)$$

or

$$(P_1T_2) = (P_2T_1)$$

or

$$P / T = \text{constant}$$

If the temperature of a gas increases, the pressure increases.

40. C is correct.

The balanced chemical equation:

$$N_2 + 3\ H_2 \rightarrow 2\ NH_3$$

Use the balanced coefficients from the written equation, apply dimensional analysis to solve for the volume of H_2 needed to produce 12.5 L NH_3:

$$V_{H2} = V_{NH3} \times (\text{mol } H_2 / \text{mol } NH_3)$$

$$V_{H2} = (12.5\ \text{L}) \times (3\ \text{mol} / 2\ \text{mol})$$

$$V_{H2} = 18.8\ \text{L}$$

Practice Set 3: Questions 41–60

41. C is correct.

Dalton's law (i.e., the law of partial pressures) states that *the pressure exerted by a mixture of gases equals the sum of the individual gas pressures*.

Convert the masses of the gases into moles:

Moles of O_2:

16 g of $O_2 \div 32$ g/mole $= 0.5$ mole

Moles of N_2:

14 g of $N_2 \div 28$ g/mole $= 0.5$ mole

Mole of CO_2:

88 g of $CO_2 \div 44$ g/mole $= 2$ moles

Total moles:

(0.5 mol + 0.5 mole + 2 mol) = 3 moles

The pressure of 1 atm (or 760 mmHg) has 38 mmHg (1 torr = 1 mmHg) contributed as H_2O vapor.

The partial pressure of CO_2:

mole fraction × (total pressure of the gas mixture – H_2O vapor)

(2 moles CO_2 / 3 moles total gas)] × (760 mmHg – 38 mmHg)

partial pressure of CO_2 = 481 mmHg

42. D is correct.

Colligative properties are solutions that depend on the ratio of the number of solute particles to the number of solvent molecules in a solution and not on the type of chemical species present.

Colligative properties include lowering of vapor pressure, the elevation of boiling point, depression of freezing point, and increased osmotic pressure

Dissolving a solute into a solvent alters the solvent's freezing point, melting point, boiling point, and vapor pressure.

43. B is correct.

Boyle's law (i.e., pressure-volume law) states that pressure and volume are inversely proportional:

$$(P_1V_1) = (P_2V_2)$$

or

$$P \times V = \text{constant}$$

If the volume of a gas increases, its pressure decreases proportionally.

44. D is correct.

Gases form homogeneous mixtures, regardless of the identities or relative proportions of the component gases. There is a relatively large distance between gas molecules (as opposed to solids or liquids where the molecules are much closer together).

When pressure is applied to gas, its volume readily decreases, and thus gases are highly compressible.

There are no attractive forces between gas molecules, which is why gas molecules can move about freely.

45. D is correct.

Solids have a definite shape and volume. For example, a granite block does not change its shape or volume regardless of the container in which it is placed.

Molecules in a solid are very tightly packed due to the strong intermolecular attractions, which prevent the molecules from moving around.

46. A is correct.

Standard temperature and pressure (STP) have a temperature of 273.15 K (0 °C, 32 °F) and a pressure of 10^5 Pa (100 kPa, 750.06 mmHg, 1 bar, 14.504 psi, 0.98692 atm).

The mm of Hg was defined as the pressure generated by a column of mercury one millimeter high. The pressure of mercury depends on temperature and gravity.

This variation in mmHg and torr is a difference in units of about 0.000015%.

In general,

1 torr = 1 mm of Hg = 0.0013158 atm.

750.06 mmHg = 0.98692 atm.

47. C is correct.

Intermolecular forces act between neighboring molecules. Examples include hydrogen bonding, dipole-dipole, dipole-induced dipole, and van der Waals (i.e., London dispersion) forces.

Stronger force results in a higher boiling point.

CH_3COOH is a carboxylic acid that can form two hydrogen bonds. Therefore, it has the highest boiling point.

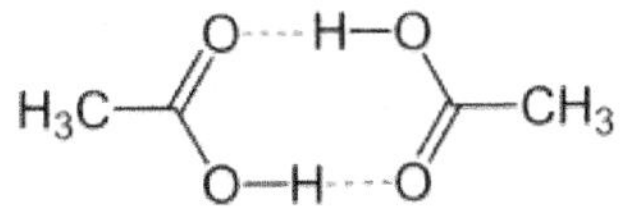

Ethanoic acid with the two hydrogen bonds indicated on the structure

48. C is correct.

Molecules in solids have the most attraction to their neighbors, followed by liquids (significant motion between the individual molecules) and then gas.

A molecule in an ideal gas has no attraction to other gas molecules. For a gas experiencing low pressure, the particles are far enough apart for no attractive forces between the individual gas molecules.

49. C is correct.

Ideal gas law:

$$PV = nRT$$

where P is pressure, V is volume, n is the number of molecules, R is the ideal gas constant, and T is the temperature of the gas.

Set the initial and final P, V and T conditions equal:

$$(P_1V_1 / T_1) = (P_2V_2 / T_2)$$

Solve for the final volume of N_2:

$$(P_2V_2 / T_2) = (P_1V_1 / T_1)$$

$$V_2 = (T_2\ P_1V_1) / (P_2T_1)$$

$$V_2 = [(295\ K) \times (750.06\ mmHg) \times (0.190\ L\ N_2)] / [(660\ mmHg) \times (298.15\ K)]$$

$$V_2 = 0.214\ L$$

50. B is correct.

Volatility is the tendency of a substance to vaporize (phase change from liquid to vapor).

Volatility is related to a substance's vapor pressure. A substance with higher vapor pressure vaporizes more readily than a substance with a lower vapor pressure at a given temperature.

Molecules with weak intermolecular attraction can increase their kinetic energy by transferring less heat due to a smaller molecular mass. The increase in kinetic energy is required for individual molecules to move from the liquid to the gaseous phase.

51. B is correct.

Barometer and manometer are used to measure pressure. They are designed to measure atmospheric pressure, while a manometer can measure lower pressure than atmospheric pressure.

A manometer has both ends of the tube open to the outside (while some may have one end closed), whereas a barometer is a type of closed-end manometer with one end of the glass tube closed and sealed with a vacuum.

The difference in mercury height on both necks indicates the capacity of a manometer.

$$820 \text{ mm} - 160 \text{ mm} = 660 \text{ mm}$$

Historically, the pressure unit of torr equals 1 mmHg or the rise/dip of 1 mm of mercury in a manometer.

Because the manometer uses mercury, the height difference (660 mm) equals its measuring capacity in torr (660 torrs).

52. C is correct.

The conditions of the ideal gases are the same, so the number of moles (i.e., molecules) is equal.

At STP, the temperature is the same, so the kinetic energy of the molecules is the same.

However, the molar mass of oxygen and nitrogen are different. Therefore, the density is different.

53. A is correct.

Vapor pressure is the pressure exerted by a vapor in equilibrium with condensed phases (i.e., solid or liquid) in a closed system at a given temperature.

The compound exists as a liquid if the external pressure > compound's vapor pressure.

A substance boils when vapor pressure = external pressure.

54. D is correct.

Gas molecules have a large amount of space between them; therefore, they can be pushed together, and gases are thus very compressible.

Molecules in solids and liquids are close together; therefore, they cannot get significantly closer and are thus nearly incompressible.

55. C is correct.

Charles' law (i.e., the law of volumes) explains how, at constant pressure, gases behave when the temperature changes:

$$V \propto T$$

or $$V / T = \text{constant}$$

or $$(V_1 / T_1) = (V_2 / T_2)$$

Volume and temperature are proportional. Therefore, an increase in one results in an increase in the other.

Gay-Lussac's law (i.e., pressure-temperature law) states that pressure is proportional to temperature:

$$P \alpha T$$

or

$$(P_1 / T_1) = (P_2 / T_2)$$

or

$$(P_1T_2) = (P_2T_1)$$

or $$P / T = \text{constant}$$

If the pressure of a gas increases, the temperature increases.

Dalton's law (i.e., the law of partial pressures) states that *the pressure exerted by a mixture of gases is equal to the sum of the individual gas pressures*

Boyle's law (i.e., pressure-volume law) states that pressure and volume are inversely proportional:

$$(P_1V_1) = (P_2V_2)$$

or $$P \times V = \text{constant}$$

If the volume of a gas increases, its pressure decreases proportionally.

Avogadro's law is an experimental gas law relating the volume of a gas to the amount of substance of gas present. *Equal volumes of gases at the same temperature and pressure have the same number of molecules*.

56. B is correct.

Sublimation is the direct state change from a solid to a gas, skipping the intermediate liquid phase.

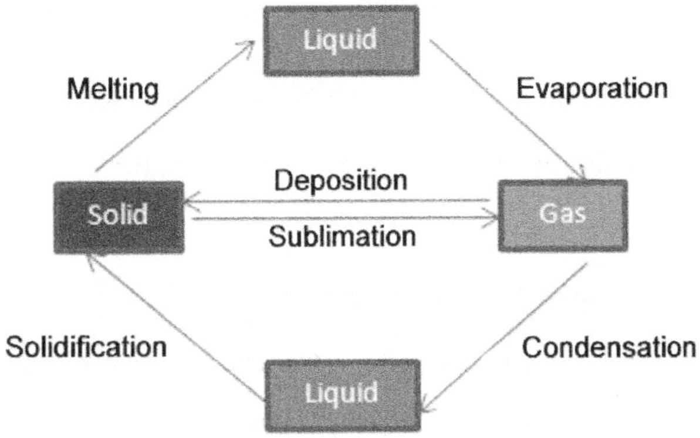

Interconversion of states of matter

An example of a compound that undergoes sublimation is solid carbon dioxide (i.e., dry ice). CO_2 changes phases from solid to gas (i.e., bypasses the liquid phase) and is often used as a cooling agent.

57. B is correct.

Graham's law of effusion states that the rate of effusion (i.e., escaping through a small hole) of a gas is inversely proportional to the square root of the molar mass of its particles.

Rate 1 / Rate 2 = √(molar mass gas 2 / molar mass gas 1)

The diffusion rate is the inverse root of the molecular weights of the gases.

Therefore, the rate of effusion is:

$O_2 / H_2 = \sqrt{(2 / 32)}$

rate of diffusion = 1:4

58. B is correct.

To calculate the number of molecules, calculate the moles of gas using the ideal gas law:

PV = nRT

n = PV / RT

Because the gas constant R is in L·atm K^{-1} mol^{-1}, pressure must be converted into atm:

320 mmHg × (1 / 760 atm/mmHg) = 320 mmHg / 760 atm

(Leave it in this fraction form because the answer choices are in this format.)

Convert the temperature to Kelvin:

10 °C + 273 = 283 K

Substitute those values into the ideal gas equation:

n = PV / RT

n = (320 mmHg / 760 atm) × 6 L / (0.0821 L·atm K^{-1} mol^{-1} × 283 K)

This expression represents the number of gas moles present in the container.

Calculate the number of molecules:

number of molecules = moles × Avogadro's number

number of molecules = (320 mmHg / 760 atm) × 6 / (0.0821 × 283) × 6.02×10^{23}

number of molecules = (320 / 760)·(6)·(6×10^{23}) / (0.0821)·(283)

59. A is correct.

At STP (standard conditions for temperature and pressure), the pressure and temperature are the same regardless of the gas.

The ideal gas law:

$$PV = nRT$$

where P is pressure, V is volume, n is the number of molecules, R is the ideal gas constant, and T is the temperature of the gas.

Therefore, one molecule of each gas occupies the same volume at STP.

Since CO_2 molecules have the largest mass, CO_2 gas has a greater mass in the same volume, and thus it has the greatest density.

60. C is correct.

Hydrogens, bonded directly to F, O or N, participate in hydrogen bonds. The hydrogen is partially positive (i.e., delta plus: $\partial+$) due to the bond to these electronegative atoms. The lone pair of electrons on the F, O or N interacts with the partial positive ($\partial+$) hydrogen to form a hydrogen bond.

Practice Set 4: Questions 61–80

61. D is correct.

Calculate the moles of each gas:

moles = mass / molar mass

moles H_2 = 9.50 g / (2 × 1.01 g/mole)

moles H_2 = 4.70 moles

moles Ne = 14.0 g / (20.18 g/mole)

moles Ne = 0.694 moles

Calculate the mole fraction of H_2:

Mole fraction of H_2 = moles of H_2 / moles in mixture

Mole fraction of H_2 = 4.70 moles / (4.70 moles + 0.694 moles)

Mole fraction of H_2 = 4.70 moles / (5.394 moles)

Mole fraction of H_2 = 0.87 moles

62. A is correct.

Vapor pressure is the pressure exerted by a vapor in equilibrium with its condensed phases (i.e., solid or liquid) in a closed system at a given temperature.

Boiling occurs when the vapor pressure of a liquid equals atmospheric pressure.

Vapor pressure increases as the temperature increases.

Liquid A boils at a lower temperature than B because the vapor pressure of Liquid A is closer to the atmospheric pressure.

63. B is correct.

Colligative properties of solutions depend on the ratio of the number of solute particles to the number of solvent molecules in a solution and not on the type of chemical species present.

Colligative properties include lowering vapor pressure, the elevation of boiling point, depression of freezing point, and increased osmotic pressure.

Freezing point (FP) depression:

$$\Delta FP = -iKm$$

where i = the number of particles produced when the solute dissociates, K = freezing point depression constant and m = molality (moles/kg solvent).

$-iKm = -2$ K

$-(1)\cdot(40)\cdot(x) = -2$

$x = -2 / -1(40)$

$x = 0.05$ molal

Assume that compound x does not dissociate.

0.05 mole compound (x) / kg camphor = 25 g / kg camphor

Therefore, if 0.05 mole = 25 g

1 mole = 500 g

64. C is correct.

Boiling occurs when the vapor pressure of a liquid equals atmospheric pressure.

Vapor pressure is the pressure exerted by a vapor in equilibrium with its condensed phases (i.e., solid or liquid) in a closed system at a given temperature.

Atmospheric pressure is the pressure exerted by the weight of air in the atmosphere.

Vapor pressure is inversely correlated with the strength of the intermolecular force.

In liquid form, the molecules are more likely to stick together due to stronger intermolecular forces. Few participate in the liquid-vapor equilibrium; therefore, the molecule would boil at a higher temperature.

65. D is correct.

Van der Waals equation describes factors that must be accounted for when the ideal gas law calculates values for nonideal gases.

The terms that affect the pressure and volume of the ideal gas law are intermolecular forces and the volume of nonideal gas molecules.

66. B is correct.

Ideal gas law:

$PV = nRT$

where P is pressure, V is volume, n is the number of molecules, R is the ideal gas constant, and T is the temperature of the gas.

Units of R can be calculated by rearranging the expression:

$R = PV/nT$

$R = \text{atm}\cdot\text{L/mol}\cdot\text{K}$

67. A is correct.

Ideal gas law:

$$PV = nRT$$

where P is pressure, V is volume, n is the number of molecules, R is the ideal gas constant, and T is the temperature of the gas.

Set the initial and final P/V/T conditions equal:

$$(P_1V_1 / T_1) = (P_2V_2 / T_2)$$

STP condition is the temperature of 0 °C (273 K) and a pressure of 1 atm.

Solve for the final temperature:

$$(P_2V_2 / T_2) = (P_1V_1 / T_1)$$

$$T_2 = (P_2V_2T_1) / (P_1V_1)$$

$$T_2 = [(0.80 \text{ atm}) \times (0.155 \text{ L}) \times (273 \text{ K})] / [(1.00 \text{ atm}) \times (0.120 \text{ L})]$$

$$T_2 = 282.1 \text{ K}$$

Then convert temperature units to degrees Celsius:

$$T_2 = (282.1 \text{ K} - 273 \text{ K})$$

$$T_2 = 9.1 \text{ °C}$$

68. C is correct.

Atmospheric pressure is the pressure exerted by the weight of air in the atmosphere.

Boiling occurs when the vapor pressure of the liquid is higher than the atmospheric pressure.

At standard atmospheric pressure and 22 °C, the vapor pressure of water is less than the atmospheric pressure, and it does not boil.

However, when a vacuum pump is used, the atmospheric pressure is reduced until it has a lower vapor pressure than water, allowing water to boil at a much lower temperature.

69. D is correct.

The kinetic theory of gases describes a gas as many small particles in constant rapid motion. These particles collide with each other and with the walls of the container.

Their average kinetic energy depends only on the absolute temperature of the system.

$$\text{temperature} = \tfrac{1}{2}mv^2$$

At high temperatures, the particles are moving greater velocity, and at a temperature of absolute zero (i.e., 0 K), there is no movement of gas particles.

Therefore, as temperature decreases, kinetic energy decreases, and so does the velocity of the gas molecules.

70. A is correct.

Dalton's law (i.e., the law of partial pressures) states that *the pressure exerted by a mixture of gases equals the sum of the individual gas pressures.*

The partial pressure of molecules in a mixture is proportional to their molar ratios.

Use the coefficients of the reaction to determine the molar ratio.

Based on that information, calculate the partial pressure of O_2:

(coefficient O_2) / (sum of coefficients in a mixture) × total pressure

$[1 / (2 + 1)] \times 1{,}250$ torr

417 torr = partial pressure of O_2

71. D is correct.

$2\ Na\ (s) + Cl_2\ (g) \rightarrow 2\ NaCl\ (s)$

In its elemental form, Na exists as a solid.

In its elemental form, chlorine exists as a gas.

72. B is correct.

Charles' law (i.e., the law of volumes) explains how, at constant pressure, gases behave when the temperature changes:

$V \alpha T$

or $V / T = \text{constant}$

or $(V_1 / T_1) = (V_2 / T_2)$

Volume and temperature are proportional.

Therefore, an increase in one term increases the other.

73. A is correct.

Gay-Lussac's law (i.e., pressure-temperature law) states that pressure is proportional to temperature:

$P \alpha T$

or $(P_1 / T_1) = (P_2 / T_2)$

or $(P_1T_2) = (P_2T_1)$

or $P / T = \text{constant}$

If the pressure of a gas increases, the temperature is increased proportionally.

Quadrupling the temperature increases the pressure fourfold.

Boyle's law (i.e., pressure-volume law) states that pressure and volume are inversely proportional:

$(P_1V_1) = (P_2V_2)$

or $P \times V = \text{constant}$

Reducing the volume by half increases pressure twofold.

Therefore, the increase in pressure would be by a factor of $4 \times 2 = 8$.

74. C is correct.

Scientists found that the relationships between pressure, temperature, and volume of a sample of gas hold for all gases, and the gas laws were developed.

Boyle's law (i.e., pressure-volume law) states that pressure and volume are inversely proportional:

$(P_1V_1) = (P_2V_2)$

or $P \times V = \text{constant}$

If the volume of a gas increases, its pressure decreases proportionally.

Charles' law (i.e., the law of volumes) explains how, at constant pressure, gases behave when the temperature changes:

$(V_1 / T_1) = (V_2 / T_2)$

Gay-Lussac's law (i.e., pressure-temperature law) states that pressure is proportional to temperature:

$P \propto T$

or $(P_1 / T_1) = (P_2 / T_2)$ or $(P_1T_2) = (P_2T_1)$

or $P / T = \text{constant}$

If the pressure of a gas increases, the temperature increases.

75. A is correct.

Sublimation is the direct state change from a solid to a gas, skipping the intermediate liquid phase. An example of a compound that undergoes sublimation is solid carbon dioxide (i.e., dry ice). CO_2 changes phases from solid to gas (i.e., bypasses the liquid phase) and is often used as a cooling agent.

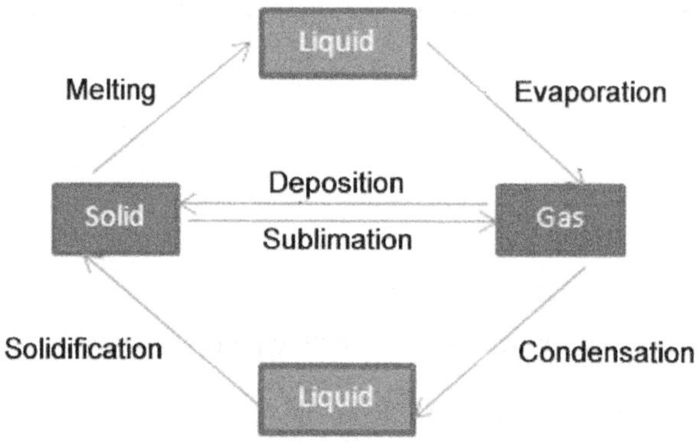

Interconversion of states of matter

76. B is correct.

Graham's law of effusion states that the rate of effusion (i.e., escaping through a small hole) of a gas is inversely proportional to the square root of the molar mass of its particles.

Rate 1 / Rate 2 = √(molar mass gas 2 / molar mass gas 1)

Set the rate of effusion of krypton over the rate of effusion of methane:

$Rate_{Kr}$ / $Rate_{CH4}$ = √ [(M_{Kr}) / (M_{CH4})]

Solve for the ratio of effusion rates:

$Rate_{Kr}$ / $Rate_{CH4}$ = √ [(83.798 g/mol) / (16.04 g/mol)]

$Rate_{Kr}$ / $Rate_{CH4}$ = 2.29

Larger gas molecules must effuse at a slower rate; the effusion rate of Kr gas molecules must be slower than methane molecules:

$Rate_{Kr}$ = [$Rate_{CH4}$ / (2.29)]

$Rate_{Kr}$ = [(631 m/s) / (2.29)]

$Rate_{Kr}$ = 276 m/s

77. C is correct.

Observe information about gas C. The temperature, pressure, and volume of the gas are indicated.

Substitute these values into the ideal gas law equation to determine the number of moles:

PV = nRT

n = PV / RT

Convert units to the units indicated in the gas constant:

volume = 668.5 mL × (1 L / 1,000 mL)

volume = 0. 669 L

pressure = 745.5 torr × (1 atm / 760 torr)

pressure = 0.981 atm

temperature = 32.0 °C + 273.15 K

temperature = 305.2 K

n = PV / RT

n = (0.981 atm × 0.669 L) / (0.0821 L·atm K^{-1} mol^{-1} × 305.2 K)

n = 0.026 mole

The Law of mass conservation states that the mass of products = the mass of reactants.

Based on this law, the mass of gas C can be determined:

mass product = mass reactant

mass A = mass B + mass C

5.2 g = 3.8 g + mass C

mass C = 1.4 g

Calculate the molar mass of C:

molar mass = 1.4 g / 0.026 mole

molar mass = 53.9 g / mole

78. D is correct.

Liquids take the shape of their container (i.e., they have an indefinite shape). Consider a liter of water poured into a cylindrical bucket or a square box. In both cases, it takes up the shape of the container.

Liquids have a definite volume. The volume of water in the given example is 1 L, regardless of its container.

79. D is correct.

Intermolecular forces act between neighboring molecules. Examples include hydrogen bonding, dipole-dipole, dipole-induced dipole, and van der Waals (i.e., London dispersion) forces.

Hydrogens, bonded directly to F, O or N, participate in hydrogen bonds. The hydrogen is partially positive (i.e., delta plus: $\partial+$) due to the bond to these electronegative atoms. The lone pair of electrons on the F, O or N interacts with the partial positive ($\partial+$) hydrogen to form a hydrogen bond.

80. D is correct.

When the balloon is placed in a freezer, the temperature of the helium in the balloon decreases because the surroundings are colder than the balloon, and heat is transferred from the balloon to the surrounding air until a thermal equilibrium is reached.

When the temperature of a gas decreases, the molecules move more slowly and become closer, causing the volume of the balloon to decrease.

From the ideal gas law,

PV = nRT

If temperature decreases, the volume must decrease (if pressure remains constant).

Stoichiometry – Detailed Explanations

==

Practice Set 1: Questions 1–20

==

1. C is correct.

One mole of an ideal gas occupies 22.71 L at STP.

3 mole × 22.71 L = 68.13 L

2. C is correct.

Use the mnemonic OIL RIG: Oxidation Is Loss, Reduction Is Gain (of electrons).

Oxidation is the loss of electrons, while reduction is the gain of electrons.

An oxidizing agent undergoes reduction, while a reducing agent undergoes oxidation.

Check the oxidation numbers of each atom.

The oxidation number for S is +6 as a reactant and +4 as a product, which means that it is reduced.

If a substance is reduced in a redox reaction, it is the oxidizing agent because it causes the other reagent (HI) to be oxidized.

3. C is correct. Balanced equation (combustion): $2\ C_6H_{14} + 19\ O_2 \rightarrow 12\ CO_2 + 14\ H_2O$

4. D is correct.

An oxidation number is a charge on an atom that is not present in the elemental state when the element is neutral.

Electronegativity refers to the attraction an atom has for additional electrons.

5. C is correct.

CH_3COOH has 2 carbons, 4 hydrogens, and 2 oxygens: $C_2O_2H_4$

Reduce to the smallest coefficients by dividing by two: COH_2

6. B is correct.

Oxidation numbers of Cl in each compound:

$NaClO_2$	+3
$Al(ClO_4)_3$	+7
$Ca(ClO_3)_2$	+5
$LiClO_3$	+5

7. A is correct.

Formula mass is a synonym for molecular mass/molecular weight (MW).

Molecular mass = (atomic mass of C) + (2 × atomic mass of O)

Molecular mass = (12.01 g/mole) + (2 × 16.00 g/mole)

Molecular mass = 44.01 g/mole

Note: 1 amu = 1 g/mole

8. C is correct.

Cadmium is located in group II, so its oxidation number is +2.

Sulfur tends to form sulfide ions (oxidation number –2).

The net oxidation number is zero, and therefore, the compound is CdS.

9. B is correct.

Balanced reaction (single replacement): Cl_2 (*g*) + 2 NaI (*aq*) → I_2 (*s*) + 2 NaCl (*aq*)

10. C is correct.

Use the mnemonic OIL RIG: Oxidation Is Loss, Reduction Is Gain (of electrons).

Oxidation is the loss of electrons, while reduction is the gain of electrons.

An oxidizing agent undergoes reduction, while a reducing agent undergoes oxidation.

When there are 2 reactants in a redox reaction, one is oxidized, and the other is reduced. The oxidized species is the reducing agent and vice versa.

Determine the oxidation number of all atoms:

I changes its oxidation state from –1 as a reactant to 0 as a product.

I is oxidized; therefore, the compound that contains I (NaI) is the reducing agent. Species in their elemental state have an oxidation number of 0.

Since the reaction is spontaneous, NaI is the *strongest* reducing agent because it can reduce another species spontaneously without needing external energy for the reaction to proceed.

11. C is correct.

To determine the remainder of a reactant, calculate how many moles of that reactant are required to produce the specified amount of product.

Use the coefficients to calculate the moles of the reactant:

moles N_2 = (1 / 2) × 18 moles = 9 moles N_2

Therefore, the remainder is:

(14.5 moles N_2) – (9 moles N_2) = 5.5 moles N_2 remaining

12. D is correct.

$CaCO_3 \rightarrow CaO + CO_2$ is a decomposition reaction because one complex compound is being broken down into two or more parts. However, the redox part is incorrect because the reactants and products are compounds (no single elements), and there is no change in the oxidation state (i.e., no gain or loss of electrons).

A: $AgNO_3 + NaCl \rightarrow AgCl + NaNO_3$ is correctly classified as a double-replacement reaction (sometimes referred to as double-displacement) because parts of two ionic compounds are exchanged to make two new compounds. It is a non-redox reaction because the reactants and products are compounds (no single elements), and there is no change in oxidation state (i.e., no gain or loss of electrons).

B: $Cl_2 + F_2 \rightarrow 2\ ClF$ is correctly classified as a synthesis reaction because two species are combining to form a more complex chemical compound as the product. It is a redox reaction because the F atoms in F_2 are reduced (i.e., gain electrons), and the Cl atoms in Cl_2 are oxidized (i.e., lose electrons), forming a covalent compound.

C: $H_2O + SO_2 \rightarrow H_2SO_3$ is correctly classified as a synthesis reaction because two species are combining to form a more complex chemical compound as the product. It is a non-redox reaction because the reactants and products are compounds (no single elements), and there is no change in oxidation state (i.e., no gain or loss of electrons).

13. C is correct.

Hydrogen is being oxidized (or ignited) to produce water.

The balanced equation that describes the ignition of hydrogen gas in the air to produce water is:

$$\tfrac{1}{2}\ O_2 + H_2 \rightarrow H_2O$$

This equation suggests that for every mole of water produced in the reaction, 1 mole of hydrogen and a ½ mole of oxygen gas are needed.

The maximum amount of water produced in the reaction is determined by the amount of the limiting reactant available. This requires identifying which reactant is the limiting reactant and can be done by comparing the number of moles of oxygen and hydrogen.

From the question, there are 10 grams of oxygen gas and 1 gram of hydrogen gas. This is equivalent to 0.3125 moles of oxygen and 0.5 moles of hydrogen.

Because the reaction requires twice as much hydrogen as oxygen gas, the limiting reactant is hydrogen gas (0.3215 moles of O_2 requires 0.625 moles of hydrogen, but only 0.5 moles of H_2 are available).

Since one equivalent of water is produced for every equivalent of hydrogen that is burned, the amount of water produced is:

$$(18\ \text{grams/mol}\ H_2O) \times (0.5\ \text{mol}\ H_2O) = 9\ \text{grams of}\ H_2O$$

14. D is correct.

Calculate the moles of Cl^- ion:

Moles of Cl^- ion = number of Cl^- atoms / Avogadro's number

Moles of Cl^- ion = 6.8×10^{22} atoms / 6.02×10^{23} atoms/mole

Moles of Cl^- ion = 0.113 moles

In a solution, $BaCl_2$ dissociates:

$BaCl_2 \rightarrow Ba^{2+} + 2\ Cl^-$

The moles of from Cl^- previous calculation can be used to calculate the moles of Ba^{2+}:

Moles of Ba^{2+} = (coefficient of Ba^{2+} / coefficient Cl^-) × moles of Cl^-

Moles of Ba^{2+} = (½) × 0.113 moles

Moles of Ba^{2+} = 0.0565 moles

Calculate the mass of Ba^{2+}:

Mass of Ba^{2+} = moles of Ba^{2+} × atomic mass of Ba

Mass of Ba^{2+} = 0.0565 moles × 137.33 g/mole

Mass of Ba^{2+} = 7.76 g

15. D is correct.

Na	Br	O_3
+1	x	(3×-2)
+1	x	-6

The sum of charges in a neutral molecule is zero:

1 + oxidation number of Br:

$1 + x + (-6) = 0$

$-5 + x = 0$

$x = +5$

oxidation number of Br = +5

16. A is correct.

Balanced equation:

$2\ Al + Fe_2O_3 \rightarrow 2\ Fe + Al_2O_3$

The sum of the coefficients of the products = 3.

17. B is correct.

2 H_2O_2 (*s*) → 2 H_2O (*l*) + O_2 (*g*) is incorrectly classified. The decomposition part of the classification is correct because one complex compound (H_2O_2) is being broken down into two or more parts (H_2O and O_2).

However, it is a redox reaction because oxygen is being lost from H_2O_2, one of the three indicators of a redox reaction (i.e., electron loss/gain, hydrogen loss/gain, oxygen loss/gain). Additionally, one of the products is a single element (O_2), which is a clue that it is a redox reaction.

$AgNO_3$ (*aq*) + KOH (*aq*) → KNO_3 (*aq*) + AgOH (*s*) is correctly classified as a non-redox reaction because the reactants and products are compounds (no single elements), and there is no change in oxidation state (i.e., no gain or loss of electrons). It is a precipitation reaction because the chemical reaction occurs in an aqueous solution, and one of the products formed (AgOH) is insoluble, making it a precipitate.

$Pb(NO_3)_2$ (*aq*) + 2 Na (*s*) → Pb (*s*) + 2 $NaNO_3$ (*aq*) is correctly classified as a redox reaction because the nitrate (NO_3^-), which dissociates, is oxidized (loses electrons), the Na^+ is reduced (gains electrons). The two combine to form the ionic compound $NaNO_3$. It is a single-replacement reaction because one element is substituted for another element in a compound, making a new compound ($2NaNO_3$) and an element (Pb).

HNO_3 (*aq*) + LiOH (*aq*) → $LiNO_3$ (*aq*) + H_2O (*l*) is correctly classified as a non-redox reaction because the reactants and products are compounds (no single elements), and there is no change in oxidation state (i.e., no gain or loss of electrons). It is a double-replacement reaction because parts of two ionic compounds are exchanged to make two new compounds.

18. C is correct.

At STP, 1 mole of gas has a volume of 22.4 L.

Calculate moles of O_2:

moles of O_2 = (15.0 L) / (22.4 L/mole)

moles of O_2 = 0.67 mole

Calculate the moles of O_2, using the fact that 1 mole of O_2 has 6.02×10^{23} O_2 molecules:

molecules of O_2 = 0.67 mole × (6.02×10^{23} molecules/mole)

molecules of O_2 = 4.03×10^{23} molecules

19. D is correct.

20. D is correct.

Use the mnemonic OIL RIG: Oxidation Is Loss, Reduction Is Gain (of electrons).

Oxidation is the loss of electrons, while reduction is the gain of electrons.

An oxidizing agent undergoes reduction, while a reducing agent undergoes oxidation.

Co^{2+} is the starting reactant because it loses electrons and produces an ion with a higher oxidation number.

Practice Set 2: Questions 21–40

21. D is correct.

Double replacement reaction:

HCl has a molar mass of 36.5 g/mol:

365 grams HCl = 10 moles HCl

10 moles are 75% of the total yield.

10 moles PCl_3 / 0.75 = 13.5 moles of HCl.

Apply the mole ratios:

Coefficients: 3 HCl = PCl_3

13.5 / 3 = moles of PCl_3

13.5 moles of HCl is produced by 4.5 moles of PCl_3

22. A is correct.

All elements in their free state have an oxidation number of zero.

23. B is correct.

Balanced chemical equation:

$C_3H_8 + 5\ O_2 \rightarrow 3\ CO_2 + 4\ H_2O$

By using the coefficients from the balanced equation, apply dimensional analysis to solve for the volume of H_2O produced from the 2.6 L C_3H_8 reacted:

$V_{H2O} = V_{C3H8} \times (\text{mol } H_2O / \text{mol } C_3H_8)$

$V_{H2O} = (2.6\text{ L}) \times (4\text{ mol} / 1\text{ mol})$

$V_{H2O} = 10.4\text{ L}$

24. C is correct.

Use the mnemonic OIL RIG: Oxidation Is Loss, Reduction Is Gain (of electrons).

Oxidation is the loss of electrons, while reduction is the gain of electrons.

An oxidizing agent undergoes reduction, while a reducing agent undergoes oxidation.

The oxidation number of Hg decreases (i.e., reduction) from +2 as a reactant ($HgCl_2$) to +1 as a product (Hg_2Cl_2). This means that Hg gained one electron.

25. C is correct.

Balanced equation (synthesis):

$$4\ P\ (s) + 5\ O_2\ (g) \rightarrow 2\ P_2O_5\ (s)$$

26. D is correct.

27. A is correct.

Formula mass is a synonym for molecular mass/molecular weight (MW).

Start by calculating the mass of the formula unit (CH_2O)

$$CH_2O = (12.01\ \text{g/mol} + 2.02\ \text{g/mol} + 16.00\ \text{g/mol}) = 30.03\ \text{g/mol}$$

Divide the molar mass with the formula unit mass:

$$180\ \text{g/mol} / 30.03\ \text{g/mol} = 5.994$$

Round to the closest whole number: 6

Multiply the formula unit by 6:

$$(CH_2O)_6 = C_6H_{12}O_6$$

28. B is correct.

Calculate the moles of LiI present. The molecular weight of LiI is 133.85 g/mol.

Moles of LiI:

$$6.45\ \text{g} / 133.85\ \text{g/mole} = 0.0482\ \text{moles}$$

Each mole of LiI contains 6.02×10^{23} molecules.

Number of molecules:

$$0.0482 \times 6.02 \times 10^{23} = 2.90 \times 10^{22}\ \text{molecules}$$

1 molecule is equal to 1 formula unit.

29. D is correct.

A combustion engine creates various nitrogen oxide compounds, often referred to as NO_x gases because each gas would have a different value of x in its formula.

30. C is correct.

Balanced equation (synthesis):

$$4\ P\ (s) + 3\ O_2\ (g) \rightarrow 2\ P_2O_3\ (s)$$

31. A is correct.

Use the mnemonic OIL RIG: Oxidation Is Loss, Reduction Is Gain (of electrons).

Oxidation is the loss of electrons, while reduction is the gain of electrons.

An oxidizing agent undergoes reduction, while a reducing agent undergoes oxidation.

Because the forward reaction is spontaneous, the reverse reaction is not spontaneous.

Sn cannot reduce Mg^{2+}; therefore, Sn is the weakest reducing agent.

32. C is correct.

Balancing Redox Equations

From the balanced equation, the coefficient for the proton can be determined. Balancing a redox equation requires balancing the atoms in the equation, but the charges must be balanced.

The equations must be separated into two different half-reactions. One of the half-reactions addresses the oxidizing component, and the other addresses the reducing component.

Magnesium is being oxidized; therefore, the unbalanced oxidation half-reaction is:

$$Mg\ (s) \rightarrow Mg^{2+}\ (aq)$$

Furthermore, the nitrogen is being reduced; therefore, the unbalanced reduction half-reaction is:

$$NO_3^-\ (aq) \rightarrow NO_2\ (aq)$$

Each half-reaction needs to be balanced for each atom at this stage, and the net electric charge on each side of the equations must be balanced.

Order of operations for balancing half-reactions:

1) Balance atoms except for oxygen and hydrogen.
2) Balance the oxygen atom count by adding water.
3) Balance the hydrogen atom count by adding protons.
    a) If in basic solution, add equal amounts of hydroxide to each side to cancel the protons.
4) Balance the electric charge by adding electrons.
5) If necessary, multiply the coefficients of one half-reaction equation by a factor that cancels the electron count when both equations are combined.
6) Cancel ions or molecules that appear on each side of the overall equation.

After determining the balanced overall redox reaction, stoichiometry indicates the moles of protons that are involved.

For magnesium:

$$Mg\ (s) \rightarrow Mg^{2+}\ (aq)$$

The magnesium is balanced with a coefficient of 1. There are no hydrogen or oxygen atoms present in the equation. The magnesium cation has a +2 charge, so 2 moles of electrons should be added to the right side to balance the charge.

The balanced half-reaction for oxidation:

$$Mg\ (s) \rightarrow Mg^{2+}\ (aq) + 2\ e^-$$

For nitrogen:

$$NO_3^-\ (aq) \rightarrow NO_2\ (aq)$$

The equation is balanced for nitrogen because one nitrogen atom appears on each side of the reaction. The nitrate reactant has three oxygen atoms, while the nitrite product has two oxygen atoms.

To balance the oxygen, one mole of water should be added to the right side:

$$NO_3^-\ (aq) \rightarrow NO_2\ (aq) + H_2O$$

Adding water to the right side of the equation introduces hydrogen atoms to that side. Therefore, the hydrogen atom count needs to be balanced. Water possesses two hydrogen atoms; therefore, two protons need to be added to the left side of the reaction:

$$NO_3^-\ (aq) + 2\ H^+ \rightarrow NO_2\ (aq) + H_2O$$

The reaction is occurring in acidic conditions. If the reaction were basic, OH^- would need to be added to each side to cancel the protons.

Next, the net charge needs to be balanced. The left side has a net charge of +1 (+2 from the protons and –1 from the electron), while the right side is neutral. Therefore, one electron should be added to the left side:

$$NO_3^-\ (aq) + 2\ H^+ + e^- \rightarrow NO_2\ (aq) + H_2O$$

When half-reactions are recombined, the electrons in the overall reaction must cancel.

The reduction half-reaction contributes one electron to the left side of the overall equation, while the oxidation half-reaction contributes two electrons to the product side.

Therefore, the coefficients of the reduction half-reaction should be doubled:

$$2 \times [NO_3^-\ (aq) + 2\ H^+ + e^- \rightarrow NO_2\ (aq) + H_2O]$$

$$= 2\ NO_3^-\ (aq) + \mathbf{4\ H^+} + 2\ e^- \rightarrow 2\ NO_2\ (aq) + 2\ H_2O$$

The answer is 4 at this step.

Combining both half-reactions gives:

$$Mg\ (s) + 2\ NO_3^-\ (aq) + \mathbf{4\ H^+} + 2\ e^- \rightarrow Mg^{2+}\ (aq) + 2\ e^- + 2\ NO_2\ (aq) + 2\ H_2O$$

Cancel the electrons that appear on each side of the reaction in the following balanced net equation:

$$Mg\ (s) + 2\ NO_3^-\ (aq) + \mathbf{4\ H^+} \rightarrow Mg^{2+}\ (aq) + 2\ NO_2\ (aq) + 2\ H_2O$$

This equation is now fully balanced for mass, oxygen, hydrogen and electric charge.

33. B is correct.

To obtain moles, divide sample mass by the molecular mass.

Because all options have the same mass, a comparison of molecular mass provides the answer without performing the actual calculation.

The molecule with the smallest molecular mass has the greatest number of moles.

34. A is correct.

The law of constant composition states that *samples of a given chemical compound have the same chemical composition by mass*. This is true for all compounds, including ethyl alcohol.

This law is referred to as the *law of definite proportions* or *Proust's Law* because this observation was first made by the French chemist Joseph Proust.

35. A is correct.

The coefficients in balanced reactions refer to moles (or molecules) but not grams.

1 mole of N_2 gas reacts with 3 moles of H_2 gas to produce 2 moles of NH_3 gas.

36. C is correct.

Calculate the number of moles of Al_2O_3:

$$2\ Al / 1\ Al_2O_3 = 0.2\ \text{moles}\ Al / x\ \text{moles}\ Al_2O_3$$

$$x\ \text{moles}\ Al_2O_3 = 0.2\ \text{moles}\ Al \times 1\ Al_2O_3 / 2\ Al$$

$$x = 0.1\ \text{moles}\ Al_2O_3$$

Therefore:

$$(0.1\ \text{mole}\ Al_2O_3)\cdot(102\ \text{g/mole}\ Al_2O_3) = 10.2\ \text{g}\ Al_2O_3$$

37. D is correct.

Because Na is a metal, Na is more electropositive than H. Therefore, the positive charge is on Na (i.e., +1), which means that the charge on H is –1.

38. A is correct.

$PbO + C \rightarrow Pb + CO$ is a single replacement reaction because only one element is transferred from one reactant to another.

However, it is a redox reaction because the lead (Pb^{2+}), which dissociates, is reduced (gains electrons), the oxygen (O^{2-}) is oxidized (loses electrons). Additionally, one of the products is a single element (Pb), which is a clue that it is a redox reaction.

C: the reaction is double-replacement and not combustion because the combustion reactions must have hydrocarbon and O_2 as reactants. Double-replacement reactions often result in the formation of a solid, water, and gas.

39. B is correct.

The sum of the oxidation numbers in a neutral molecule must be zero.

The oxidation number of each H is +1, and the oxidation number of each O is –2.

If x denotes the oxidation number of S in H_2SO_4, then:

$$2(+1) + x + 4(-2) = 0$$

$$+2 + x + -8 = 0$$

$$x = +6$$

In H_2SO_4, the oxidation sate of sulfur is +6.

40. A is correct.

To calculate the number of molecules, start by calculating the moles of gas using the ideal gas equation:

$$PV = nRT$$

Rearrange to isolate the number of moles:

$$n = PV / RT$$

Because the gas constant R is in L·atm K^{-1} mol^{-1}, the pressure must be converted into atm:

$$780 \text{ torr} \times (1 \text{ atm} / 760 \text{ torr}) = 1.026 \text{ atm}$$

Convert the volume to liters:

$$500 \text{ mL} \times 0.001 \text{ L/mL} = 0.5 \text{ L}$$

Substitute into the ideal gas equation:

$$n = PV / RT$$

$$n = 1.026 \text{ atm} \times 0.5 \text{ L} / (0.08206 \text{ L·atm K}^{-1} \text{mol}^{-1} \times 320 \text{ K})$$

$$n = 0.513 \text{ L·atm} / (26.259 \text{ L·atm mol}^{-1})$$

$$n = 0.0195 \text{ mol}$$

===

Practice Set 3: Questions 41–60

===

41. B is correct.

Use the mnemonic OIL RIG: Oxidation Is Loss, Reduction Is Gain (of electrons).

The oxidizing and reducing agents are reactants, not products, in a redox reaction.

The oxidizing agent is the species that gets reduced (i.e., gains electrons) and causes oxidation of the other species.

The reducing agent is the species that gets oxidized (i.e., loses electrons) and reduces the other species.

42. D is correct.

Balancing a redox equation in acidic solution by the half-reaction method involves each of the steps described.

43. D is correct.

Formula mass is a synonym for molecular mass/molecular weight (MW).

MW of $C_6H_{12}O_6$ = (6 × atomic mass C) + (12 × atomic mass H) + (6 × atomic mass O)

MW of $C_6H_{12}O_6$ = (6 × 12.01 g/mole) + (12 × 1.01 g/mole) + (6 × 16.00 g/mole)

MW of $C_6H_{12}O_6$ = (72.06 g/mole) + (12.12 g/mole) + (96.00 g/mole)

MW of $C_6H_{12}O_6$ = 180.18 g/mole

44. A is correct.

The molecular mass of oxygen is 16 g/mole.

The atomic mass unit (amu) or dalton (Da) is the standard unit for indicating mass on an atomic or molecular scale (atomic mass). One amu is approximately the mass of one nucleon (a single proton or neutron) and is numerically equivalent to 1 g/mol.

45. D is correct.

Balanced equation (combustion): $2\ C_2H_6 + 7\ O_2 \rightarrow 4\ CO_2 + 6\ H_2O$

46. B is correct.

Use the mnemonic OIL RIG: Oxidation Is Loss, Reduction Is Gain (of electrons).

Oxidation is the loss of electrons, while reduction is the gain of electrons.

An oxidizing agent undergoes reduction, while a reducing agent undergoes oxidation.

Evaluate the oxidation number in each atom.

S changes from +6 as a reactant to –2 as a product, reducing it because the oxidation number decreases.

47. C is correct.

Balanced equation (double replacement):

$$C_2H_5OH\ (g) + 3\ O_2\ (g) \rightarrow 2\ CO_2\ (g) + 3\ H_2O\ (g)$$

48. A is correct.

Note: 1 amu (atomic mass unit) = 1 g/mole.

Moles of C = mass of C / atomic mass of C

Moles of C = (4.50 g) / (12.01 g/mole)

Moles of C = 0.375 mole

49. C is correct.

According to the law of mass conservation, there should be equal amounts of carbon in the product (CO_2) and reactant (the hydrocarbon sample).

Start by calculating the amount of carbon in the product (CO_2).

First, calculate the mass % of carbon in CO_2:

Molecular mass of CO_2 = atomic mass of carbon + (2 × atomic mass of oxygen)

Molecular mass of CO_2 = 12.01 g/mole + (2 × 16.00 g/mole)

Molecular mass of CO_2 = 44.01 g/mole

Mass % of carbon in CO_2 = (mass of carbon / molecular mass of CO_2) × 100%

Mass % of carbon in CO_2 = (12.01 g/mole / 44.01 g/mole) × 100%

Mass % of carbon in CO_2 = 27.3%

Calculate the mass of carbon in the CO_2:

Mass of carbon = mass of CO_2 × mass % of carbon in CO_2

Mass of carbon = 8.98 g × 27.3%

Mass of carbon = 2.45 g

The mass of carbon in the starting reactant (the hydrocarbon sample) should be 2.45 g.

Mass % carbon in hydrocarbon = (mass carbon in sample / total mass sample) × 100%

Mass % of carbon in hydrocarbon = (2.45 g / 6.84 g) × 100%

Mass % of carbon in hydrocarbon = 35.8%

50. B is correct.

The number of each atom on the reactants' side must be the same and equal in number to the number of atoms on the product side of the reaction.

51. D is correct.

Using ½ reactions, the balanced reaction is:

$$6\ (Fe^{2+} \rightarrow Fe^{3} + 1\ e^{-})$$

$$Cr_2O_7^{2-} + 14\ H^+ + 6\ e^- \rightarrow 2\ Cr^{3+} + 7\ H_2O$$

$$Cr_2O_7^{2-} + 14\ H^+ + 6\ Fe^{2+} \rightarrow 2\ Cr^{3+} + 6\ Fe^{3+} + 7\ H_2O$$

The sum of coefficients in the balanced reaction is 36.

52. A is correct.

Balanced reaction: $4\ RuS\ (s) + 9\ O_2 + 4\ H_2O \rightarrow 2\ Ru_2O_3\ (s) + 4\ H_2SO_4$

Calculate the moles of H_2SO_4:

$49\ g \times 1\ mol / 98\ g = 0.5$ mole H_2SO_4

If 0.5 mole H_2SO_4 is produced, set a ratio for O_2 consumed.

$9\ O_2 / 4\ H_2SO_4 = x$ moles $O_2 / 0.5$ mole H_2SO_4

x moles $O_2 = 9\ O_2 / 4\ H_2SO_4 \times 0.5$ mole H_2SO_4

$x = 1.125$ moles $O_2 \times 22.4$ liters / 1 mole

$x = 25.2$ liters of O_2

53. C is correct.

Identify the limiting reactant:

Al is the limiting reactant because 2 atoms combine with 1 molecule of ferric oxide to produce the products. If the reaction starts with equal numbers of moles of each reactant, Al is depleted.

Use moles of Al and coefficients to calculate moles of other products/reactants.

Coefficient of Al = coefficient of Fe

Moles of iron produced = 0.20 moles.

54. B is correct.

S_2 O_8

$2x$ (8×-2)

The sum of charges on the molecule is –2:

$2x + (8 \times -2) = -2$

$2x + (-16) = -2$

$2x = +14$

oxidation number of S = +7

55. D is correct.

Na is in group IA, so its oxidation number is +1.

Oxygen is usually −2, with some exceptions, such as peroxide (H_2O_2), where it is −1.

Cr is a transition metal and could have more than one possible oxidation number.

To determine Cr's oxidation number, use the known oxidation numbers:

$2(+1) + Cr + 4(-2) = 0$

$2 + Cr - 8 = 0$

$Cr = 6$

56. B is correct.

The net charge of $H_2SO_4 = 0$

Hydrogen has a common oxidation number of +1, whereas each oxygen atom has an oxidation number of −2.

H_2	S	O_4
(2×1)	x	(4×-2)

$2 + x + -8 = 0$

$x = +6$

In H_2SO_4, the oxidation of sulfur is +6.

57. C is correct.

Use the mnemonic OIL RIG: <u>O</u>xidation <u>I</u>s <u>L</u>oss, <u>R</u>eduction <u>I</u>s <u>G</u>ain (of electrons).

Oxidation is the loss of electrons, while reduction is the gain of electrons.

An oxidizing agent undergoes reduction, while a reducing agent undergoes oxidation.

The oxidation number for Sn is +2 as a reactant and +4 as a product; an increase in oxidation number means that Sn^{2+} is oxidized in this reaction.

58. D is correct.

Na_2Cl_2 and 2 NaCl are not the same.

Na_2Cl_2 is a molecule of 2 Na and 2 Cl.

NaCl is a compound that consists of 1 Na and 1 Cl.

59. D is correct.

I: represents the volume of 1 mol of an ideal gas at STP

II: mass of 1 mol of PH_3

III: number of molecules in 1 mol of PH_3

60. D is correct.

Use the mnemonic OIL RIG: Oxidation Is Loss, Reduction Is Gain (of electrons).

Oxidation is the loss of electrons, while reduction is the gain of electrons.

An oxidizing agent undergoes reduction, while a reducing agent undergoes oxidation.

Co^{2+} is the starting reactant because it loses electrons and produces an ion with a higher oxidation number.

==

Practice Set 4: Questions 61–80

==

61. B is correct.

Use the mnemonic OIL RIG: Oxidation Is Loss, Reduction Is Gain (of electrons).

The oxidizing and reducing agents are reactants, not products, in a redox reaction.

The oxidizing agent is the species that is reduced (i.e., gains electrons).

The reducing agent is the species that is oxidized (i.e., loses electrons).

62. D is correct.

Calculations:

% mass Cl = (molecular mass Cl) / (molecular mass of compound) × 100%

% mass Cl = 4(35.5 g/mol) / [12 g/mol + 4(35.5 g/mol)] × 100%

% mass Cl = (142 g/mol) / (154 g/mol) × 100%

% mass Cl = 0.922 × 100% = 92%

63. D is correct.

I: ionic charge of reactants must equal the ionic charge of products, which means that no electrons can "disappear" from the reaction; they can only be transferred from one atom to another. Therefore, the total charge will always be the same.

II: atoms of each reactant must equal atoms of product, which is true for chemical reactions because atoms in chemical reactions cannot be created or destroyed.

III: electrons gained by one atom must be lost by another.

64. B is correct.

Determine the molecular weight of each compound.

Note, unlike % by mass, there is no division.

$Cl_2C_2H_4$ = 2(35.5 g/mol) + 2(12 g/mol) + 4(1 g/mol)

$Cl_2C_2H_4$ = 71 g/mol + 24 g/mol + 4 g/mol

$Cl_2C_2H_4$ = 99 g/mol

65. C is correct.

Formula mass is a synonym for molecular mass/molecular weight (MW).

Start by calculating the mass of the formula unit (CH):

$$CH = (12.01\ g/mol + 1.01\ g/mol) = 13.02\ g/mol$$

Divide the molar mass by the formula unit mass:

$$78\ g/mol\ /\ 13.02\ g/mol = 5.99$$

Round to the closest whole number: 6

Multiply the formula unit by 6:

$$(CH)_6 = C_6H_6$$

66. D is correct.

There is 1 oxygen atom from GaO plus 6 oxygen atoms from $(NO_3)_2$, so the total is 7.

67. B is correct.

Ethanol (C_2H_5OH) undergoes combination with oxygen to produce carbon dioxide and water.

68. C is correct.

Balanced reaction (combustion):

$$2\ C_3H_7OH + 9\ O_2 \rightarrow 6\ CO_2 + 8\ H_2O$$

69. C is correct.

Balanced equation (double replacement):

$$Co_2O_3\ (s) + 3\ CO\ (g) \rightarrow 2\ Co\ (s) + 3\ CO_2\ (g)$$

70. B is correct.

The molar volume (V_m) is the volume occupied by one mole of a substance (i.e., element or compound) at a given temperature and pressure.

71. A is correct.

Use the mnemonic OIL RIG: Oxidation Is Loss, Reduction Is Gain (of electrons).

Oxidation is the loss of electrons, while reduction is the gain of electrons.

An oxidizing agent undergoes reduction, while a reducing agent undergoes oxidation.

Because the forward reaction is spontaneous, the reverse reaction is not spontaneous.

Mg^{2+} cannot oxidize Sn; therefore, Mg^{2+} is the weakest oxidizing agent.

72. D is correct.

H_2 is the limiting reactant.

24 moles of H_2 should produce: $24 \times (2 / 3) = 16$ moles of NH_3

If 13.5 moles are produced, the yield:

13.5 moles / 16 moles × 100% = 84%

73. B is correct.

Balanced reaction:

$H_2 + \frac{1}{2} O_2 \rightarrow H_2O$

Multiply the equations by 2 to remove the fraction:

$2 H_2 + O_2 \rightarrow 2 H_2O$

Find the limiting reactant by calculating the moles of each reactant:

Moles of hydrogen = mass of hydrogen / (2 × atomic mass of hydrogen)

Moles of hydrogen = 25 g / (2 × 1.01 g/mole)

Moles of hydrogen = 12.38 moles

Moles of oxygen = mass of oxygen / (2 × atomic mass of oxygen)

Moles of oxygen = 225 g / (2 × 16.00 g/mole)

Moles of oxygen = 7.03 moles

Divide the number of moles of each reactant by its coefficient.

The reactant with a smaller number of moles is the limiting reactant.

Hydrogen = 12.38 moles / 2

Hydrogen = 6.19 moles

Oxygen = 7.03 moles / 1

Oxygen = 7.03 moles

Because hydrogen has a smaller number of moles after the division, hydrogen is the limiting reactant, and hydrogen will be depleted in the reaction.

$2 H_2 + O_2 \rightarrow 2 H_2O$

Hydrogen and water have coefficients of 2; therefore, these have the same number of moles.

There are 12.38 moles of hydrogen, which means this reaction produces 12.38 moles of water.

Molecular mass of H_2O = (2 × molecular mass of hydrogen) + atomic mass of oxygen

Molecular mass of H_2O = (2 × 1.01 g/mole) + 16.00 g/mole

Molecular mass of H_2O = 18.02 g/mole

Mass of H_2O = moles of H_2O × molecular mass of H_2O

Mass of H_2O = 12.38 moles × 18.02 g/mole

Mass of H_2O = 223 g

74. C is correct.

Li	Cl	O_2
+1	x	(2 × –2)

The sum of charges in a neutral molecule is zero:

$1 + x + (2 \times -2) = 0$

$1 + x + (-4) = 0$

$x = -1 + 4$

oxidation number of Cl = +3

75. D is correct.

Using half-reactions, the balanced reaction is:

To balance hydrogen ↓ To balance oxygen ↓

$$4[5\ e^- + 8\ H^+ + MnO_4^- \rightarrow Mn^{2+} + 4\ (H_2O)]$$

$$5[(H_2O) + C_3H_7OH \rightarrow C_2H_5CO_2H + 4\ H^+ + 4\ e^-]$$

$$12\ H^+ + 4\ MnO_4^- + 5\ C_3H_7OH \rightarrow 11\ H_2O + 4\ Mn^{2+} + 5\ C_2H_5CO_2H$$

Multiply by a common multiple of 4 and 5:

The sum of the product's coefficients: 11 + 4 + 5 = 20.

76. A is correct.

PbO (*s*) + C (*s*) → Pb (*s*) + CO (*g*) is a single-replacement reaction because one element (i.e., oxygen) is being transferred from one reactant to another.

77. D is correct.

The reactants in a chemical reaction are on the left of the reaction arrow; the products are on the right.

In this reaction, the reactants are $C_6H_{12}O_6$, H_2O and O_2.

There is a distinction between the terms *reactant* and *reagent.*

A reactant is a substance consumed in the course of a chemical reaction.

A reagent is a substance (e.g., solvent) added to a system to cause a chemical reaction.

78. C is correct.

Determine the number of moles of He:

$4 \text{ g} \div 4.0 \text{ g/mol} = 1 \text{ mole}$

He has 2 electrons since He has atomic number 2 (# electrons = # protons for neutral atoms).

Therefore, 1 mole of He × 2 electrons / mole = 2 mole of electrons

79. D is correct.

In this molecule, Br has the oxidation number of –1 because it is a halogen, and the gaining of one electron results in a complete octet for bromine.

The sum of charges in a neutral molecule = 0

0 = (oxidation state of Fe) + (3 × oxidation state of Br)

0 = (oxidation state of Fe) + (3 × –1)

0 = oxidation state of Fe – 3

oxidation state of Fe = +3

80. B is correct.

A mole is a unit of measurement used to express amounts of a chemical substance.

The number of molecules in a mole is 6.02×10^{23}, which is Avogadro's number.

However, Avogadro's number relates to the number of molecules, not the amount of substance.

Molar mass refers to the mass per mole of a substance.

Formula mass is a term that is sometimes used to mean *molecular mass* or *molecular weight*, and it refers to the mass of a particular molecule.

Notes for active learning

Thermochemistry – Detailed Explanations

Practice Set 1: Questions 1–20

1. A is correct.

A reaction's enthalpy is specified by ΔH (not ΔG). ΔH determines whether a reaction is exothermic (releases heat to surroundings) or endothermic (absorbs heat from surroundings).

To predict spontaneity of reaction, use the Gibbs free energy equation: $\Delta G = \Delta H° - T\Delta S$

The reaction is spontaneous (i.e., exergonic) if ΔG is negative.

The reaction is nonspontaneous (i.e., endergonic) if ΔG is positive.

2. B is correct.

Heat is energy, and the energy from the heat is transferred to the gas molecules, which increases their movement (i.e., kinetic energy).

Temperature is a measure of the average kinetic energy of the molecules.

3. D is correct.

Heat capacity is the amount of heat required to increase the temperature of *the whole sample* by 1 °C.

Specific heat is the heat required to increase the temperature of *1 gram* of sample by 1 °C.

Heat = mass × specific heat × change in temperature:

$$q = m \times c \times \Delta T$$

$$q = 21.0\ g \times 0.382\ J/g{\cdot}°C \times (68.5\ °C - 21.0\ °C)$$

$$q = 21.0\ g \times 0.382\ J/g{\cdot}°C \times (47.5\ °C)$$

$$q = 381\ J$$

4. C is correct. This is a theory/memorization question. However, it can be solved by comparing the options.

Kinetic energy is correlated with temperature, so the formula that involves temperature would be a good choice.

Another approach is to analyze the units; apply the units to the variables and evaluate them. For example, choices A and B have moles, pressure (Pa, bar, or atm), and area (m^2), so it is not possible for them to result in energy (J) when multiplied.

Option D has molarity (moles/L) and volume (L or m^3); they cannot result in energy (J) when multiplied.

Option C is nRT: mol × (J/mol K) × K; units but J cancel, so this is a plausible formula for kinetic energy.

For questions using formulas, dimensional analysis limits the answer choices on this type of question.

5. A is correct.

ΔH refers to enthalpy (or heat).

Endothermic reactions have heat as a reactant.

Exothermic reactions have heat as a product.

Exothermic reactions release heat and cause the temperature of the immediate surroundings to rise (i.e., a net loss of energy).

Endothermic reactions absorb heat and cool the surroundings (i.e., a net gain of energy).

Endothermic reactions absorb energy to break strong bonds to form a less stable state (i.e., positive enthalpy).

Exothermic reactions release energy when forming stronger bonds to produce a more stable state (i.e., negative enthalpy).

6. D is correct.

Endothermic reactions absorb energy to break strong bonds to form a less stable state (i.e., positive enthalpy).

Exothermic reactions release energy when forming stronger bonds to produce a more stable state (i.e., negative enthalpy).

The spontaneity of a reaction is determined by the calculation of ΔG using:

$$\Delta G = \Delta H - T\Delta S$$

The reaction is spontaneous when the value of ΔG is negative.

In this problem, the reaction is endothermic, which means that ΔH is positive.

The reaction decreases S, which means that ΔS value is negative.

Substituting these values into the equation:

$$\Delta G = \Delta H - T(-\Delta S)$$

$$\Delta G = \Delta H + T\Delta S\text{, with } \Delta H \text{ and } \Delta S \text{ positive values}$$

For this problem, the value of ΔG is positive.

The reaction does not occur because ΔG needs to be negative for a spontaneous reaction to proceed, the products are more stable than the reactants.

7. D is correct.

Chemical energy is potential energy stored in molecular bonds.

Electrical energy is the kinetic energy associated with the motion of electrons (e.g., in wires, circuits, lightning). The potential energy stored in a flashlight battery is electrical energy.

Heat is the spontaneous transfer of energy from a hot object to a cold one with no displacement or deformation of the objects (i.e., no work done).

8. A is correct.

Entropy is higher for less organized states (i.e., more random or disordered).

9. C is correct.

ΔH refers to enthalpy (or heat).

Endothermic reactions have heat as a reactant.

Exothermic reactions have heat as a product.

Exothermic reactions release heat and cause the temperature of the immediate surroundings to rise (i.e., a net loss of energy).

Endothermic reactions absorb heat and cool the surroundings (i.e., a net gain of energy).

Endothermic reactions absorb energy to break strong bonds to form a less stable state (i.e., positive enthalpy). Exothermic reactions release energy when forming stronger bonds to produce a more stable state (i.e., negative enthalpy).

The reaction is nonspontaneous (i.e., endergonic) if the products are less stable than the reactants and ΔG is positive.

The reaction is spontaneous (i.e., exergonic) if the products are more stable than the reactants and ΔG is negative.

10. D is correct.

This cannot be determined when the given $\Delta G = \Delta H - T\Delta S$, terms cancel.

The system is at equilibrium when $\Delta G = 0$, but in this question, each side of the equation cancel:

$$\underbrace{X - RY}_{\Delta G \text{ term}} = \underbrace{X - RY}_{\substack{\Delta H - T\Delta S \\ \text{term}}}, \quad \text{so } 0 = 0$$

11. C is correct.

Gibbs free energy:

$$\Delta G = \Delta H - T\Delta S$$

Stable molecules are spontaneous because they have a negative (or relatively low) ΔG.

ΔG is most negative (most stable) when ΔH is the smallest and ΔS is largest.

12. A is correct.

ΔH refers to enthalpy (or heat).

Endothermic reactions have heat as a reactant.

Exothermic reactions have heat as a product.

The reaction is nonspontaneous (i.e., endergonic) if the products are less stable than the reactants and ΔG is positive.

The reaction is spontaneous (i.e., exergonic) if the products are more stable than the reactants and ΔG is negative.

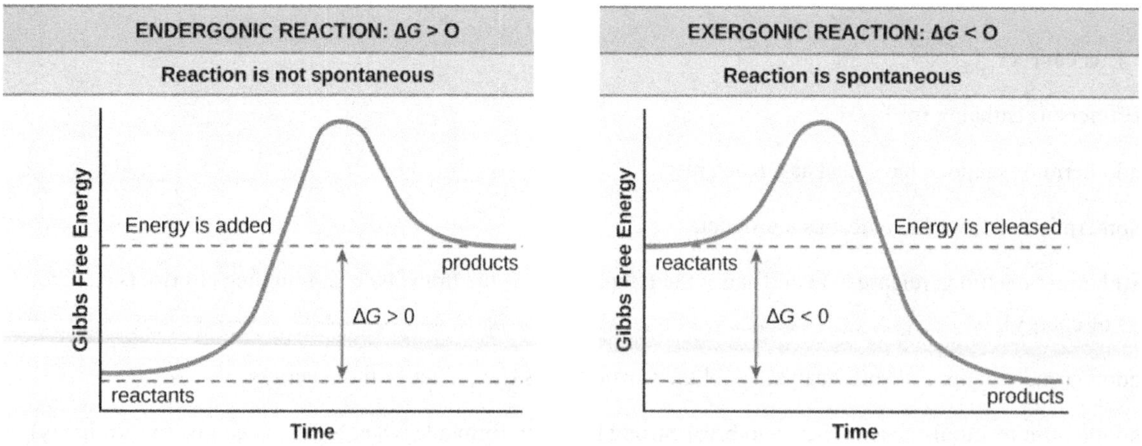

Exothermic reactions release heat and cause the temperature of the immediate surroundings to rise (i.e., a net loss of energy).

Endothermic reactions absorb heat and cool the surroundings (i.e., a net gain of energy).

Endothermic reactions absorb energy to break strong bonds to form a less stable state (i.e., positive enthalpy).

Exothermic reactions release energy when forming stronger bonds to produce a more stable state (i.e., negative enthalpy).

13. B is correct.

Gibbs free energy: $\Delta G = \Delta H - T\Delta S$

Entropy (ΔS) determines the favorability of chemical reactions.

If entropy is large, the reaction is more likely to proceed.

14. D is correct.

In a chemical reaction, bonds within reactants are broken down, and new bonds are formed to create products. Therefore, in bond dissociation problems,

$$\Delta H \text{ reaction} = \text{sum of bond energy in reactants} - \text{sum of bond energy in products}$$

For $H_2C{=}CH_2 + H_2 \rightarrow CH_3{-}CH_3$:

$$\Delta H_{reaction} = \text{sum of bond energy in reactants} - \text{sum of bond energy in products}$$

$$\Delta H_{reaction} = [(C{=}C) + 4(C{-}H) + (H{-}H)] - [(C{-}C) + 6(C{-}H)]$$

$$\Delta H_{reaction} = [612\text{ kJ} + (4 \times 412\text{ kJ}) + 436\text{ kJ}] - [348\text{ kJ} + (6 \times 412\text{ kJ})]$$

$$\Delta H_{reaction} = -124\text{ kJ}$$

Remember that this is the opposite of ΔH_f problems, where:

$$\Delta H_{reaction} = (\text{sum of } \Delta H_f \text{ products}) - (\text{sum of } \Delta H_f \text{ reactants})$$

15. C is correct.

Bond dissociation energy is the energy required to break a bond between two gaseous atoms and is useful in estimating the enthalpy change in a reaction.

16. B is correct.

The ΔH value is positive, which means the reaction is endothermic and absorbs energy from the surroundings.

ΔH value is expressed as the energy released/absorbed per mole (for species with a coefficient of 1).

If the coefficient is 2 then it is the energy transferred per 2 moles. For example, 2 moles of NO are produced from 1 mole for each reagent for the stated reaction. Therefore, 43.2 kcal is produced for 2 moles of NO.

17. A is correct.

State functions depend only on the initial and final states of the system and are independent of the paths taken to reach the final state.

Common thermodynamic state functions include internal energy, enthalpy, entropy, pressure, temperature, and volume.

Work and heat relate to the change in energy of a system when it moves from one state to another, depending on how a system changes between states.

18. B is correct.

The temperature (i.e., average kinetic energy) of a substance remains constant during a phase change (e.g., solid to liquid or liquid to gas).

For example, the heat (i.e., energy) breaks bonds between the ice molecules and changes phases into the liquid phase. Since the average kinetic energy of the molecules does not change at the moment of the phase change (i.e., melting), the temperature of the molecules does not change.

19. C is correct.

In a chemical reaction, bonds within reactants are broken, and new bonds form to create products.

Bond dissociation:

$$\Delta H_{reaction} = (\text{sum of bond energy in reactants}) - (\text{sum of bond energy in products})$$

$$O{=}C{=}O + 3\ H_2 \rightarrow CH_3{-}O{-}H + H{-}O{-}H$$

$$\Delta H_{reaction} = (\text{sum of bond energy in reactants}) - (\text{sum of bond energy in products})$$

$$\Delta H_{reaction} = [2(C{=}O) + 3(H{-}H)] - [3(C{-}H) + (C{-}O) + (O{-}H) + 2(O{-}H)]$$

$$\Delta H_{reaction} = [(2 \times 743\text{ kJ}) + (3 \times 436\text{ kJ})] - [(3 \times 412\text{ kJ}) + 360\text{ kJ} + 463\text{ kJ} + (2 \times 463\text{ kJ})]$$

$$\Delta H_{reaction} = -191\text{ kJ}$$

This is the reverse of ΔH_f problems, where:

$$\Delta H_{reaction} = (\text{sum of } \Delta H_{f\ product}) - (\text{sum of } \Delta H_{f\ reactant})$$

20. C is correct.

Entropy indicates how a system is organized; increased entropy means less order.

The entropy of the universe is increasing because a system moves towards more disorder unless energy is added to the system.

The second law of thermodynamics states that the entropy of interconnected systems, without the addition of energy, always increases.

Practice Set 2: Questions 21–40

21. A is correct.

Gibbs free energy:

$$\Delta G = \Delta H - T\Delta S$$

The value that is the largest positive value is the most endothermic while the value that is the largest negative value is the most exothermic.

ΔH refers to enthalpy (or heat).

Endothermic reactions have heat as a reactant.

Exothermic reactions have heat as a product.

Exothermic reactions release heat and cause the temperature of the immediate surroundings to rise (i.e., a net loss of energy).

Endothermic reactions absorb heat and cool the surroundings (i.e., a net gain of energy).

Endothermic reactions absorb energy to break strong bonds to form a less stable state (i.e., positive enthalpy). Exothermic reactions release energy when forming stronger bonds to produce a more stable state (i.e., negative enthalpy).

22. C is correct.

Potential Energy **Stored energy and the energy of position (gravitational)**	**Kinetic Energy** **Energy of motion: motion of waves, electrons, atoms, molecules, and substances**
Chemical Energy Chemical energy is the energy stored in the bonds of atoms and molecules. Examples of stored chemical energy: biomass, petroleum, natural gas, propane, coal.	**Thermal Energy** Thermal energy is internal energy in substances; it is the vibration and movement of atoms and molecules within substances. Geothermal energy is an example of thermal energy.

23. A is correct.

By convention, the ΔG for an element in its standard state is 0.

24. B is correct.

Test strategy: when given a choice among similar explanations, carefully evaluate the most descriptive one.

However, check the statement because sometimes the most descriptive option is not accurate.

Here, the longest option is accurate, so that would be the best choice among all options.

25. C is correct.

ΔG is always negative for spontaneous reactions.

26. C is correct.

An insulator reduces conduction (i.e., transfer of thermal energy through matter).

Air and vacuum are excellent insulators. Storm windows, which have air wedged between two glass panes, work by utilizing this conduction principle.

27. A is correct.

ΔH refers to enthalpy (or heat).

Endothermic reactions have heat as a reactant.

Exothermic reactions have heat as a product.

Exothermic reactions release heat and cause the temperature of the immediate surroundings to rise (i.e., a net loss of energy).

Endothermic reactions absorb heat and cool the surroundings (i.e., a net gain of energy).

The reaction is nonspontaneous (i.e., endergonic) if the products are less stable than the reactants and ΔG is positive.

The reaction is spontaneous (i.e., exergonic) if the products are more stable than the reactants and ΔG is negative.

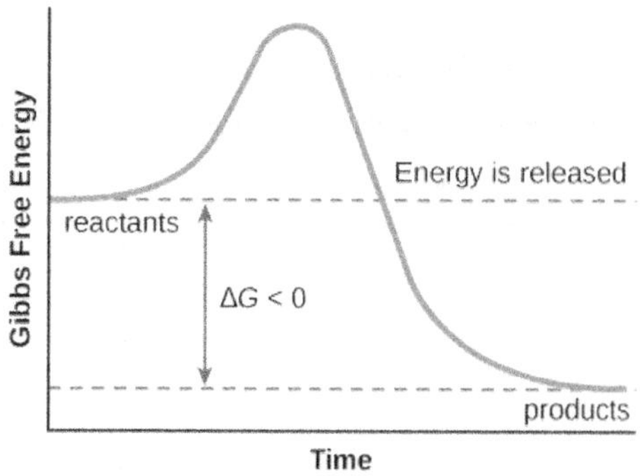

Exergonic reactions are spontaneous and have reactants with more energy than the products.

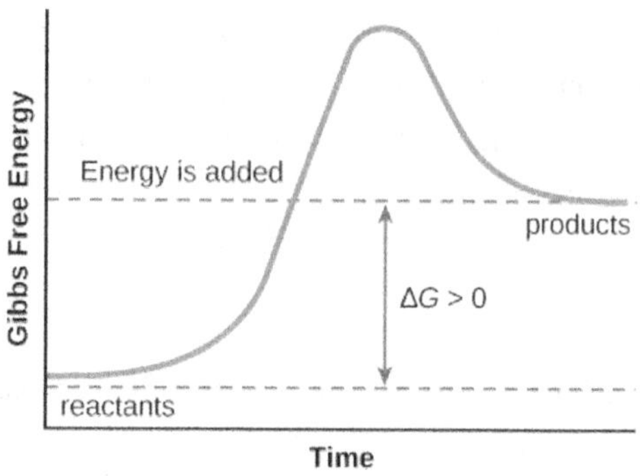

Endergonic reactions are nonspontaneous and have products with more energy than the reactants.

28. D is correct.

Heat capacity is the amount of heat required to increase the temperature of *the whole sample* by 1 °C.

Specific heat is the heat required to increase the temperature of *1 gram* of sample by 1 °C.

q = mass × heat of condensation

q = 16 g × 1,380 J/g

q = 22,080 J

29. A is correct.

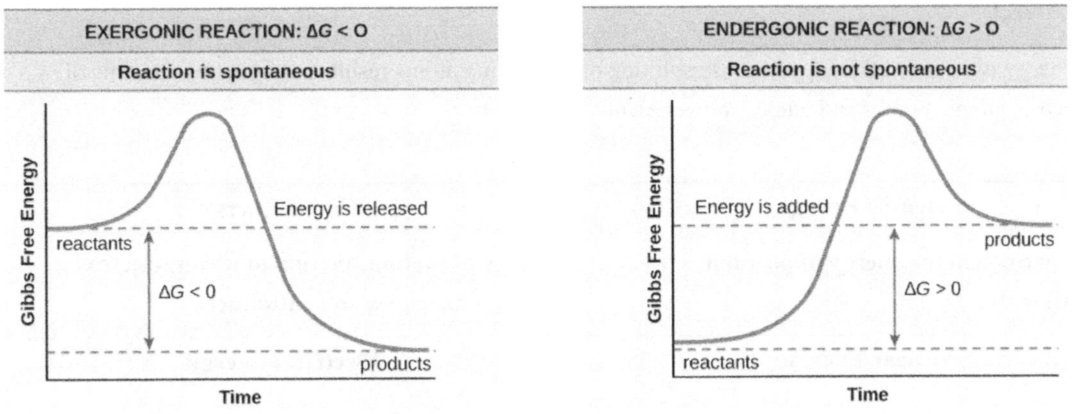

$\Delta G = \Delta H - T\Delta S$ refers to exergonic when ΔG is negative and endergonic when ΔG is positive.

If ΔS is 0 then ΔG = ΔH because the TΔS term cancels because it equals zero).

ΔH refers to exothermic when ΔH is negative and endothermic when ΔH is positive.

30. D is correct.

The heat of formation is defined as the heat required or released when creating one mole of the substance from its elements.

If the given reaction is reversed and divided by two, it would be the formation reaction of NH_3:

$\frac{1}{2} N_2 + 3/2\ H_2 \rightarrow NH_3$

ΔH of this reaction would be reversed (i.e., plus to minus sign) and divided by two:

$\Delta H = -(92.4\ \text{kJ/mol}) / 2$

$\Delta H = -46.2\ \text{kJ/mol}$

31. B is correct.

Gibbs free energy:

$\Delta G = \Delta H - T\Delta S$

With a positive ΔH and negative ΔS, ΔG is positive, so the reaction is nonspontaneous.

32. C is correct.

Entropy (ΔS) measures the degree of disorder in a system.

When a change causes the components of a system to transform into a higher energy phase (solid → liquid, liquid → gas), the system's entropy increases.

33. D is correct.

During phase changes, only the mass and heat of specific phase changes are required to calculate the energy released or absorbed. Because water is turning into ice, the specific heat required is the heat of solidification.

34. B is correct.

Nuclear energy is considered energy because splitting or combining atoms results in enormous amounts of energy used in atomic bombs and nuclear power plants.

Potential Energy **Stored energy and the energy of position (gravitational)**	**Kinetic Energy** **Energy of motion: motion of waves, electrons, atoms, molecules, and substances**
Nuclear Energy Nuclear energy is the energy stored in the nucleus of an atom. It is the energy that holds the nucleus. The nucleus of the uranium atom is an example of nuclear energy.	**Electrical Energy** Electrical energy is the movement of electrons. Lightning and electricity are examples of electrical energy.

35. D is correct.

Entropy is a measurement of disorder. Gases contain more energy than liquids of the same element and have a higher degree of randomness.

Entropy is the unavailable energy that cannot be converted into mechanical work.

36. A is correct.

There are two different methods available to solve this problem. Both methods are based on the definition of ΔH formation as energy consumed/released when one mole of a molecule is produced from its elemental atoms.

Method 1.

Rearrange and add the equations for the formation reaction of C_2H_5OH.

Information provided by the problem:

$C_2H_5OH + 3\ O_2 \rightarrow 2\ CO_2 + 3\ H_2O \quad \Delta H = 327$ kcal

$H_2O \rightarrow H_2 + \frac{1}{2}\ O_2 \quad \Delta H = 68.3$ kcal

$C + O_2 \rightarrow CO_2 \quad \Delta H = -94.1$ kcal

Place C_2H_5OH on the product side. For the other two reactions, arrange them so the elements are on the left and the molecules are on the right (remember that the goal is to create a formation reaction: elements forming a molecule).

When reversing the direction of a reaction, change the positive/negative sign of ΔH.

Multiplying the whole reaction by a coefficient would multiply ΔH by the same ratio.

$2\ CO_2 + 3\ H_2O \rightarrow C_2H_5OH + 3\ O_2 \qquad \Delta H = -327\ \text{kcal}$

$3\ H_2 + 3/2\ O_2 \rightarrow 3\ H_2O \qquad \Delta H = -204.9\ \text{kcal}$

$\underline{2\ C + 2\ O_2 \rightarrow 2\ CO_2 \qquad \Delta H = -188.2\ \text{kcal}}$

$2\ C + 3\ H_2 + 1/2\ O_2 \rightarrow C_2H_5OH \qquad \Delta H = -720.1\ \text{kcal}$

ΔH formation of C_2H_5OH is –720.1 kcal.

Method 2.

Heat of formation (ΔH_f) data can be used to calculate ΔH of a reaction:

$\Delta H_{reaction}$ = the sum of $\Delta H_{f\ product}$ – sum of $\Delta H_{f\ reactant}$

Apply the formula on the first equation:

$C_2H_5OH + 3\ O_2 \rightarrow 2\ CO_2 + 3\ H_2O \quad \Delta H = 327\ \text{kcal}$

$\Delta H_{reaction}$ = the sum of $\Delta H_{f\ product}$ – sum of $\Delta H_{f\ reactant}$

$327\ \text{kcal} = [2(\Delta H_f\ CO_2) + 3(\Delta H_f\ H_2O)] - [(\Delta H_f\ C_2H_5OH) + 3(\Delta H_f\ O_2)]$

Oxygen (O_2) is an element, so $\Delta H_f = 0$

For CO_2 and H_2O, ΔH_f information can be obtained from the other 2 reactions:

$H_2O \rightarrow H_2 + ½\ O_2 \qquad \Delta H = 68.3\ \text{kcal}$

$C + O_2 \rightarrow CO_2 \qquad \Delta H = -94.1\ \text{kcal}$

The CO_2 reaction is a formation reaction – creating one mole of a molecule from its elements. Therefore, ΔH_f of CO_2 = –94.1 kcal

If the H_2O reaction is reversed, it will be a formation reaction.

$H_2 + ½\ O_2 \rightarrow H_2O \qquad \Delta H = -68.3\ \text{kcal}$

Using those values, calculate ΔH_f of C_2H_5OH:

$327\ \text{kcal} = [2(\Delta H_f\ CO_2) + 3(\Delta H_f\ H_2O)] - [(\Delta H_f\ C_2H_5OH) + 3(\Delta H_f\ O_2)]$

$327\ \text{kcal} = [2(-94.1\ \text{kcal}) + 3(-68.3\ \text{kcal})] - [(\Delta H_f\ C_2H_5OH) + 3(0\ \text{kcal})]$

$327\ \text{kcal} = [(-188.2\ \text{kcal}) + (-204.9\ \text{kcal})] - (\Delta H_f\ C_2H_5OH)$

$327\ \text{kcal} = (-393.1\ \text{kcal}) - (\Delta H_f\ C_2H_5OH)$

$\Delta H_f\ C_2H_5OH = -720.1\ \text{kcal}$

37. B is correct.

$S + O_2 \rightarrow SO_2 + 69.8$ kcal

The reaction indicates that 69.8 kcal of energy is released (i.e., on the product side).

All species in this reaction have the coefficient of 1, which means that the ΔH value is calculated by reacting 1 mole of each reagent to create 1 mole of product.

The atomic mass of sulfur is 32.1 g, which means that this reaction uses 32.1 g of sulfur.

38. D is correct.

The disorder of the system is decreasing as more complex and ordered molecules are forming from the reaction of H_2 and O_2 gases.

3 moles of gas are combining to form 2 moles of gas.

Additionally, two molecules are joining to form one molecule.

39. D is correct.

In thermodynamics, an isolated system is enclosed by rigid, immovable walls through which neither matter nor energy passes.

Temperature is a measure of energy and is not a form of energy. Therefore, the temperature cannot be exchanged between the system and the surroundings.

A closed system can exchange energy (as heat or work) with its surroundings, but not matter.

An isolated system cannot exchange energy (as heat or work) or matter with the surroundings.

An open system can exchange energy and matter with the surroundings.

40. D is correct.

State functions depend only on the system's initial and final states and are independent of the paths taken to reach the final state.

Common thermodynamic state functions include internal energy, enthalpy, entropy, pressure, temperature, and volume.

Practice Set 3: Questions 41–60

---

41. A is correct.

When comparing various fuels in varying forms (e.g., solid, liquid, or gas), it is easiest to compare energy released in terms of the mass because matter has mass regardless of its state (as opposed to volume, which is convenient for a liquid or gas, but not for a solid).

Moles are more complicated because they have to be converted to mass before a direct comparison can be made between the fuels.

42. C is correct.

Convection is heat carried by fluids (i.e., liquids and gases). Heat is prevented from leaving the system through convection when the lid is placed on the cup.

43. B is correct.

For a reaction to be spontaneous, the entropy ($\Delta S_{system} + \Delta S_{surrounding}$) has to be positive (i.e., greater than 0).

44. D is correct.

The explosion of gases: chemical → heat energy

Heat converting into steam, which in turn moves a turbine: heat → mechanical energy

The generator creates electricity: mechanical → electrical energy

Potential Energy **Stored energy and the energy of position (gravitational)**	**Kinetic Energy** **Energy of motion: motion of waves, electrons, atoms, molecules, and substances**
Chemical Energy Chemical energy is the energy stored in the bonds of atoms and molecules. Examples of stored chemical energy: biomass, petroleum, natural gas, propane, coal. **Nuclear Energy** Nuclear energy is the energy stored in the nucleus of an atom. It is the energy holding the nucleus. The nucleus of the uranium atom is an example of nuclear energy.	**Radiant Energy** Radiant energy is electromagnetic energy that travels in transverse waves. Radiant energy includes visible light, x-rays, gamma rays, and radio waves. Solar energy is an example of radiant energy. **Thermal Energy** Thermal energy is internal energy in substances; it is the vibration and movement of atoms and molecules within substances. Geothermal energy is an example of thermal energy.

Stored Mechanical Energy	Motion
Stored mechanical energy is energy stored in objects by the application of a force. Compressed springs and stretched rubber bands are examples of stored mechanical energy. **Gravitational Energy** Gravitational Energy is the energy of place or position. Water in a reservoir behind a hydropower dam is an example of gravitational potential energy. When the water is released to spin the turbines, it becomes kinetic energy.	The movement of objects or substances from one place to another is motion. Wind and hydropower are examples of motion. **Sound** Sound is the movement of energy through substances in longitudinal (compression/rarefaction) waves. **Electrical Energy** Electrical energy is the movement of electrons. Lightning and electricity are examples of electrical energy.

45. A is correct.

Relationship between enthalpy (ΔH) and internal energy (ΔE):

Enthalpy = Internal energy + work (for gases, work = PV)

$$\Delta H = \Delta E + \Delta(PV)$$

Solving for ΔE:

$$\Delta E = \Delta H - \Delta(PV)$$

According to ideal gas law:

$$PV = nRT$$

Substitute ideal gas law to the previous equation:

$$\Delta E = \Delta H - \Delta(nRT)$$

R and T are constant, which leaves Δn as the variable.

The reaction is $C_2H_2\ (g) + 2H_2\ (g) \rightarrow C_2H_6\ (g)$.

There are three gas molecules on the left and one on the right, which means $\Delta n = 1 - 3 = -2$.

In this problem, the temperature is not provided. However, the presence of degree symbols (ΔG°, ΔH°, ΔS°) indicates that those are standard values, which are measured at 25 °C or 298.15 K.

Double-check the units before performing calculations; ΔH is in kilojoules, while the gas constant (R) is 8.314 J/mol K. Convert ΔH to joules before calculating.

Use the Δn and T values to calculate ΔE:

$$\Delta E = \Delta H - \Delta nRT$$

$$\Delta E = -311{,}500\ \text{J/mol} - (-2 \times 8.314\ \text{J/mol K} \times 298.15\ \text{K})$$

$$\Delta E = -306{,}542\ \text{J} \approx -306.5\ \text{kJ}$$

46. D is correct.

A closed system can exchange energy (as heat or work) but not matter with its surroundings.

An isolated system cannot exchange energy (as heat or work) or matter with the surroundings.

An open system can exchange energy and matter with the surroundings.

47. C is correct.

An increase in entropy indicates an increase in disorder; find a reaction that creates a higher number of molecules on the product side compared to the reactant side.

48. C is correct.

When enthalpy is calculated, work done by gases is not considered; however, the energy of the reaction includes the work done by gases.

The largest difference between the energy of the reaction and enthalpy would be for reactions with the largest difference in the number of gas molecules between the product and reactants.

49. B is correct.

ΔH refers to enthalpy (or heat).

Endothermic reactions have heat as a reactant.

Exothermic reactions have heat as a product.

Endothermic reactions absorb energy to break strong bonds to form a less stable state (i.e., positive enthalpy).

Exothermic reactions release energy when forming stronger bonds to produce a more stable state (i.e., negative enthalpy).

The reaction is nonspontaneous (i.e., endergonic) if the products are less stable than the reactants and ΔG is positive.

The reaction is spontaneous (i.e., exergonic) if the products are more stable than the reactants and ΔG is negative.

50. B is correct.

All thermodynamic functions in $\Delta G = \Delta H - T\Delta S$ refer to the system.

51. C is correct.

To predict spontaneity of reaction, use the Gibbs free energy equation:

$$\Delta G = \Delta H° - T\Delta S$$

The reaction is spontaneous if ΔG is negative.

Substitute the given values to the equation:

$$\Delta G = -113.4 \text{ kJ/mol} - [T \times (-145.7 \text{ J/K mol})]$$

$\Delta G = -113.4$ kJ/mol + (T × 145.7 J/K mol)

Important: note that the units aren't identical, ΔH° is in kJ and ΔS° is in J. Convert kJ to J (1 kJ = 1,000 J):

$\Delta G = -113{,}400$ J/mol + (T × 145.7 J/K mol)

It can be predicted that the value of ΔG would be negative if the value of T is small.

If T goes higher, ΔG approaches a positive value, and the reaction would be nonspontaneous.

52. C is correct.

Enthalpy:

U + PV

53. C is correct.

State functions depend only on the initial and final states of the system and are independent of the paths taken to reach the final state. Common thermodynamic state functions include internal energy, enthalpy, entropy, pressure, temperature, and volume.

An *extensive property* is a property that changes when the size of the sample changes. Examples include mass, volume, length, and total charge.

Entropy has no absolute zero value, and a substance at zero Kelvin has zero entropy (i.e., no motion).

54. C is correct.

In an exothermic reaction, the bonds formed are stronger than the bonds broken.

55. D is correct.

Calculate the value of $\Delta H_f = \Delta H_{f\ product} - \Delta H_{f\ reactant}$

$\Delta H_f = -436.8\ \text{kJ mol}^{-1} - (-391.2\ \text{kJ mol}^{-1})$

$\Delta H_f = -45.6\ \text{kJ mol}^{-1}$

To determine spontaneity, calculate the Gibbs free energy:

$\Delta G = \Delta H° - T\Delta S$

The reaction is spontaneous if ΔG is negative:

$\Delta G = -45.6\ \text{kJ} - T\Delta S$

Typical ΔS values are around 100–200 J.

If the ΔH value is –45.6 kJ or –45,600 J, the value of ΔG would still be negative unless TΔS is less than –45,600.

Therefore, the reaction would be spontaneous over a broad range of temperatures.

56. A is correct.

Enthalpy is an extensive property: it varies with the quantity of the matter.

Reaction 1 : $P_4 + 6\ Cl_2 \rightarrow 4\ PCl_3 \quad \Delta H = -1{,}289$ kJ

Reaction 2 : $3\ P_4 + 18\ Cl_2 \rightarrow 12\ PCl_3 \quad \Delta H = ?$

The only difference between reactions 1 and 2 is the coefficients.

In reaction 2, coefficients are three times reaction 1. Reaction 2 consumes three times as many reactants as reaction 1, and reaction 2 releases three times as much energy as reaction 1.

$\Delta H = 3 \times -1{,}289$ kJ $= -3{,}867$ kJ

57. C is correct.

ΔH refers to enthalpy (or heat).

Endothermic reactions have heat as a reactant.

Exothermic reactions have heat as a product.

Endothermic reactions absorb energy to break strong bonds to form a less stable state (i.e., positive enthalpy). Since endothermic reactions consume energy, there should be heat on the reactants' side.

Exothermic reactions release energy when forming stronger bonds to produce a more stable state (i.e., negative enthalpy).

The reaction is nonspontaneous (i.e., endergonic) if the products are less stable than the reactants and ΔG is positive.

The reaction is spontaneous (i.e., exergonic) if the products are more stable than the reactants and ΔG is negative.

58. D is correct.

According to the second law of thermodynamics, the entropy gain of the universe must be positive (i.e., increased disorder). The entropy of a system, however, can be negative if the surroundings experience an increase in entropy greater than the negative entropy change experienced by the system.

59. B is correct.

In thermodynamics, an isolated system is enclosed by rigid, immovable walls through which neither matter nor energy passes. Bell jars are designed not to let air or other materials in or out. Insulated means that heat cannot get in or out.

Evaporation indicates a phase change (vaporization) from liquid to gaseous phase; however, it does not imply that the matter has escaped its system.

A closed system can exchange energy (as heat or work) with its surroundings, but not matter.

An isolated system cannot exchange energy (as heat or work) or matter with the surroundings.

An open system can exchange energy and matter with the surroundings.

60. A is correct.

The Law of Conservation of Energy states that an isolated system's total energy (e.g., potential or kinetic) remains constant and is conserved. Energy can be neither created nor destroyed but is transformed from one form to another. For instance, chemical energy can be converted to kinetic energy in the explosion of a firecracker.

The Law of Conservation of Mass states that for a system closed to transfers of matter and energy, the mass of the system must remain constant over time.

Law of definite proportions (Proust's law) states that a chemical compound contains the same proportion of elements by mass. The law of definite proportions forms the basis of stoichiometry (i.e., from the known amounts of separate reactants, the amount of the product can be calculated).

Avogadro's law is an experimental gas law relating the volume of a gas to the amount of substance of gas present. It states that equal volumes of gases have the same number of molecules at the same temperature and pressure.

Boyle's law is an experimental gas law that describes how the pressure of a gas tends to increase as the volume of a gas decreases. It states that the absolute pressure exerted by a given mass of an ideal gas is inversely proportional to the volume it occupies if the temperature and amount of gas remain unchanged within a closed system.

Practice Set 4: Questions 61–80

---

61. C is correct.

ΔH refers to enthalpy (or heat).

Endothermic reactions have heat as a reactant.

Exothermic reactions have heat as a product.

Exothermic reactions release heat and cause the temperature of the immediate surroundings to rise (i.e., a net loss of energy).

Endothermic reactions absorb heat and cool the surroundings (i.e., a net gain of energy).

Endothermic reactions absorb energy to break strong bonds to form a less stable state (i.e., positive enthalpy). Exothermic reactions release energy when forming stronger bonds to produce a more stable state (i.e., negative enthalpy).

The reaction is nonspontaneous (i.e., endergonic) if the products are less stable than the reactants and ΔG is positive.

The reaction is spontaneous (i.e., exergonic) if the products are more stable than the reactants and ΔG is negative.

62. C is correct.

Temperature is a measure of the average kinetic energy of the molecules.

Kinetic energy is proportional to temperature.

In most thermodynamic equations, temperatures are expressed in Kelvin, so convert the Celsius temperatures to Kelvin:

$$25\ °C + 273.15 = 298.15\ K$$

$$50\ °C + 273.15 = 323.15\ K$$

Calculate the kinetic energy using simple proportions:

$$KE = (323.15\ K / 298.15\ K) \times 500\ J$$

$$KE = 540\ J$$

63. C is correct.

The equation relates the change in entropy (ΔS) at different temperatures.

Entropy increases at higher temperatures (i.e., increased kinetic energy).

64. B is correct.

A change is exothermic when energy is released to the surroundings.

Energy loss occurs when a substance changes to a more rigid phase (e.g., gas → liquid or liquid → solid).

65. A is correct.

Potential Energy **Stored energy and the energy of position (gravitational)**	**Kinetic Energy** **Energy of motion: motion of waves, electrons, atoms, molecules, and substances**
Chemical Energy Chemical energy is the energy stored in the bonds of atoms and molecules. Examples of stored chemical energy: biomass, petroleum, natural gas, propane, coal.	**Electrical Energy** Electrical energy is the movement of electrons. Lightning and electricity are examples of electrical energy.

66. B is correct.

At equilibrium, there is no potential, and therefore neither direction of the reaction is favored.

Reaction potential (E) indicates the tendency of a reaction to be spontaneous in either direction.

Because the solution is in equilibrium, no further reactions occur in either direction. Therefore, $E = 0$.

67. D is correct.

The transformation from solid to gas is accompanied by an increase in entropy (i.e., disorder).

The molecules of gas have greater kinetic energy than the molecules of solids.

68. C is correct.

The reaction is nonspontaneous (i.e., endergonic) if products are less stable than reactants and ΔG is positive.

The reaction is spontaneous (i.e., exergonic) if the products are more stable than the reactants and ΔG is negative.

69. A is correct.

The spontaneity of a reaction is determined by evaluating the Gibbs free energy, or ΔG.

$$\Delta G = \Delta H - T\Delta S$$

A reaction is spontaneous if $\Delta G < 0$ (or ΔG is negative).

For ΔG to be negative, ΔH has to be less than $T\Delta S$.

A reaction is nonspontaneous if $\Delta G > 0$ (or ΔG is positive and ΔH is greater than $T\Delta S$).

70. B is correct.

The reaction is nonspontaneous (i.e., endergonic) if products are less stable than reactants and ΔG is positive.

The reaction is spontaneous (i.e., exergonic) if products are more stable than the reactants and ΔG is negative.

Entropy (ΔS) is determined by the relative number of molecules (compounds):

The reactants have one molecule (with a coefficient total of 2). The products have two molecules (with a coefficient total of 7).

The number of molecules has increased from reactants to products. Therefore, the entropy (i.e., disorder) has increased.

71. D is correct.

ΔG indicates the energy made available to do non P–V work. This is a useful quantity for processes like batteries and living cells that do not have expanding gases.

72. C is correct.

The internal energy of a system is the energy contained within the system, excluding the kinetic energy of motion of the system as a whole and the potential energy of the system as a whole due to external force fields.

The internal energy of a system can be changed by transfers of matter and by work and heat transfer. When impermeable container walls prevent matter transfer, the system is said to be closed. The First law of thermodynamics states that the increase in internal energy equals the heat added plus the work done on the system by its surroundings. If the container walls do not pass energy or matter, the system is isolated, and its internal energy remains constant.

Bond energy is internal potential energy (PE), while thermal energy is internal kinetic energy (KE).

73. D is correct.

Endothermic reactions absorb energy to break strong bonds to form a less stable state (i.e., positive enthalpy). Exothermic reactions release energy when forming stronger bonds to produce a more stable state (i.e., negative enthalpy).

To predict spontaneity of reaction, use the Gibbs free energy equation:

$$\Delta G = \Delta H° - T\Delta S$$

The reaction is nonspontaneous (i.e., endergonic) if the products are less stable than the reactants and ΔG is positive.

The reaction is spontaneous (i.e., exergonic) if the products are more stable than the reactants and ΔG is negative.

It is the amount of energy that is determinative. Some bonds are stronger than others, so there is a net gain or net loss of energy when formed.

74. A is correct.

The starting material is ice at −25 °C, which means its temperature must be raised to 0 °C before melting.

The first term in the heat calculation is:

specific heat of ice (A) × mass × change of temperature

75. C is correct.

The heat of formation of water vapor is like the highly exothermic process for the heat of hydrogen combustion.

Heat is required to convert H_2O (*l*) into vapor; therefore, the heat of formation of water vapor must be less exothermic than H_2O (*l*).

Solid → liquid → gas: endothermic because the reaction absorbs energy to break strong bonds to form a less stable state (i.e., positive enthalpy).

Gas → liquid → solid: exothermic because the reaction releases energy when forming stronger bonds to produce the more stable state (i.e., negative enthalpy).

76. D is correct.

77. B is correct.

Einstein established the equivalence of mass and energy; the energy equivalent of mass is given by $E = mc^2$.

Energy can be converted to mass, and mass can be converted to energy if the total energy, including mass-energy, is conserved.

In nuclear fission, the mass of the parent nucleus is greater than the sum of the masses of the daughter nuclei. This deficit in mass is converted into energy.

78. D is correct.

A closed system can exchange energy (as heat or work) but not matter with its surroundings.

An isolated system cannot exchange energy (as heat or work) or matter with the surroundings.

An open system can exchange energy and matter with the surroundings.

79. A is correct.

Gases have the highest entropy (i.e., disorder or randomness).

Solutions have higher entropy than pure phases since the molecules are more scattered.

The liquid phase has less entropy than the aqueous phase.

80. A is correct.

In a chemical reaction, bonds within reactants are broken, and new bonds form.

Therefore, in bond dissociation problems:

$\Delta H_{reaction}$ = sum of bond energy in reactants – the sum of bond energy in products.

This is the opposite of ΔH_f problems, where:

$\Delta H_{reaction}$ = sum of ΔH_f product – sum of ΔH_f reactant

For $N_2 + 3\ H_2 \rightarrow 2\ NH_3$

$\Delta H_{reaction}$ = sum of bond energy in reactants – sum of bond energy in products

$\Delta H_{reaction} = [(N{\equiv}N) + 3(H–H)] – [2 \times 3(N–H)]$

$\Delta H_{reaction} = [(946\ kJ) + (3 \times 436\ kJ) – (2 \times (3 \times 389\ kJ)]$

$\Delta H_{reaction} = –80\ kJ$

Kinetics and Equilibrium – Detailed Explanations

Practice Set 1: Questions 1–20

1. A is correct.

General formula for the equilibrium constant of a reaction:

$aA + bB \leftrightarrow cC + dD$

$K_{eq} = ([C]^c \times [D]^d) / ([A]^a \times [B]^b)$

For the reaction above:

$K_{eq} = [C] / [A]^2 \times [B]^3$

2. B is correct.

Two gases having the same temperature have the same average molecular kinetic energy.

Therefore,

$(\frac{1}{2})m_A v_A^2 = (\frac{1}{2})m_B v_B^2$

$m_A v_A^2 = m_B v_B^2$

$(m_B / m_A) = (v_A / v_B)^2$

$(v_A / v_B) = 2$

$(m_B / m_A) = 2^2 = 4$

The only gases listed with a mass ratio of about 4 to 1 are iron (55.8 g/mol) and nitrogen (14 g/mol).

3. A is correct.

The slowest reaction would be the reaction with the highest activation energy and the lowest temperature.

4. A is correct.

The rate law is calculated experimentally by comparing trials and determining how changes in the initial concentrations of the reactants affect the rate of the reaction.

rate = $k[A]^x \cdot [B]^y$

where k is the rate constant, and the exponents x and y are the partial reaction orders (i.e., determined experimentally). They are not equal to the stoichiometric coefficients.

To determine the order of reactant A, find two experiments where the concentrations of B are identical, and concentrations of A are different.

Use data from experiments 2 and 3 to calculate order of A:

$Rate_3 / Rate_2 = ([A]_3 / [A]_2)^{\text{order of A}}$

$0.500 / 0.500 = (0.060 / 0.030)^{\text{order of A}}$

$1 = (2)^{\text{order of A}}$

Order of A = 0

5. C is correct.

The balanced equation:

$C_2H_6O + 3\ O_2 \rightarrow 2\ CO_2 + 3\ H_2O$

The rate of reaction/consumption is proportional to the coefficients.

The rate of carbon dioxide production is twice the rate of ethanol consumption:

$2 \times 4.0\ M\ s^{-1} = 8.0\ M\ s^{-1}$

6. D is correct.

General formula for the equilibrium constant of a reaction:

$a\text{A} + b\text{B} \leftrightarrow c\text{C} + d\text{D}$

$K_{eq} = ([C]^c \times [D]^d) / ([A]^a \times [B]^b)$

For equilibrium constant calculations, only include species in aqueous or gas phases.

$K_{eq} = 1 / [Cl_2]$

7. B is correct.

K_{eq} = [products] / [reactants]

If the K_{eq} is less than 1 (e.g., 6.3×10^{-14}), then the numerator (i.e., products) is smaller than the denominator (i.e., reactants), and fewer products have formed relative to the reactants.

If the reaction favors reactants compared to products, the equilibrium lies to the left.

8. A is correct.

The higher the energy of activation, the slower the reaction is. The height of this barrier is independent of the determination of spontaneous (ΔG is negative with products more stable than reactants) or nonspontaneous reactions (ΔG is positive with products less stable than reactants).

9. D is correct.

The main function of catalysts is lowering the reaction's activation energy (energy barrier), thus increasing its rate (k).

Temperature measures the average kinetic energy (i.e., KE = ½mv^2) of the molecules.

Except for zero-order chemical reactions, increased concentration of reactants increases the probability that the reactants collide with sufficient energy and orientation to overcome the energy of the activation barrier and proceed toward products.

10. D is correct.

Every reaction has activation energy: the amount of energy required by the reactant to start reacting.

On the graph, activation energy can be estimated by calculating the distance between the initial energy level (i.e., reactant) and the peak (i.e., transition state). Activation energy is *not* the difference between energy levels of the initial and final state (reactant and product).

A catalyst provides an alternative pathway for the reaction to proceed to product formation. It lowers the energy of activation (i.e., relative energy between reactants and transition state) and, therefore, speeds the reaction rate.

Catalysts do not affect the Gibbs free energy (ΔG: stability of products vs. reactants) or the enthalpy (ΔH: bond breaking in reactants or bond making in products).

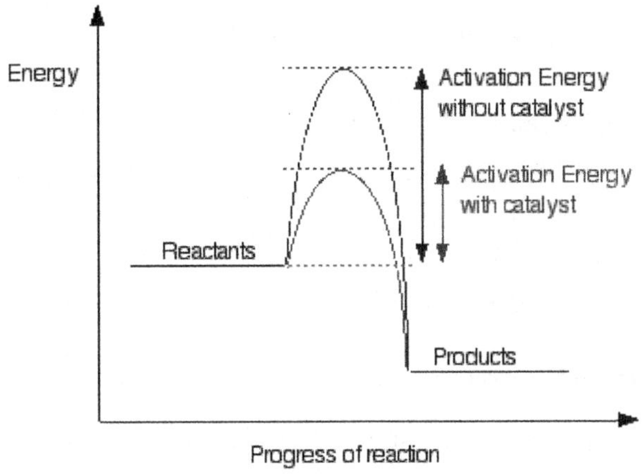

11. D is correct.

The activation energy is the energy barrier that must be overcome to transform reactant(s) into the product(s).

A reaction with higher activation energy (energy barrier) has a decreased rate (k).

12. B is correct.

The activation energy is the energy barrier that must be overcome to transform reactant(s) into the product(s).

A reaction with lower activation energy (energy barrier) has an increased rate (k).

13. D is correct.

The activation energy is the energy barrier that must be overcome to transform reactant(s) into the product(s).

A reaction with higher activation energy (energy barrier) has a decreased rate (k).

If the graphs are on the same scale, the height of the activation energy (R to the highest peak on the graph) is greatest in graph d.

14. B is correct.

If the graphs are on the same scale, the height of the activation energy (R to the highest peak on the graph) is the smallest in graph b.

The main function of catalysts is lowering the reaction's activation energy (energy barrier), thus increasing its rate (*k*).

Although catalysts decrease the amount of energy required to reach the rate-limiting transition state, they do *not* decrease the relative energy of the products and reactants. Therefore, a catalyst does not affect ΔG.

A catalyst provides an alternative pathway for the reaction to proceed to product formation. It lowers the energy of activation (i.e., relative energy between reactants and transition state) and, therefore, speeds the reaction rate.

Catalysts do not affect the Gibbs free energy (ΔG: stability of products vs. reactants) or the enthalpy (ΔH: bond breaking in reactants or bond making in products).

Catalysts do not affect ΔG (relative levels of R and P). Graph D shows an endergonic reaction with products less stable than reactants. All other graphs show an exergonic reaction with the same relative difference between the more stable products and the less stable reactants; only the activation energy is different.

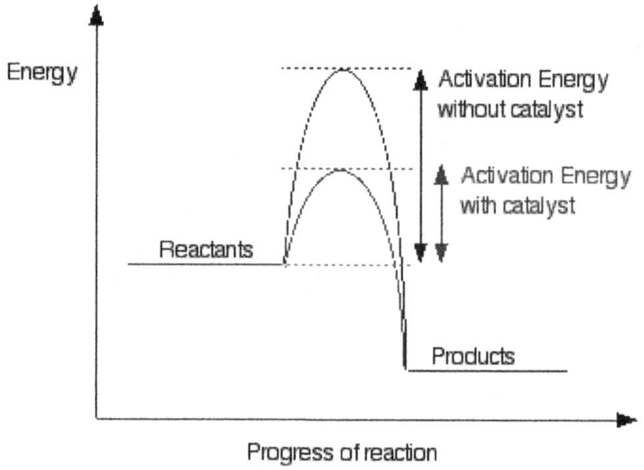

15. B is correct.

A reaction is at equilibrium when the forward reaction rate equals the rate of the reverse reaction. Achieving equilibrium is common for reactions.

Catalysts, by definition, are regenerated during a reaction and are not consumed by the reaction.

16. D is correct.

The molecules must collide with sufficient energy, the frequency of collision, and the proper orientation to overcome the activation energy barrier.

17. B is correct.

A catalyst lowers the energy of activation, which increases the rate of the reaction.

The main function of catalysts is lowering the reaction's activation energy (energy barrier), thus increasing its rate (*k*).

Although catalysts decrease the amount of energy required to reach the rate-limiting transition state, they do *not* decrease the relative energy of the products and reactants. Therefore, a catalyst does not affect ΔG.

18. D is correct.

When changing the reaction conditions, Le Châtelier's principle states that equilibrium shifts to counteract the change. If the concentration of reactants or products changes, the position of the equilibrium changes.

Adding reactants or removing products shifts the equilibrium to the right.

According to Le Châtelier's principle, adding or removing a solid does not affect the equilibrium position.

However, solid NaOH dissociates into Na^+ and OH^- when placed in an aqueous solution. The OH^- combines with H^+ to form water. Reducing H^+ as a product drives the reaction to the right.

19. A is correct.

When changing the reaction conditions, Le Châtelier's principle states that equilibrium shifts to counteract the change. If the reaction temperature, pressure, or volume changes, the position of equilibrium changes.

Adding reactants or removing products shifts the equilibrium to the right.

Heat (i.e., energy-related to temperature) is a reactant, and increasing its value drives product formation.

Decreasing the pressure on the reaction vessel drives the reaction toward reactants (i.e., to the left). The relative molar concentration of the reactants (i.e., 2 + 6) is greater than the products (i.e., 4 + 3).

The other choices listed drive the reaction toward reactants (i.e., to the left).

20. C is correct.

First, calculate the concentration/molarity of each reactant:

H_2 : 0.20 moles / 4.00 L = 0.05 M

X_2 : 0.20 moles / 4.00 L = 0.05 M

HX: 0.800 moles / 4.00L = 0.20 M

Use the concentrations to calculate Q:

$$Q = [HX]^2 / [H_2]\cdot[X_2]$$

$$Q = (0.20)^2 / (0.05 \times 0.05)$$

$$Q = 16$$

Because $Q < K_c$ (i.e., 16 < 24.4), the reaction shifts to the right.

For Q to increase and match K_c, the numerator (i.e., $[HX]^2$), which is the product (right side) of the equation, needs to be increased.

Practice Set 2: Questions 21–40

21. B is correct.

General formula for the equilibrium constant of a reaction:

$aA + bB \leftrightarrow cC + dD$

$K_{eq} = ([C]^c \times [D]^d) / ([A]^a \times [B]^b)$

For equilibrium constant (K_c) calculation, only include species in aqueous or gas phases:

$K_c = [CO_2]\,[H_2O]^2 / [CH_4]$

22. A is correct.

The rate law is calculated by determining how changes in concentrations of the reactants affect the initial rate of the reaction.

rate = $k[A]^x \cdot [B]^y$

where k is the rate constant, and the exponents x and y are the partial reaction orders; they are not equal to the stoichiometric coefficients.

Whenever the fast (i.e., second) step follows the slow (i.e., first) step, the fast step is assumed to reach equilibrium, and the equilibrium concentrations are used for the rate law of the slow step.

23. D is correct.

The main function of catalysts is lowering the reaction's activation energy (energy barrier), thus increasing its rate (k).

Although catalysts decrease the amount of energy required to reach the rate-limiting transition state, they do *not* decrease the relative energy of the products and reactants. Therefore, a catalyst does not affect ΔG.

A catalyst provides an alternative pathway for the reaction to proceed to product formation. It lowers the activation energy (i.e., relative energy between reactants and transition state) and speeds the reaction rate.

Catalysts do not affect the Gibbs free energy (ΔG: stability of products vs. reactants) or the enthalpy (ΔH: bond breaking in reactants or bond making in products).

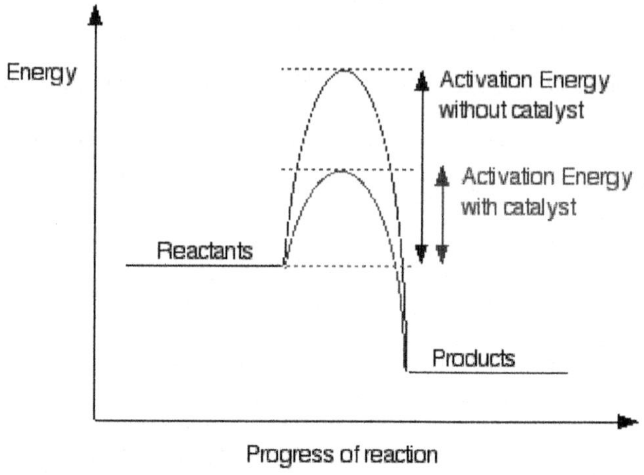

24. B is correct.

Equilibrium reactions vary between conditions, so any proportion of product/reactant mass is possible.

The question refers to the state *after* a reaction reached equilibrium.

For a reaction that favors product formation ($K_{eq} > 1$), the amount of product would be more than reactants.

For a reaction that favors reactants ($K_{eq} < 1$), the amount of product would be less than reactants.

When $K_{eq} = 1$, the amounts of products and reactants would be equal.

25. C is correct.

The rate law is calculated experimentally by comparing trials and determining how changes in the initial concentrations of the reactants affect the rate of the reaction.

$$\text{rate} = k[A]^x \cdot [B]^y$$

where k is the rate constant, and the exponents x and y are the partial reaction orders (i.e., determined experimentally). They are not equal to the stoichiometric coefficients.

Rate laws cannot be determined from the balanced equation (i.e., used to determine equilibrium) unless the reaction occurs in a single step.

From the data, when the concentration doubles, the rate quadrupled. Since 4 is 2^2, the rate law is second order, therefore the rate = $k[H_2]^2$.

26. D is correct.

Lowering temperature shifts the equilibrium to the right because heat is a product (lowering the temperature is like removing the product). Increasing the temperature (i.e., heating the system) has the opposite effect and shifts the equilibrium to the left (i.e., toward products).

Removal of H_2 (i.e., reactant) shifts the equilibrium to the left.

Addition of NH_3 (i.e., products) shifts the equilibrium to the left.

A catalyst lowers the energy of activation but does not affect the position of the equilibrium.

27. B is correct.

The order of the reaction must be determined experimentally.

A reaction's order can be identical to its coefficients. For example, the coefficients correlate to the rate law in a single-step reaction or during the slow step of a multi-step reaction.

28. A is correct.

General formula for the equilibrium constant of a reaction: $aA + bB \leftrightarrow cC + dD$

$$K_{eq} = ([C]^c \times [D]^d) / ([A]^a \times [B]^b)$$

For equilibrium constant calculation, only include species in aqueous or gas phases: $K_{eq} = 1 / [CO] \times [H_2]^2$

29. D is correct.

Two peaks in this reaction indicate two energy-requiring steps with one intermediate (i.e., C) and each peak (i.e., B and D) as an activated complex (i.e., transition states). The activated complex (i.e., transition state) is undergoing bond breaking/bond making events.

30. A is correct.

The activation energy for the slow step of a reaction is the distance from the starting material (or an intermediate) to the activated complex (i.e., transition state) with the absolute highest energy (i.e., highest point on the graph).

31. C is correct.

The activation energy for the slow step of a reverse reaction is the distance from an intermediate (or product) to the activated complex (i.e., transition state) with the absolute highest energy (i.e., highest point on the graph). The slow step may not have the greatest magnitude for activation energy (e.g., E→ D on the graph).

32. B is correct.

The change in energy (or ΔH) is the difference between the energy contents of reactants and products.

33. A is correct.

As the temperature (i.e., average kinetic energy) *increases*, the particles move faster (i.e., increased kinetic energy) and collide more frequently per unit time. This *increases* the *reaction rate*.

34. A is correct.

When changing the reaction conditions, Le Châtelier's principle states that equilibrium shifts to counteract the change. If the reaction temperature, pressure, or volume is changed, the position of equilibrium changes.

According to Le Châtelier's principle, endothermic ($+\Delta H$) reactions increase the formation of products at higher temperatures.

Since each side of the reaction has 2 gas molecules, a change in pressure does not affect equilibrium.

35. B is correct.

All chemical reactions eventually reach equilibrium, the state at which the reactants and products are present in concentrations with no further tendency to change with time. The rate of production of each product (i.e., forward reaction) equals the rate of their consumption by the reverse reaction.

36. D is correct.

Chemical equilibrium refers to a dynamic process whereby the *rate* at which a reactant molecule is transformed into a product is the same as the *rate* for a product molecule to be transformed into a reactant.

All chemical reactions eventually reach equilibrium, the state at which the reactants and products are present in concentrations with no further tendency to change with time.

37. A is correct.

Expression for equilibrium constant:

$$K = [H_2O]^2 \cdot [Cl_2]^2 / [HCl]^4 \cdot [O_2]$$

Solve for $[Cl_2]$:

$$[Cl_2]^2 = (K \times [HCl]^4 \cdot [O_2]) / [H_2O]^2$$

$$[Cl_2]^2 = [46.0 \times (0.150)^4 \times 0.395] / (0.625)^2$$

$$[Cl_2]^2 = 0.0235$$

$$[Cl_2] = 0.153 \text{ M}$$

38. A is correct.

To determine the amount of each compound at equilibrium, consider the chemical reaction written in the form:

$$a\text{A} + b\text{B} \leftrightarrow c\text{C} + d\text{D}$$

The equilibrium constant (K_c) is defined as:

$$K_c = ([\text{C}]^c \times [\text{D}]^d) / ([\text{A}]^a \times [\text{B}]^b)$$

or

$$K_c = [\text{products}] / [\text{reactants}]$$

If the K_{eq} is greater than 1, the numerator (i.e., products) is larger than the denominator (i.e., reactants), and more products have formed relative to the reactants.

If the reaction favors products compared to reactants, the equilibrium lies to the right.

39. D is correct.

Gases (as opposed to liquids and solids) are most sensitive to changes in pressure.

There are three moles of hydrogen gas as a reactant and three moles of water vapor as a product.

Therefore, changes in pressure result in proportionate changes to the forward and reverse reactions.

40. C is correct.

Increased pressure or increased concentration of reactants increases the probability that the reactants collide with sufficient energy and orientation to overcome the energy of activation barrier and proceed toward products.

Practice Set 3: Questions 41–60

---

41. A is correct.

General formula for the equilibrium constant of a reaction:

$aA + bB \leftrightarrow cC + dD$

$K_{eq} = ([C]^c \times [D]^d) / ([A]^a \times [B]^b)$

For equilibrium constant calculation, only include species in aqueous or gas phases:

$K_{eq} = [NO]^4 \times [H_2O]^6 / [NH_3]^4 \times [O_2]^5$

42. D is correct.

The high levels of CO_2 cause a person to hyperventilate to reduce the number of CO_2 (reactant).

Hyperventilating has the effect of driving the reaction toward products.

Removing products (i.e., HCO_3^- and H^+) or intermediates (H_2CO_3) drives the reaction toward products.

43. B is correct.

The units of the rate constants:

Zero-order reaction: M sec^{-1}

First order reaction: sec^{-1}

Second order reaction: L $mole^{-1}$ sec^{-1}

44. D is correct.

A reaction proceeds when the reactant(s) have sufficient energy to overcome activation energy and form products. Increasing the temperature increases the molecule's kinetic energy, the frequency of collision, and the probability that the reactants will overcome the barrier (energy of activation) to form products.

45. B is correct.

Low activation energy increases the rate because the energy required for the reaction to proceed is lower.

High temperature increases the rate because faster-moving molecules have a greater probability of collision, facilitating the reaction.

Combined, lower activation energy and a higher temperature result in the highest relative rate for the reaction.

46. C is correct.

The rate law is calculated by comparing trials and determining how changes in the initial concentrations of the reactants affect the rate of the reaction.

rate = $k[A]^x \cdot [B]^y$

where k is the rate constant, and the exponents x and y are the partial reaction orders (i.e., determined experimentally). They are not equal to the stoichiometric coefficients.

Start by identifying two reactions where XO concentration is constant, and O_2 is different: experiments 1 and 2. When the concentration of O_2 is doubled, the rate is also doubled. This indicates that the order of the reaction for O_2 is 1.

Now, determine the order for XO.

Find 2 reactions where the concentration of O_2 is constant and XO is different: experiments 2 and 3. When the concentration of XO is tripled, the rate is multiplied by a factor of 9. $3^2 = 9$, which means the order of the reaction with respect to XO is 2.

Therefore, the expression of rate law is:

rate = $k[XO]^2 \cdot [O_2]$

47. B is correct.

General formula for the equilibrium constant of a reaction:

$a\text{A} + b\text{B} \leftrightarrow c\text{C} + d\text{D}$

$K_{eq} = ([C]^c \times [D]^d) / ([A]^a \times [B]^b)$

For equilibrium constant calculation, only include species in aqueous or gas phases.

$K_{eq} = 1 / [CO_2]$

48. D is correct.

K_{eq} = [products] / [reactants]

If the numerator (i.e., products) is smaller than the denominator (i.e., reactants), fewer products have formed relative to the reactants. K_{eq} is less than 1.

49. C is correct.

The activation energy is the energy barrier for a reaction to proceed. The forward reaction is measured from the reactants to the highest energy level in the reaction.

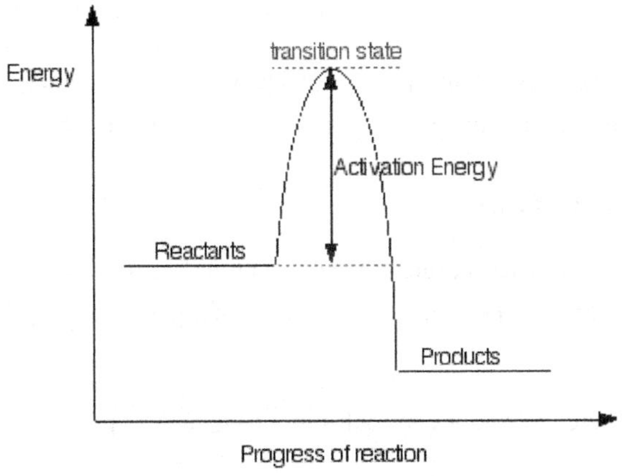

A reaction mechanism with low energy of activation proceeds faster than a reaction with high energy of activation.

50. D is correct.

As the average kinetic energy (i.e., KE = $\frac{1}{2}mv^2$) *increases*, the particles move faster, collide more frequently per unit time and possess greater energy when they collide. This *increases* the *reaction rate*. Hence the *reaction rate* of most *reactions increases* with *increasing temperature*.

For reversible reactions, it is common for the rate law to depend on the concentration of the products. The overall rate is negative (except for autocatalytic reactions). As the concentration of products decreases, their collision frequency decreases, and the reverse reaction rate decreases. Therefore, the overall rate of the reaction increases.

The primary function of catalysts is lowering the reaction's activation energy, thus increasing its rate (k).

51. C is correct.

General formula for the equilibrium constant of a reaction:

$$aA + bB \leftrightarrow cC + dD$$

$$K_{eq} = ([C]^c \times [D]^d) / ([A]^a \times [B]^b)$$

For the ionization equilibrium (K_i) constant calculation, only include species in aqueous or gas phases (which, in this case, is all):

$$K_i = [H^+]\cdot[HS^-] / [H_2S]$$

52. B is correct.

As the temperature (i.e., average kinetic energy) *increases*, the particles move faster, collide more frequently per unit time and possess greater energy when they collide. This *increases* the *reaction rate*. Hence the *reaction rate* of most *reactions increases* with *increasing temperature*.

53. B is correct.

Reactions are spontaneous when Gibbs free energy (ΔG) is negative, and the reaction is described as exergonic; the products are more stable than the reactants.

Reactions are nonspontaneous when Gibbs free energy (ΔG) is positive, and the reaction is described as endergonic; the products are less stable than the reactants.

54. C is correct.

Le Châtelier's principle states that the equilibrium position shifts to counteract the change when changing the reaction conditions. If the reaction temperature, pressure, or volume is changed, the position of equilibrium changes.

There are 2 moles on each side, so changes in pressure (or volume) do not shift the equilibrium.

55. A is correct.

All of the choices are possible events of chemical reactions, but they do not need to happen for a reaction to occur, except that reactant particles must collide (i.e., make contact) with each other.

56. D is correct.

The enthalpy or heat of reaction (ΔH) is not a function of temperature.

Since the reaction is exothermic, Le Châtelier's principle states that increasing the temperature decreases the forward reaction.

57. C is correct.

When changing the reaction conditions, Le Châtelier's principle states that equilibrium shifts to counteract the change. If the reaction's concentration changes, the position of the equilibrium changes.

According to Le Châtelier's Principle, adding or removing a solid at equilibrium does not affect equilibrium. However, adding solid $KC_2H_3O_2$ is equivalent to adding K^+ and $C_2H_3O_2^-$ (i.e., product) as it dissociates in an aqueous solution. The increased concentration of products shifts the equilibrium toward reactants (i.e., to the left).

Increasing the pH decreases the $[H^+]$ and shifts the reaction to products (i.e., right).

58. B is correct.

When changing the reaction conditions, Le Châtelier's principle states that equilibrium shifts to counteract the change. If the reaction temperature, pressure, or volume is changed, the position of equilibrium changes.

The $K_{eq} = 2.8 \times 10^{-21}$, which indicates a low concentration of products compared to reactants:

$$K_{eq} = [\text{products}] / [\text{reactants}]$$

59. D is correct.

Chemical equilibrium refers to a dynamic process whereby the rate at which a reactant molecule is transformed into a product is the same as the rate for a product molecule to be transformed into a reactant.

The rate of the forward reaction is equal to the rate of the reverse reaction.

60. C is correct.

Increased pressure or increased concentration of reactants increases the probability that the reactants collide with sufficient energy and orientation to overcome the energy of activation barrier and proceed toward products.

Practice Set 4: Questions 61–80

---

61. A is correct.

To determine the order with respect to W, compare the data for trials 2 and 4.

The concentrations of X and Y do not change when comparing trials 2 and 4, but the concentration of W changes from 0.015 to 0.03, which corresponds to an increase by a factor of 2.

However, the rate did not increase (0.08 remains 0.08); therefore, the order with respect to W is 0.

The order for X is determined by comparing the data for trials 1 and 3.

The concentrations for W and Z are constant, and the concentration for X increases by a factor of 3 (from 0.05 to 0.15). The rate of the reaction increased by a factor of 9 (0.04 to 0.36).

Since $3^2 = 9$, the order of the reaction with respect to X is two.

The order with respect to Y is found by comparing the data from trials 1 and 5. [W] and [X] do not change, but [Y] goes up by a factor of 4. The rate from trials 1 to 5 goes up by a factor of 2.

Since $4^{1/2} = 2$, the order of the reaction with respect to Y is ½.

The overall order is found by the sum of the orders for X, Y, and Z: 0 + 2 + ½ = 2½.

62. B is correct.

Since the order with respect to W is zero, the rate of formation of Z does not depend on the concentration of W.

63. D is correct.

The rate law is calculated by comparing trials and determining how changes in the initial concentrations of the reactants affect the rate of the reaction.

$$\text{rate} = k[A]^x \cdot [B]^y$$

where k is the rate constant, and the exponents x and y are the partial reaction orders (i.e., determined experimentally). They are not equal to the stoichiometric coefficients.

Use the partial orders determined in question **61** to express the rate law:

$$\text{rate} = k[W]^0[X]^2[Z]^{½}$$

Substitute the values for trial #1:

$$0.04 = k\,[0.01]^0[0.05]^2[0.04]^{½}$$
$$0.04 = (k)\cdot(1)\cdot(2.5 \times 10^{-3})\cdot(2 \times 10^{-1})$$

$$k = 80$$

64. A is correct.

When changing the conditions of a reaction, Le Châtelier's principle states that the position of the equilibrium shifts to counteract the change. If the reaction temperature, pressure, or volume changes, the position of the equilibrium changes.

Decreasing the temperature of the system favors the production of more heat. It shifts the reaction towards the exothermic side. Because $\Delta H > 0$ (i.e., heat is a reactant), the reaction is endothermic. Therefore, decreasing the temperature shifts the equilibrium toward the reactants.

65. D is correct.

General formula for the equilibrium constant of a reaction:

$a\text{A} + b\text{B} \leftrightarrow c\text{C} + d\text{D}$

$K_{eq} = ([\text{C}]^c \times [\text{D}]^d) / ([\text{A}]^a \times [\text{B}]^b)$

For the equilibrium constant (K_i), only include species in aqueous phases:

$K_i = [H^+] \times [H_2PO_4^-] / [H_3PO_4]$

66. C is correct.

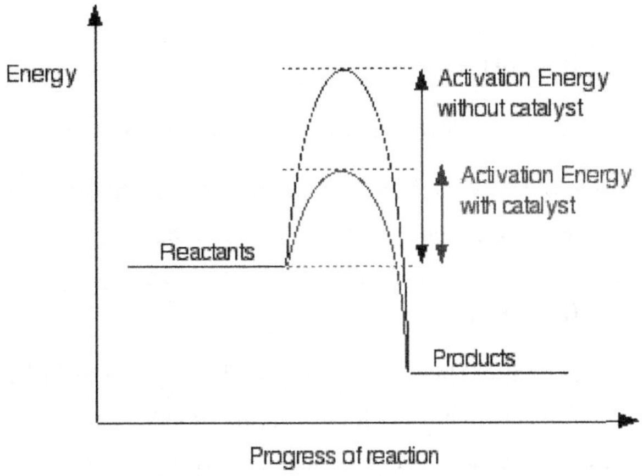

Catalysts speed up reactions and therefore increase the rate at which equilibrium is reached. However, they do not alter the thermodynamics of a reaction and therefore do not change free energy ΔG.

A catalyst provides an alternative pathway for the reaction to proceed to product formation. It lowers the energy of activation (i.e., relative energy between reactants and transition state) and increases the reaction rate.

Catalysts do not affect the Gibbs free energy (ΔG: stability of products vs. reactants) or the enthalpy (ΔH: bond breaking in reactants or bond making in products).

67. D is correct.

General formula for the equilibrium constant of a reaction:

$a\text{A} + b\text{B} \leftrightarrow c\text{C} + d\text{D}$

$K_{eq} = ([\text{C}]^c \times [\text{D}]^d) / ([\text{A}]^a \times [\text{B}]^b)$

For the ionization equilibrium constant (K_i) calculation, only include species in aqueous or gas phases (which, in this case, is all of them):

$K_i = [H^+]\cdot[HSO_3^-] / [H_2SO_3]$

68. D is correct.

K_{eq} = [products] / [reactants]

If the K_{eq} is greater than 1, then the numerator (i.e., products) is greater than the denominator (i.e., reactants), and more products have formed relative to the reactants.

69. B is correct.

The activation energy is the energy barrier for a reaction to proceed.

For the forward reaction, it is measured from the energy of the reactants to the highest energy level in the reaction.

70. A is correct.

Lowering the temperature decreases the reaction rate because the molecules involved in the reaction move and collide more slowly, so the reaction occurs at a slower speed.

Increasing the concentration of reactants due to the increased probability that the reactants collide with sufficient energy and orientation to overcome the energy of the activation barrier and proceed toward products.

The primary function of catalysts is lowering the reaction's activation energy, thus increasing its rate (k).

71. B is correct.

The activation energy is the energy barrier that must be overcome to transform reactant(s) into the product(s).

The molecules must collide with the proper orientation and energy (i.e., kinetic energy is the average temperature) to overcome the energy of the activation barrier.

72. B is correct.

The activation energy is the energy barrier that must be overcome to transform reactant(s) into the product(s). The molecules must collide with the proper orientation and energy (i.e., kinetic energy is the average temperature) to overcome the energy of the activation barrier.

Solids describe molecules with limited relative motion.

Liquids are molecules with more relative motion than solids, while gases exhibit the most relative motion.

73. B is correct.

The rate law is calculated experimentally by comparing trials and determining how changes in the initial concentrations of the reactants affect the rate of the reaction.

$$\text{rate} = k[A]^x \cdot [B]^y$$

where k is the rate constant, and the exponents x and y are the partial reaction orders (i.e., determined experimentally). They are not equal to the stoichiometric coefficients.

Comparing Trials 1 and 3, [A] increased by a factor of 3, as did the reaction rate; thus, the reaction is first order with respect to A.

Comparing Trials 1 and 2, [B] increased by a factor of 4 and the reaction rate increased by a factor of $16 = 4^2$. Thus, the reaction is second order with respect to B.

Therefore, the rate = $k[A] \cdot [B]^2$

74. D is correct.

The molecules must collide with sufficient energy, the frequency of collision, and the proper orientation to overcome the activation energy barrier.

75. D is correct.

When changing the reaction conditions, Le Châtelier's principle states that equilibrium shifts to counteract the change. If the reaction temperature, pressure, or volume is changed, the position of equilibrium changes.

Removing NH_3 and adding N_2 shifts the equilibrium to the right.

Removing N_2 or adding NH_3 shifts the equilibrium to the left.

76. B is correct.

Increasing [$HC_2H_3O_2$] and [H^+] affects equilibrium because both species are part of the equilibrium equation.

K_{eq} = [products] / [reactants]

$K_{eq} = [C_2H_3O_2^-] / [HC_2H_3O_2] \cdot [H^+]$

Adding solid $NaC_2H_3O_2$: in aqueous solutions, $NaC_2H_3O_2$ dissociates into Na^+ (*aq*) and $C_2H_3O_2^-$ (*aq*). Therefore, the overall concentration of $C_2H_3O_2^-$ (*aq*) increases, and it affects the equilibrium.

Adding solid $NaNO_3$: this is a salt and does not react with reactant or product molecules; therefore, it will not affect the equilibrium.

Adding solid NaOH: the reactant is an acid because it dissociates into hydrogen ions and anions in aqueous solutions. It reacts with bases (e.g., NaOH) to create water and salt. Thus, NaOH reduces the concentration of the reactant, which affects the equilibrium.

Increasing [$HC_2H_3O_2$] and [H^+] increases reactants and drives the reaction toward product formation.

77. D is correct.

Calculate the value of the equilibrium expression: hydrogen iodide concentration decreases for equilibrium to be reached.

K = [products] / [reactants]

$K = [HI]^2 / [H_2]·[I_2]$

To determine the state of a reaction, calculate its reaction quotient (Q). Q has the same calculation method as equilibrium constant (K); the difference is K must be calculated at the point of equilibrium, whereas Q can be calculated at any time.

If $Q > K$, reaction shifts towards reactants (more reactants, fewer products)

If $Q = K$, reaction is at equilibrium

If $Q < K$, reaction shifts towards products (more products, less reactants)

$Q = [HI]^2 / [H_2]·[I_2]$

$Q = (3)^2 / 0.4 \times 0.6$

$Q = 37.5$

Because $Q > K$ (i.e., (37.5 > 35), the reaction shifts toward reactants (larger denominator), and the amount of product (HI) decreases.

78. C is correct.

Decreasing the concentration of reactants (or decreasing pressure) decreases the reaction rate because there is a decreased probability that two reactant molecules collide with sufficient energy to overcome the energy of activation and form products.

79. B is correct.

Chemical equilibrium expression is dependent on the stoichiometry of the reaction, more specifically, the coefficients of each species involved in the reaction.

The mechanism refers to the pathway of product formation (e.g., S_N1 or S_N2) but does not affect the relative energies of the reactants and products (i.e., a position for the equilibrium).

The rate refers to the time needed to achieve equilibrium but does not affect the equilibrium position.

80. D is correct.

K = [products] / [reactants]

$K = [CH_3OH] / [H_2O]^2·[CO]$

None of the stated changes tend to decrease the *magnitude* of the equilibrium constant K.

For exothermic reactions ($\Delta H < 0$), decreasing the temperature increases the magnitude of K.

Changes in volume, pressure, or concentration do not affect the magnitude of K.

Notes for active learning

Notes for active learning

Solution Chemistry – Detailed Explanations

Practice Set 1: Questions 1–20

1. B is correct.

Solubility is proportional to pressure. Use simple proportions to compare solubility in different pressures:

$$P_1 / S_1 = P_2 / S_2$$

$$S_2 = P_2 / (P_1 / S_1)$$

$$S_2 = 4.5 \text{ atm} / (1.00 \text{ atm} / 1.90 \text{ cc}/100 \text{ mL})$$

$$S_2 = 8.55 \text{ cc} / 100 \text{ mL}$$

2. C is correct.

The correct interpretation of molarity is moles of solute per liter of solvent.

3. A is correct.

NaCl is a charged, ionic salt (Na^+ and Cl^-) and therefore is extremely water-soluble.

Hexanol can hydrogen bond with water due to the hydroxyl group, but the long hydrocarbon chain reduces its solubility, and the chains interact via hydrophobic interactions forming micelles.

Aluminum hydroxide has poor solubility in water and requires the addition of Brønsted-Lowry acids to dissolve completely.

4. A is correct.

Spectator ions appear on each side of the ionic equation.

Ionic equation of the reaction:

$$Ba^{2+} (aq) + 2\,Cl^- (aq) + 2\,K^+ (aq) + CrO_4^- (aq) \rightarrow BaCrO_4 (s) + 2\,K^+ (aq) + 2\,Cl^- (aq)$$

5. D is correct.

Higher pressure exerts more pressure on the solution, which allows more molecules to dissolve in the solvent.

The lower temperature increases gas solubility.

As the temperature increases, solvent and solute molecules move faster, and it is more difficult for the solvent molecules to bond with the solute.

6. A is correct.

Solutes dissolve in the solvent to form a solution.

7. B is correct.

When the compound is dissolved in water, the attached hydrate/water crystals dissociate and become part of the water solvent.

Therefore, the only ions left are 1 Co^{2+} and 2 NO_3^-.

8. D is correct.

The mass of a solution is calculated by taking the mass of the solution and container and subtracting the mass of the container.

9. A is correct.

An insoluble compound is incapable of being dissolved (especially with reference to water).

Hydroxide salts of Group I elements are soluble, while hydroxide salts of Group II elements (Ca, Sr and Ba) are slightly soluble.

Salts containing nitrate ions (NO_3^-) are generally soluble.

Most sulfate salts are soluble. Important exceptions: $BaSO_4$, $PbSO_4$, Ag_2SO_4 and $SrSO_4$.

Hydroxide salts of transition metals and Al^{3+} are insoluble.

Thus, $Fe(OH)_3$, $Al(OH)_3$ and $Co(OH)_2$ are not soluble.

10. C is correct.

% v/v solution = (volume of acetone / volume of solution) × 100%

% v/v solution = {25 mL / (25 mL + 75 mL)} × 100%

% v/v solution = 25%

11. C is correct.

The van der Waals forces in the hydrocarbons are relatively weak, while hydrogen bonds between H_2O molecules are strong intermolecular bonds.

Bonds between polar H_2O and nonpolar octane are weaker than those between two polar H_2O molecules (i.e., hydrogen bonds).

12. A is correct.

To find the K_{sp}, take the molarities (or concentrations) of the products (cC and dD) and multiply them.

For K_{sp}, only aqueous species are included in the calculation.

If any of the products have coefficients, raise the product to that coefficient power and multiply the concentration by that coefficient:

$$K_{sp} = [C]^c \cdot [D]^d$$

$$K_{sp} = [Cu^{2+}]^3 \cdot [PO_4^{3-}]^2$$

The reactant (aA) is solid and is not included in the K_{sp} equation.

Solids are not included when calculating equilibrium constant expressions because their concentrations do not change the expression. Any change in their concentrations is insignificant and is thus omitted.

13. C is correct.

The *like dissolves like* rule applies when a solvent is miscible with a solute that has similar properties.

A polar solute (e.g., ethanol) is miscible with a polar solvent (e.g., water).

14. C is correct.

In a net ionic equation, substances that do not dissociate in aqueous solutions are written in their molecular form (not broken down into ions).

Gases, liquids, and solids are written in molecular form.

Some solutions do not dissociate into ions in water or do so in minimal amounts (e.g., weak acids such as HF, CH_3COOH). They are also written in molecular form.

15. B is correct.

A supersaturated solution contains more of the dissolved material than could be dissolved by the solvent under normal conditions. Increased heat allows for a solution to become supersaturated. The term also refers to the vapor of a compound with a higher partial pressure than the vapor pressure of that compound.

A supersaturated solution forms a precipitate if seed crystals are added as the solution reaches a lower energy state when solutes precipitate from the solution.

16. D is correct.

Mass % = mass of solute / mass of solution

Rearrange that equation to solve for mass of solution:

mass of solution = mass of solute / mass %

mass of solution = 122 g / 7.50%

mass of solution = 122 g / 0.075

mass of solution = 1,627 g

17. B is correct.

Determine moles of NH_3:

Moles of NH_3 = mass of NH_3 / molecular weight of NH_3

Moles of NH_3 = 15.0 g / (14.01 g/mol + 3 × 1.01 g/mol)

Moles of NH_3 = 15.0 g / (17.04 g/mol)

Moles of NH_3 = 0.88 mol

Determine the volume of the solution (solvent + solute):

Volume of solution = mass / density

Volume of solution = (250 g + 15 g) / 0.974 g/mL

Volume of solution = 272.1 mL

Convert volume to liters:

Volume = 272.1 mL × 0.001 L / mL

Volume = 0.2721 L

Divide moles by the volume to calculate molarity:

Molarity = moles / liter of solution

Molarity = 0.88 mol / 0.2721 L

Molarity = 3.23 M

18. A is correct.

Solutions with the highest concentration of ions have the highest boiling point.
Calculate the concentration of ions in each solution:

0.2 M $Al(NO_3)_3$ = 0. 2 M × 4 ions = 0.8 M

0. 2 M $MgCl_2$ = 0. 2 M × 3 ions = 0.6 M

0. 2 M glucose = 0. 2 M × 1 ion = 0.2 M (glucose does not dissociate into ions in solution)

0. 2 M Na_2SO_4 = 0. 2 M × 3 ions = 0.6 M

Water = 0 M

19. A is correct.

Calculate the number of ions in each option:

A: $Li_3PO_4 \rightarrow 3\ Li^+ + PO_4^{3-}$ (4 ions)

B: $Ca(NO_3)_2 \rightarrow Ca^{2+} + 2\ NO_3^-$ (3 ions)

C: $MgSO_4 \rightarrow Mg^{2+} + SO_4^{2-}$ (2 ions)

D: $(NH_4)_2SO_4 \rightarrow 2\ NH_4^+ + SO_4^{2-}$ (3 ions)

20. D is correct.

An electrolyte is a substance that produces an electrically conducting solution when dissolved in a polar solvent (e.g., water).

The dissolved electrolyte separates into positively-charged cations and negatively-charged anions.

Strong electrolytes dissociate entirely (or almost completely) because the resulting ions are stable in the solution.

Practice Set 2: Questions 21–40

---

21. B is correct

Note that carbonate salts help remove 'hardness' in water.

Generally, SO_4, NO_3 and Cl salts tend to be soluble in water, while CO_3 salts are less soluble.

22. A is correct.

A *spectator ion* exists in the same form on the reactant and product sides of a chemical reaction.

Balanced equation:

$$Pb(NO_3)_2\ (aq) + H_2SO_4\ (aq) \rightarrow PbSO_4\ (s) + 2\ HNO_3\ (aq)$$

23. C is correct.

A polar solute is miscible with a polar solvent.

Ascorbic acid is polar and is therefore miscible in water.

24. B is correct.

Apply the formula for the molar concentration of a solution:

$$M_1V_1 = M_2V_2$$

Substitute the given volume and molar concentrations of HCl, solve for the final volume of HCl of the resulting dilution:

$$V_2 = [M_1V_1] / (M_2)$$

$$V_2 = [(2.00\ M\ HCl) \times (0.125\ L\ HCl)] / (0.400\ M\ HCl)$$

$$V_2 = 0.625\ L\ HCl$$

Solve for the volume of water needed to be added to the initial volume of HCl in order to obtain the final diluted volume of HCl:

$$V_{H_2O} = V_2 - V_1$$

$$V_{H_2O} = (0.625\ L\ HCl) - (0.125\ L\ HCl)$$

$$V_{H_2O} = 0.500\ L = 500\ mL$$

25. A is correct.

The *–ate* ending indicates the species with more oxygen than species ending in *–ite*.

However, it does not indicate a specific number of oxygen molecules.

26. C is correct.

Solubility is the property of a solid, liquid, or gaseous substance (i.e., solute); it dissolves in a solid, liquid, or gaseous solvent to form a solution (i.e., solute in the solvent).

The solubility of a substance depends on the physical and chemical properties of the solute and solvent and the temperature, pressure, and pH of the solution.

Gaseous solutes (e.g., oxygen) exhibit complex behavior with temperature. As the temperature rises, gases usually become less soluble in water but more soluble in organic solvents.

27. A is correct.

A saturated solution contains the maximum amount of dissolved material in the solvent under normal conditions. Increased heat allows for a solution to become supersaturated.

A saturated solution refers to the vapor of a compound with a higher partial pressure than the vapor pressure of that compound.

A saturated solution forms a precipitate as more solute is added to the solution.

28. D is correct.

Immiscible refers to the property (of solutions) of when two or more substances (e.g., oil and water) are mixed and eventually separate into two layers.

Miscible is when two liquids are mixed, but they do not necessarily interact chemically.

In contrast to miscibility, *soluble* means the substance (solid, liquid, or gas) can *dissolve* in another solid, liquid, or gas. In other words, a substance *dissolves* when it becomes incorporated into another substance. Also, in contrast to miscibility, solubility involves a *saturation point* at which a substance cannot dissolve further, and a mass, the *precipitate*, begins to form.

29. D is correct.

In this case, hydration means dissolving in water rather than reacting with water.

Therefore, the compound dissociates into its ions (without involving water in the actual chemical reaction).

30. A is correct.

Depending on the solubility of a solute, there are three possible results:

1) a dilute solution has less solute than the maximum amount that it can dissolve;

2) a saturated solution has the same amount as its solubility;

3) a precipitate forms if there is more solute than can be dissolved, so the excess solute separates from the solution (i.e., crystallization).

Precipitation lowers the concentration of the solute to the saturation level to increase the solution's stability.

Gas solubility is inversely proportional to temperature and proportional to pressure.

31. D is correct.

Like dissolves like means that polar substances tend to dissolve in polar solvents and nonpolar substances in nonpolar solvents.

Molecules that can form hydrogen bonds with water are soluble.

Salts are ionic compounds.

The anion and cation bond with the polar molecule of water and are soluble.

32. B is correct.

Like dissolves like means that polar substances tend to dissolve in polar solvents and nonpolar substances in nonpolar solvents.

Methanol (CH_3OH) is a polar molecule. Therefore, methanol is soluble in water because it can make a hydrogen bond with water.

33. B is correct.

To find the K_{sp}, take the molarities or concentrations of the products (cC and dD) and multiply them.

For K_{sp}, only aqueous species are included in the calculation.

If the products have coefficients in front, raise the product to that coefficient power and multiply the concentration by that coefficient:

$$K_{sp} = [C]^c \cdot [D]^d$$

$$K_{sp} = [Au^{3+}] \cdot [Cl^-]^3$$

The reactant (aA) is solid and is not included in the K_{sp} equation.

Solids are not included when calculating equilibrium constant expressions because their concentrations do not change the expression. Any change in their concentrations is insignificant and is thus omitted.

34. B is correct.

Break down the molecules into their constituent ions:

$$CaCO_3 + 2\ H^+ + 2\ NO_3^- \rightarrow Ca^{2+} + 2\ NO_3^- + CO_2 + H_2O$$

Remove species that appear on each side of the reaction:

$$CaCO_3 + 2\ H^+ \rightarrow Ca^{2+} + CO_2 + H_2O$$

35. D is correct.

According to the problem, there are 0.950 moles of nitrate ion in a $Fe(NO_3)_3$ solution.

Because there are three nitrates (NO_3) ions for each $Fe(NO_3)_3$ molecule, the moles of $Fe(NO_3)_3$ is:

(1 mole / 3 mole) × 0.950 moles = 0.317 moles

Calculate the volume of solution:

Volume of solution = moles of solute / molarity

Volume of solution = 0.317 moles / 0.550 mol/L

Volume of solution = 0.576 L

Convert the volume into milliliters:

0.576 L × 1,000 mL/L = 576 mL

36. A is correct.

mEq represents the amount in milligrams of a solute equal to 1/1,000 of its gram equivalent weight, considering the valency of the ion.

Millimolar (mM) = mEq / valence.

Divide the given concentration of Ca^{2+} by 2:

$[Ca^{2+}]$ = [48 mEq Ca^{2+}] / 2 = 24 mM Ca^{2+}

$[Ca^{2+}]$ = 24 mM Ca^{2+}

Divide the concentration of Ca^{2+} by 1,000 to obtain the concentration of Ca^{2+} in units of molarity:

$[Ca^{2+}]$ = 24 mM Ca^{2+} × [(1 M) / (1,000 mM)]

$[Ca^{2+}]$ = 0.024 M Ca^{2+}

37. A is correct.

Start by calculating the number of moles:

Moles of LiOH = mass of LiOH / molar mass of LiOH

Moles of LiOH = 36.0 g / (24.0 g/mol)

Moles of LiOH = 1.50 moles

Divide moles by volume to calculate molarity:

Molarity = moles / volume

Molarity = 1.50 moles / (975 mL × 0.001 L/mL)

Molarity = 1.54 M

38. C is correct.

The glucose content is 8.50 % (m/v).

The solute (glucose) is measured in grams, but the solution volume is measured in milliliters.

Therefore, the mass and volume of the solution are interchangeable (1 g = 1 mL).

% (m/v) of glucose = mass of glucose / volume of solution

volume of solution = mass of glucose / % (m/v) of glucose

volume of solution = 60 g / 8.50%

volume of solution = 706 mL

39. D is correct.

When AgCl dissociates, equal amounts of Ag^+ and Cl^- are produced.

If the concentration of Cl^- is B, this must be the concentration of Ag^+.

Therefore, B can be the concentration of Ag.

The concentration of silver ion can be determined by dividing the K_{sp} by the concentration of chloride ion:

$K_{sp} = [Ag^+]\cdot[Cl^-]$.

Therefore, the concentration of Ag can be A/B moles/liter.

40. B is correct.

Strong electrolytes dissociate entirely (or almost completely) in water.

Strong acids and bases dissociate almost completely, but weak acids and bases dissociate only slightly.

Practice Set 3: Questions 41–60

---

41. D is correct.

Solute-solute and solvent-solvent attractions are essential in establishing bonding amongst themselves for solvent and solute.

Once they are mixed in a solution, the solute-solvent attraction becomes the major attraction force, but the other two forces (solute-solute and solvent-solvent attractions) are still present.

42. B is correct.

Since the empirical formula for magnesium iodide is MgI_2, two moles of dissolved I^- result from each mole of dissolved MgI_2.

Therefore, if $[MgI_2] = 0.40$ M, then $[I^-] = 2(0.40\text{ M}) = 0.80$ M.

43. B is correct.

Ammonia forms hydrogen bonds, and SO_2 is polar.

44. A is correct.

The chlorite ion (chlorine dioxide anion) is ClO_2^-.

Chlorite is a compound that contains this group, with chlorine in an oxidation state of +3.

Formula	Cl^-	ClO^-	ClO_2^-	ClO_3^-	ClO_4^-
Anion name	chloride	hypochlorite	chlorite	chlorate	perchlorate
Oxidation state	−1	+1	+3	+5	+7

45. D is correct.

Water softeners cannot remove ions from the water without replacing them.

The process replaces the ions that cause scaling (precipitation) with non-reactive ions.

46. B is correct.

These compounds are rarely encountered in most chemistry problems and are part of the long list of solubility rule exceptions.

Salts containing nitrate ions (NO_3^-) are generally soluble.

47. A is correct.

Hydration involves the interaction of water molecules with the solute. The water molecules exchange bonding relationships with the solute, whereby water-water bonds break and water-solute bonds form.

When an ion is hydrated, it is surrounded and bonded by water molecules. The average number of water molecules bonding to an ion is its *hydration number*.

Hydration numbers can vary but often are 4 or 6.

48. C is correct.

Solvation describes the process whereby the solvent surrounds the solute molecules.

49. D is correct.

Miscible refers to the property (of solutions) when two or more substances (e.g., water and alcohol) are mixed without separating.

50. C is correct.

The *like dissolves like* rule applies when a solvent is miscible with a solute that has similar properties.

A nonpolar solute is immiscible with a polar solvent.

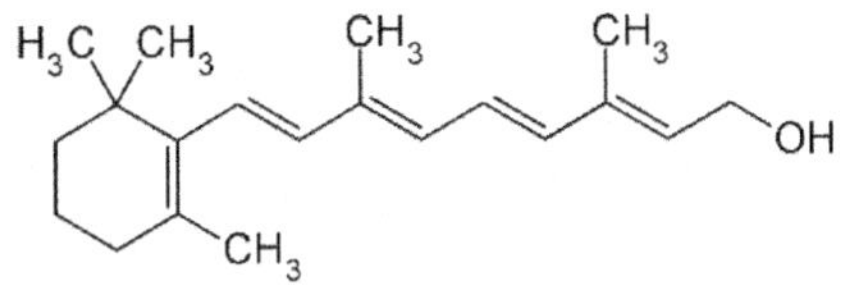

Retinol (vitamin A)

51. B is correct.

Only aqueous (*aq*) species are broken down into their ions for ionic equations.

Solids, liquids, and gases stay the same.

52. C is correct.

For K_{sp}, only aqueous species are included in the calculation.

The decomposition of $PbCl_2$:

$PbCl_2 \rightarrow Pb^{2+} + 2\ Cl^-$

$K_{sp} = [Pb^{2+}]\cdot[Cl^-]^2$

When x moles of $PbCl_2$ fully dissociate, x moles of Pb and $2x$ moles of Cl^- are produced:

$K_{sp} = (x)\cdot(2x)^2$

$K_{sp} = 4x^3$

53. B is correct.

Strong electrolytes dissociate entirely (or almost completely) in water.

Strong acids and bases dissociate nearly completely (i.e., form stable anions), but weak acids and bases dissociate only slightly (i.e., form unstable anions).

54. A is correct.

A *spectator ion* is an *ion* that exists in the same form on the reactant and product sides of a chemical reaction.

Balanced equation for potassium hydroxide and nitric acid:

$$KOH + HNO_3 \rightarrow KNO_3 + H_2O$$

Ionic equation:

$$K^+ \, {}^-OH + H^+ \, NO_3^- \rightarrow K^+ \, NO_3^- + H_2O$$

55. A is correct.

When a solution is diluted, the moles of solute (n) is constant.

However, the molarity and volume will change because n = MV:

$$n_1 = n_2$$

$$M_1V_1 = M_2V_2$$

$$V_2 = (M_1V_1) / M_2$$

$$V_2 = (0.20 \text{ M} \times 6.0 \text{ L}) / 14 \text{ M}$$

$$V_2 = 0.086 \text{ L}$$

Convert to milliliters:

$$0.086 \text{ L} \times (1{,}000 \text{ mL} / \text{L}) = 86 \text{ mL}$$

56. B is correct.

When a solution is diluted, the moles of solute are constant.

Use this formula to calculate the new molarity:

$$M_1V_1 = M_2V_2$$

$$160 \text{ mL} \times 4.50 \text{ M} = 595 \text{ mL} \times M_2$$

$$M_2 = (160 \text{ mL} / 595 \text{ mL}) \times 4.50 \text{ M}$$

$$M_2 = 1.21 \text{ M}$$

57. B is correct.

Like dissolves like means that polar substances tend to dissolve in polar solvents and nonpolar substances in nonpolar solvents.

Since benzene is nonpolar, look for a nonpolar substance.

Silver chloride is ionic, while CH_2Cl_2, H_2S and SO_2 are polar.

58. B is correct.

Molarity can vary based on temperature because it involves the volume of solution.

At different temperatures, the volume of water varies slightly, which affects the molarity.

59. D is correct.

Concentration:

solute / volume

For example: 10 g / 1 liter = 10 g/liter

2 g / 1 liter = 2 g/liter

60. C is correct.

The NaOH content is 5.0% (w/v). It means that the solute (NaOH) is measured in grams, but the solution volume is measured in milliliters.

Therefore, in this problem, the mass and volume of the solution are interchangeable (1 g = 1 mL).

mass of NaOH = % NaOH × volume of solution

mass of NaOH = 5.0% × 75.0 mL

mass of NaOH = 3.75 g

Practice Set 4: Questions 61–80

---

61. D is correct.

Adding NaCl increases [Cl^-] in solution (i.e., common ion effect), which increases the precipitation of $PbCl_2$ because the ion product increases. This increase causes lead chloride to precipitate and the concentration of free chloride in solution to decrease.

62. A is correct.

AgCl has a stronger tendency to form than $PbCl_2$ because of its smaller K_{sp}.

Therefore, as AgCl forms, an equivalent amount of $PbCl_2$ dissolves.

63. A is correct.

For the dissolution of $PbCl_2$:

$PbCl_2 \rightarrow Pb^{2+} + 2\ Cl^-$

$K_{sp} = [Pb^{2+}]\cdot[2\ Cl^-]^2 = 10^{-5}$

$K_{sp} = (x)\cdot(2x)^2 = 10^{-5}$

$4x^3 = 10^{-5}$

$x \approx 0.014$

$[Cl^-] = 2(0.014)$

$[Cl^-] = 0.028$

For the dissolution of AgCl:

$AgCl \rightarrow Ag^+ + Cl^-$

$K_{sp} = [Ag^+]\cdot[Cl^-]$

$K_{sp} = (x)\cdot(x)$

$10^{-10} = x^2$

$x = 10^{-5}$

$[Cl^-] = 10^{-5}$

64. A is correct.

The intermolecular bonding (i.e., van der Waals) in alkanes is similar.

65. C is correct.

Ideally, dilute solutions are so dilute that solute molecules do not interact.

Therefore, the mole fraction of the solvent approaches one.

66. D is correct.

Henry's law states that, at a constant temperature, the amount of gas that dissolves in a volume of liquid is directly proportional to the partial pressure of that gas in equilibrium with that liquid.

Tyndall effect is light scattering by particles in a colloid (or particles in a very fine suspension).

A colloidal suspension contains microscopically dispersed insoluble particles (i.e., colloid) suspended throughout the liquid. The colloid particles are larger than those of the solution but not large enough to precipitate due to gravity.

67. C is correct.

Miscible refers to the property (of solutions) when two or more substances (e.g., water and alcohol) are mixed without separating.

68. B is correct.

Start by calculating moles of glucose:

Moles of glucose = mass of glucose / molar mass of glucose

Moles of glucose = 10.0 g / (180.0 g/mol)

Moles of glucose = 0.0555 mol

Then, divide moles by volume to calculate molarity:

Molarity of glucose = moles of glucose / volume of glucose

Molarity of glucose = 0.0555 mol / (100 mL × 0.001 L/mL)

Molarity = 0.555 M

69. D is correct.

An insoluble compound is incapable of being dissolved (especially with reference to water).

Hydroxide salts of Group I elements are soluble.

Hydroxide salts of Group II elements (Ca, Sr and Ba) are slightly soluble.

Hydroxide salts of transition metals and Al^{3+} are insoluble. Thus, $Fe(OH)_3$, $Al(OH)_3$, $Co(OH)_2$ are not soluble.

Most sulfate salts are soluble. Important exceptions are $BaSO_4$, $PbSO_4$, Ag_2SO_4, and $SrSO_4$.

Salts containing Cl^-, Br^- and I^- are generally soluble. Exceptions are halide salts of Ag^+, Pb^{2+} and $(Hg_2)^{2+}$. $AgCl$, $PbBr_2$, and Hg_2Cl_2 are insoluble.

Chromates are frequently insoluble. Examples: $PbCrO_4$, $BaCrO_4$

70. C is correct.

The overall reaction is:

$$Zn\ (s) + 2\ HCl\ (aq) \rightarrow ZnCl_2\ (aq) + H_2\ (g)$$

Because HCl and $ZnCl_2$ are ionic, they dissociate into ions in aqueous solutions:

$$Zn\ (s) + 2\ H^+\ (aq) + 2\ Cl^-\ (aq) \rightarrow Zn^{2+}\ (aq) + 2\ Cl^-\ (aq) + H_2\ (g)$$

The common ion (Cl^-) cancels to yield the net ionic reaction:

$$Zn\ (s) + 2\ H^+\ (aq) \rightarrow Zn^{2+}\ (aq) + H_2\ (g)$$

71. D is correct.

The *like dissolves like* rule applies when a solvent is miscible with a solute that has similar properties.

A polar solute (e.g., ketone, alcohols, or carboxylic acids) is miscible with a polar (water) solvent.

The three molecules are polar and are therefore soluble in water.

72. A is correct.

Calculation of the solubility constant (K_{sp}) is similar to the equilibrium constant.

The concentration of each species is raised to the power of their coefficients and multiplied with each other.

For K_{sp}, only aqueous species are included in the calculation.

$$CaF_2\ (s) \rightarrow Ca^{2+}\ (aq) + 2F^-\ (aq)$$

The concentration of fluorine ions can be determined using the concentration of calcium ions:

$$[F^-] = 2 \times [Ca^{2+}]$$

$$[F^-] = 2 \times 0.00021\ M$$

$$[F^-] = 4.2 \times 10^{-4}\ M$$

Then, K_{sp} can be determined:

$$K_{sp} = [Ca^{2+}] \cdot [F^-]^2$$

$$K_{sp} = (2.1 \times 10^{-4}\ M) \times (4.2 \times 10^{-4}\ M)^2$$

$$K_{sp} = 3.7 \times 10^{-11}$$

73. B is correct.

To determine the number of equivalents:

(4 moles / liter) × (3 equivalents / mole) × (1/3 liter)

= 4 equivalents

74. A is correct.

An electrolyte is a substance that produces an electrically conducting solution when dissolved in a polar solvent (e.g., water).

The dissolved electrolyte separates into positively charged cations and negatively charged anions.

Electrolytes conduct electricity.

75. A is correct.

This is a frequently encountered insoluble salt in chemistry problems.

Most sulfate salts are soluble. Important exceptions are $BaSO_4$, $PbSO_4$, Ag_2SO_4, and $SrSO_4$.

76. C is correct.

When the ion product is equal to or greater than K_{sp}, precipitation of the salt occurs.

If the ion product value is less than K_{sp}, precipitation does not occur.

77. C is correct.

Gas solubility is inversely proportional to temperature, so a low temperature increases solubility.

The high pressure of O_2 above the solution increases the pressure of the solution (which improves solubility) and the amount of O_2 available for dissolving.

78. C is correct.

Molality is the number of solute moles dissolved in 1,000 grams of solvent.

The mass of the solution is 1,000 g + mass of solute.

Mass of solute (CH_3OH) = moles CH_3OH × molecular mass of CH_3OH

Mass of solute (CH_3OH) = 8.60 moles × [12.01 g/mol + (4 × 1.01 g/mol) + 16 g/mol]

Mass of solute (CH_3OH) = 8.60 moles × (32.05 g/mol)

Mass of solute (CH_3OH) = 275.63 g

Total mass of solution: 1,000 g + 275.63 g = 1,275.63 g

Volume of solution = mass / density

Volume of solution = 1,275.63 g / 0.94 g/mL

Volume of solution = 1,357.05 mL

Divide moles by volume to calculate molarity:

Molarity = number of moles / volume

Molarity = 8.60 moles / 1,357.05 mL

Molarity = 6.34 M

79. B is correct.

By definition, a 15.0% aqueous solution of KI contains 15% KI and the remainder (100 % – 15% = 85%) is water.

If there are 100 g of KI solution, it has 15 g of KI and 85 g of water.

The answer choice of 15 g KI / 100 g water is incorrect because 100 g is the mass of the solution; the actual mass of water is 85 g.

80. D is correct.

Molarity equals moles of solute divided by liters of solution.

$1\ cm^3 = 1\ mL$

Molarity:

$(0.75\ mol) / (0.075\ L) = 10\ M$

Notes for active learning

Notes for active learning

Acids and Bases – Detailed Explanations

Practice Set 1: Questions 1–20

1. A is correct.

Acids dissociate protons to form the conjugate base, while a conjugate base accepts a proton to form the acid.

2. B is correct.

$pH = -\log[H^+]$

$pH = -\log(0.10)$

$pH = 1$

3. A is correct.

Salt is a combination of an acid and a base.

Its pH is determined by the acid and base that created the salt.

Combinations include:

Weak acid and strong base: salt is basic

Strong acid and weak base: salt is acidic

Strong acid and strong base: salt is neutral

Weak acid and weak base: could be anything (acidic, basic, or neutral)

Examples of strong acids and bases that commonly appear in chemistry problems:

Strong acids: HCl, HBr, HI, H_2SO_4, $HClO_4$ and HNO_3

Strong bases: NaOH, KOH, $Ca(OH)_2$ and $Ba(OH)_2$

The hydroxides of Group I and II metals are considered strong bases. Examples include LiOH (lithium hydroxide), NaOH (sodium hydroxide), KOH (potassium hydroxide), RbOH (rubidium hydroxide), CsOH (cesium hydroxide), $Ca(OH)_2$ (calcium hydroxide), $Sr(OH)_2$ (strontium hydroxide) and $Ba(OH)_2$ (barium hydroxide).

4. C is correct.

The Brønsted-Lowry acid-base theory focuses on the ability to accept and donate protons (H^+).

A Brønsted-Lowry acid is a term for a substance that donates a proton (H^+) in an acid-base reaction, while a Brønsted-Lowry base is a term for a substance that accepts a proton.

5. C is correct.

A solution's conductivity is correlated to the number of ions present in a solution. The bulb is shining brightly implies that the solution is a conductor, which means that the solution has a high concentration of ions.

6. C is correct.

An amphoteric compound reacts as an acid (i.e., donates protons) as well as a base (i.e., accepts protons).

One type of amphoteric species is amphiprotic molecules, accepting or donating a proton (H^+). Examples of amphiprotic molecules include amino acids (i.e., an amine and carboxylic acid group) and self-ionizable compounds such as water.

7. D is correct.

Strong acids completely dissociate protons into the aqueous solution. The resulting anion is stable, which accounts for the ~100% ionization of the strong acid.

Weak acids do not entirely (or appreciably) dissociate protons into the aqueous solution. The resulting anion is unstable, which accounts for a small ionization of the weak acid.

Each of the listed strong acids forms an anion stabilized by resonance.

$HClO_4$ (perchloric acid) has a pK_a of –10.

H_2SO_4 (sulfuric acid) is a diprotic acid and has a pK_a of –3 and 1.99.

HNO_3 (nitrous acid) has a pK_a of –1.4.

8. A is correct.

To find the pH of strong acid and strong base solutions (where $[H^+] > 10^{-6}$), use the equation:

$$pH = -\log[H_3O^+]$$

Given that $[H^+] < 10^{-6}$, the pH = 4.

Note that this approach only applies to strong acids and strong bases.

9. B is correct.

The self-ionization (autoionization) of water is an ionization reaction in pure water or an aqueous solution, in which a water molecule, H_2O, deprotonates (loses the nucleus of one of its hydrogen atoms) to become a hydroxide ion (^-OH).

10. B is correct.

An electrolyte is a substance that dissociates into cations (i.e., positive ions) and anions (i.e., negative ions) when placed in solution.

NH_3 is a weak acid with a pK_a of 38. Therefore, it does not readily dissociate into H^+ and $^-NH_2$.

HCO_3^- (bicarbonate) has a pK_a of about 10.3 and is considered a weak acid because it does not dissociate completely. Therefore, it does not readily dissociate into H^+ and CO_3^{2-}.

HCN (nitrile) is a weak acid with a pK_a of 9.3.

Therefore, it does not readily dissociate into H^+ and ^-CN (i.e., cyanide).

11. B is correct.

If the $[H_3O^+] = [^-OH]$, it has a pH of 7, and the solution is neutral.

12. D is correct.

The Arrhenius acid-base theory states that acids produce H^+ ions (protons) in H_2O solution and bases produce OH^- ions (hydroxide) in H_2O solution.

13. C is correct.

The Brønsted-Lowry acid-base theory focuses on the ability to accept and donate protons (H^+).

A Brønsted-Lowry acid is a term for a substance that donates a proton in an acid-base reaction, while a Brønsted-Lowry base is a term for a substance that accepts a proton.

The definition is expressed in terms of an equilibrium expression

acid + base ↔ conjugate base + conjugate acid.

14. B is correct.

According to the Brønsted-Lowry acid-base theory:

An acid (reactant) dissociates a proton to become the conjugate base (product).

A base (reactant) gains a proton to become the conjugate acid (product).

The definition is expressed in terms of an equilibrium expression:

acid + base ↔ conjugate base + conjugate acid

15. C is correct.

The pH scale has a range from 1 to 14, with 7 being a neutral pH.

Acidic solutions have a pH below 7, while basic solutions have a pH above 7.

The pH scale is a log scale where 7 has a 50% deprotonated and 50% protonated (neutral) species (1:1 ratio).

At a pH of 6, there are 1 deprotonated : 10 protonated species. The ratio is 1:10.

At a pH of 5, there are 1 deprotonated : 100 protonated species. The ratio is 1:100.

A pH change of 1 unit changes the ratio by 10×.

Lower pH (< 7) results in more protonated species (e.g., cation), while an increase in pH (> 7) results in more deprotonated species (e.g., anion).

16. A is correct.

pI is the symbol for isoelectric point: the pH where a protein ion has zero net charge.

To calculate pI of amino acids with 2 pK_a values, take the average of the pK_a's:

$$pI = (pK_{a1} + pK_{a2}) / 2$$

$$pI = (2.2 + 4.2) / 2$$

$$pI = 3.2$$

17. D is correct.

Acidic solutions contain hydronium ions (H_3O^+). These ions are in the aqueous form because they are dissolved in water. Although chemists often write H^+ (*aq*), referring to a single hydrogen nucleus (a proton), it exists as the hydronium ion (H_3O^+).

18. C is correct.

The equivalence point is the point at which chemically equivalent quantities of acid and base have been mixed.

The moles of acid are equivalent to the moles of base.

The endpoint (related to, but not the same as the equivalence point) refers to the point at which the indicator changes color in a colorimetric titration. The endpoint can be found by an indicator, such as phenolphthalein (i.e., it turns colorless in acidic solutions and pink in basic solutions).

A buffer is an aqueous solution that consists of a weak acid and its conjugate base, or vice versa.

Buffered solutions resist changes in pH and are often used to keep the pH at a nearly constant value in many chemical applications. It does this by readily absorbing or releasing protons (H^+) and ^-OH.

When an acid is added to the solution, the buffer releases ^-OH and accepts H^+ ions from the acid.

When a base is added, the buffer accepts ^-OH ions from the base and releases protons (H^+).

Using the Henderson-Hasselbalch equation:

$$pH = pK_a + \log([A^-] / [HA]),$$

where [HA] = concentration of the weak acid, in units of molarity; $[A^-]$ = concentration of the conjugate base, in units of molarity.

$$pK_a = -\log(K_a)$$

where K_a = acid dissociation constant.

From the question, the concentration of the conjugate base equals the concentration of the weak acid. This equates to the following expression:

$$[HA] = [A^-]$$

Rearranging: $[A^-] / [HA]$, like in the Henderson-Hasselbalch equation:

$$[A^-] / [HA] = 1$$

Substituting the value into the Henderson-Hasselbalch equation:

$$pH = pK_a + \log(1)$$

$$pH = pK_a + 0$$

$$pH = pK_a$$

When $pH = pK_a$, the titration is in the buffering region.

19. D is correct.

The ionic product constant of water:

$$K_w = [H_3O^+]\cdot[^-OH]$$

$$[^-OH] = K_w / [H_3O^+]$$

$$[^-OH] = [1 \times 10^{-14}] / [7.5 \times 10^{-9}]$$

$$[^-OH] = 1.3 \times 10^{-6}$$

20. C is correct.

Consider the K_a presented in the question – which species is more acidic or basic than NH_3?

NH_4^+ is correct because it is NH_3 after absorbing one proton.

The concentration of H^+ in water is proportional to K_a. NH_4^+ is the conjugate acid of NH_3.

A: H^+ is acidic.

B: NH_2^- is NH_3 with one less proton. If a base loses a proton, it would be an even stronger base with a higher proton affinity, so it is not weaker than NH_3.

D: Water is neutral.

Practice Set 2: Questions 21–40

21. C is correct.

Acid as a reactant produces a conjugate base, while a base as a reactant produces a conjugate acid.

The conjugate base of a chemical species is that species after H^+ has dissociated.

Therefore, the conjugate base of HSO_4^- is SO_4^{2-}.

The conjugate base of H_3O^+ is H_2O.

22. A is correct.

Start by calculating the moles of $Ca(OH)_2$:

Moles of $Ca(OH)_2$ = molarity $Ca(OH)_2$ × volume of $Ca(OH)_2$

Moles of $Ca(OH)_2$ = 0.1 M × (30 mL × 0.001 L/mL)

Moles of $Ca(OH)_2$ = 0.003 mol

Use the coefficients from the reaction equation to determine moles of HNO_3:

Moles of HNO_3 = (coefficient of HNO_3) / [coefficient $Ca(OH)_2$ × moles of $Ca(OH)_2$]

Moles of HNO_3 = (2 / 1) × 0.003 mol

Moles of HNO_3 = 0.006 mol

Divide moles by molarity to calculate volume:

Volume of HNO_3 = moles of HNO_3 / molarity of HNO_3

Volume of HNO_3 = 0.006 mol / 0.2 M

Volume of HNO_3 = 0.03 L

Convert volume to milliliters:

0.03 L × 1000 mL / L = 30 mL

23. D is correct.

Acidic solutions have a pH less than 7 due to a higher concentration of H^+ ions than ^-OH ions.

Basic solutions have a pH greater than 7 due to a higher concentration of ^-OH ions relative to H^+ ions.

24. D is correct.

An amphoteric compound reacts as an acid (i.e., donates protons) as well as a base (i.e., accepts protons).

Examples of amphoteric molecules include amino acids (i.e., an amine and carboxylic acid group) and self-ionizable compounds such as water.

25. B is correct.

Strong acids (i.e., reactants) proceed towards products.

K_a = [products] / [reactants]

The molecule with the largest K_a is the strongest acid.

$pK_a = -\log K_a$

The molecule with the smallest pK_a is the strongest acid.

Strong acids dissociate a proton to produce the weakest conjugate base (i.e., most stable anion).

Weak acids dissociate a proton to produce the strongest conjugate base (i.e., least stable anion).

26. B is correct.

Balanced reaction:

$H_3PO_4 + 3\ LIOH = Li_3PO_4 + 3\ H_2O$

In the neutralization of acids and bases, the result is salt and water.

Phosphoric acid and lithium hydroxide react, so the resulting compounds are lithium phosphate and water.

27. B is correct.

The stability of the compound determines base strength. If the compound is unstable in its present state, it seeks a bonding partner (e.g., H+ or another atom) by donating its electrons for the new bond formation.

The 8 strong bases are: LiOH (lithium hydroxide), NaOH (sodium hydroxide), KOH (potassium hydroxide), $Ca(OH)_2$ (calcium hydroxide), RbOH (rubidium hydroxide), $Sr(OH)_2$, (strontium hydroxide), CsOH (cesium hydroxide) and $Ba(OH)_2$ (barium hydroxide).

28. D is correct.

The formula for pH:

$pH = -\log[H^+]$

Rearrange to solve for $[H^+]$:

$[H^+] = 10^{-pH}$

$[H^+] = 10^{-2}$ M

$[H^+] = 0.01$ M

29. C is correct.

Ionization	Dissociation
The process that produces new charged particles	The separation of charged particles existing in a compound
Involves polar covalent compounds or metals	Involves ionic compounds
Involves covalent bonds between atoms	Involves ionic bonds in compounds
Produces charged particles	Produces charged or electrically neutral particles
Irreversible	Reversible
Example: $HCl \rightarrow H^+ + Cl^-$ $Mg \rightarrow Mg^{2+} + 2\ e^-$	Example: $PbBr_2 \rightarrow Pb^{2+} + 2\ Br^-$

30. C is correct.

A base is a chemical substance with a pH greater than 7 and feels slippery because it dissolves the fatty acids and oils from the skin, reducing the friction between skin cells.

Under acidic conditions, litmus paper is red, and under basic conditions, it is blue.

Many bitter-tasting foods are alkaline because bitter compounds often contain amine groups, which are weak bases.

Acids are known to have a sour taste (e.g., lemon juice) because the sour taste receptors on the tongue detect the dissolved hydrogen (H^+) ions.

31. A is correct.

The 7 strong acids are HCl (hydrochloric acid), HNO_3 (nitric acid), H_2SO_4 (sulfuric acid), HBr (hydrobromic acid), HI (hydroiodic acid), $HClO_3$ (chloric acid), and $HClO_4$ (perchloric acid).

The 8 strong bases are: LiOH (lithium hydroxide), NaOH (sodium hydroxide), KOH (potassium hydroxide), $Ca(OH)_2$ (calcium hydroxide), RbOH (rubidium hydroxide), $Sr(OH)_2$, (strontium hydroxide), CsOH (cesium hydroxide) and $Ba(OH)_2$ (barium hydroxide).

32. C is correct.

The Arrhenius acid-base theory states that acids produce H^+ ions in H_2O solution and bases produce ^-OH ions in H_2O solution.

The Brønsted-Lowry acid-base theory focuses on the ability to accept and donate protons (H^+).

A Brønsted-Lowry acid is a term for a substance that donates a proton in an acid-base reaction, while a Brønsted-Lowry base is a term for a substance that accepts a proton.

Lewis acids are defined as electron pair acceptors, whereas Lewis bases are electron pair donors.

33. A is correct.

By the Brønsted-Lowry acid-base theory:

An acid (reactant) dissociates a proton to become the conjugate base (product).

A base (reactant) gains a proton to become the conjugate acid (product).

The definition is expressed in terms of an equilibrium expression:

acid + base ↔ conjugate base + conjugate acid

34. D is correct.

With polyprotic acids (i.e., more than one H^+ present), the pK_a indicates the pH at which the H^+ is deprotonated. If the pH goes above the first pK_a, one proton dissociates, and so on.

In this example, the pH is above the first and second pK_a, so two acid groups are deprotonated while the third acidic proton is unaffected.

35. A is correct.

It is important to identify the acid that is active in the reaction.

The parent acid is defined as the most protonated form of the buffer. The number of dissociating protons an acid can donate depends on the charge of its conjugate base.

$Ba_2P_2O_7$ is given as one of the products in the reaction. Because barium is a group 2B metal, it has a stable oxidation state of +2. Because two barium cations are present in the product, the charge of P_2O_7 ion (the conjugate base in the reaction) must be –4.

Therefore, the fully protonated form of this conjugate must be $H_4P_2O_7$, which is a tetraprotic acid because it has 4 protons that can dissociate.

36. C is correct.

The greater the concentration of H_3O^+, the more acidic the solution is.

37. B is correct.

A triprotic acid has three protons that can dissociate.

38. B is correct.

KCl and NaI are salts.

Two salts only react if one of the products precipitates.

In this example, the products (KI and NaCl) are soluble in water, so they do not react.

39. B is correct.

Sodium acetate is a basic compound because acetate is the conjugate base of acetic acid, a weak acid ("the conjugate base of a weak acid acts as a base in the water").

The addition of a base to a solution, even a buffered solution, increases the pH.

40. D is correct.

An acid anhydride is a compound that has two acyl groups bonded to the same oxygen atom.

Anhydride means *without water* and is formed via dehydration (i.e., removal of H_2O) reaction.

Practice Set 3: Questions 41–60

41. A is correct.

The Brønsted-Lowry acid-base theory focuses on the ability to accept and donate protons (H^+).

A Brønsted-Lowry acid is a term for a substance that donates a proton (H^+) in an acid-base reaction, while a Brønsted-Lowry base is a term for a substance that accepts a proton.

42. A is correct.

Strong acids (i.e., reactants) proceed towards products.

$$K_a = [\text{products}] / [\text{reactants}]$$

The molecule with the largest K_a is the strongest acid.

$$pK_a = -\log K_a$$

The molecule with the smallest pK_a is the strongest acid.

Strong acids dissociate a proton to produce the weakest conjugate base (i.e., most stable anion).

Weak acids dissociate a proton to produce the strongest conjugate base (i.e., least stable anion).

43. D is correct.

Learn the ions involved in boiler scale formations: CO_3^{2-} and the metal ions.

44. B is correct.

A buffer is an aqueous solution that consists of a weak acid and its conjugate base, or vice versa.

Buffered solutions resist changes in pH and are often used to keep the pH at a nearly constant value in many chemical applications. It does this by readily absorbing or releasing protons (H^+) and ^-OH.

When an acid is added to the solution, the buffer releases ^-OH and accepts H^+ ions from the acid.

To create a buffer solution, there needs to be a pair of a weak acid/base and its conjugate, or a salt that contains an ion from the weak acid/base.

Since the problem indicates that sulfoxylic acid (H_2SO_2), which has a pK_a 7.97, needs to be in the mixture, the other component would be an HSO_2^- ion (bisulfoxylate) of $NaHSO_2$.

45. C is correct.

Acidic salt is a salt that still contains H^+ in its anion. It is formed when a polyprotic acid is partially neutralized, leaving at least 1 H^+.

For example:

$H_3PO_4 + 2\ KOH \rightarrow K_2HPO_4 + 2\ H_2O$: (partial neutralization, K_2HPO_4 is acidic salt)

While:

$H_3PO_4 + 3\ KOH \rightarrow K_3PO_4 + 3\ H_2O$: (complete neutralization, K_3PO_4 is not acidic salt)

46. D is correct.

An acid anhydride is a compound that has two acyl groups bonded to the same oxygen atom.

Anhydride means *without water* and is formed via dehydration (i.e., removal of H_2O) reaction.

47. A is correct.

By the Brønsted-Lowry definition, an acid donates protons while a base accepts protons.

On the product side of the reaction, H_2O acts as a base (i.e., the conjugate base of H_3O^+), and HCl acts as an acid (i.e., the conjugate acid of Cl^-).

48. D is correct.

The ratio of the conjugate base to the acid must be determined from the pH of the solution and the pK_a of the acidic component in the reaction.

In the reaction, $H_2PO_4^-$ acts as the acid, and HPO_4^{2-} acts as the base, so the pK_a of $H_2PO_4^-$ should be used in the equation.

Substitute the given values into the Henderson-Hasselbalch equation:

$pH = pK_a + \log[\text{salt / acid}]$

$7.35 = 6.87 + \log[\text{salt / acid}]$

Since $H_2PO_4^-$ is acting as the acid, subtract 6.87 from each side:

$0.48 = \log[\text{salt / acid}]$

The log base is 10, so the inverse log gives:

$10^{0.48} = (\text{salt / acid})$

$(\text{salt / acid}) = 3.02$

The ratio between the conjugate base or salt and the acid is 3.02 / 1.

49. B is correct.

An electrolyte is a substance that dissociates into cations (i.e., positive ions) and anions (i.e., negative ions) when placed in solution. The light bulb is dimly lit, indicating that the solution contains only a low concentration (i.e., partial ionization) of the ions.

An electrolyte produces an electrically conducting solution when dissolved in a polar solvent (e.g., water). The dissolved ions disperse uniformly through the solvent. If an electrical potential (i.e., voltage) is applied to such a solution, the cations of the solution migrate towards the electrode (i.e., an abundance of electrons). In contrast, the anions migrate towards the electrode (i.e., a deficit of electrons).

50. B is correct.

An acid is a chemical substance with a pH of less than 7, producing H^+ ions in water. An acid can be neutralized by a base (i.e., a substance with a pH above 7) to form a salt.

Acids are known to have a sour taste (e.g., lemon juice) because the sour taste receptors on the tongue detect the dissolved hydrogen (H^+) ions.

However, acids are not known to have a slippery feel; this is characteristic of bases. Bases feel slippery because they dissolve the fatty acids and oils from the skin and reduce the friction between the skin cells.

51. C is correct.

The Arrhenius acid-base theory states that acids produce H^+ ions (protons) in H_2O solution and bases produce ^-OH ions (hydroxide) in H_2O solution.

The Brønsted-Lowry acid-base theory focuses on the ability to accept and donate protons (H^+).

A Brønsted-Lowry acid is a term for a substance that donates a proton in an acid-base reaction, while a Brønsted-Lowry base is a term for a substance that accepts a proton.

52. A is correct.

By the Brønsted-Lowry acid-base theory:

An acid (reactant) dissociates a proton to become the conjugate base (product).

A base (reactant) gains a proton to become the conjugate acid (product).

The definition is expressed in terms of an equilibrium expression:

$$\text{acid} + \text{base} \leftrightarrow \text{conjugate base} + \text{conjugate acid.}$$

53. A is correct.

By the Brønsted-Lowry acid-base theory:

An acid (reactant) dissociates a proton to become the conjugate base (product).

A base (reactant) gains a proton to become the conjugate acid (product).

The definition is expressed in terms of an equilibrium expression:

acid + base $\leftrightarrow$ conjugate base + conjugate acid.

HCl dissociates completely and therefore is a strong acid.

Strong acids produce weak (i.e., stable) conjugate bases.

54. D is correct.

A diprotic acid has two protons that can dissociate.

55. A is correct.

Protons (H^+) migrate between amino acid and solvent, depending on the pH of solvent and pK_a of functional groups on the amino acid.

Carboxylic acid groups can donate protons, while amine groups can receive protons.

For the carboxylic acid group:

If the pH of solution < pK_a : the group is protonated and neutral

If the pH of solution > pK_a : the group is deprotonated and negative

For the amine group:

If the pH of solution < pK_a : the group is protonated and positive

If the pH of solution > pK_a : the group is deprotonated and neutral

56. C is correct.

Acidic solutions contain hydronium ions (H_3O^+). These ions are in the aqueous form because they are dissolved in water.

Although chemists often write H^+ (*aq*), referring to a single hydrogen nucleus (a proton), it exists as the hydronium ion (H_3O^+).

57. A is correct.

With a K_a of 10^{-5}, the pH of a 1 M solution of the carboxylic acid, $CH_3CH_2CH2CO_2H$, would be 5 and is a weak acid.

Only $CH_3CH_2CH_2CO_2H$ can be considered a weak acid because it yields a (relatively) unstable anion.

58. C is correct.

Condition 1: pH = 4

$[H_3O^+] = 10^{-pH} = 10^{-4}$

Condition 2: pH = 7

$[H_3O^+] = 10^{-pH} = 10^{-7}$

Ratio of $[H_3O^+]$ in condition 1 and condition 2:

$10^{-4} : 10^{-7} = 1{,}000 : 1$

Solution with a pH of 4 has 1,000 times greater $[H^+]$ than a solution with a pH of 7.

Note: $[H_3O^+]$ is equivalent to $[H^+]$

59. B is correct.

The pH of a buffer is calculated using the Henderson-Hasselbalch equation:

$\text{pH} = \text{p}K_a + \log([\text{conjugate base}] / [\text{conjugate acid}])$

When [acid] = [base], the fraction is 1.

Log 1 = 0,

$\text{pH} = \text{p}K_a + 0$

If the K_a of the acid is 4.6×10^{-4} (between 10^{-4} and 10^{-3}), the pK_a (and therefore the pH) is between 3 and 4.

60. C is correct.

$K_w = [H^+]\cdot[^-OH]$ is the definition of the ionization constant for water.

==

Practice Set 4: Questions 61–80

==

61. A is correct.

Weak acids react with a strong base, converted to its weak conjugate base, creating an overall basic solution.

Strong acids do react with a strong base, creating a neutral solution.

Weak acids partially dissociate when dissolved in water; unlike strong acids, they do not readily form ions.

Strong acids are much more corrosive than weak acids.

62. C is correct.

K_w is the water ionization constant (or *water autoprotolysis constant*).

It can be determined experimentally and equals 1.011×10^{-14} at 25 °C (1.00×10^{-14} is used).

63. D is correct.

All options are correct descriptions of the reaction, but the correct choice is the most descriptive.

64. A is correct.

The Brønsted-Lowry acid-base theory focuses on the ability to accept and donate protons (H^+).

A Brønsted-Lowry acid is a term for a substance that donates a proton in an acid-base reaction, while a Brønsted-Lowry base is a term for a substance that accepts a proton.

65. A is correct.

Buffered solutions resist changes in pH and are often used to keep the pH at a nearly constant value in many chemical applications. It does this by readily absorbing or releasing protons (H^+) and ^-OH.

H_2SO_4 is a strong acid. Weak acids and their salts are good buffers.

A buffer is an aqueous solution that consists of a weak acid and its conjugate base, or vice versa.

When an acid is added to the solution, the buffer releases ^-OH and accepts H^+ ions from the acid.

When a base is added, the buffer accepts ^-OH ions from the base and releases protons (H^+).

66. D is correct.

The main use of litmus paper is to test whether a solution is acidic or basic.

Litmus paper can be used to test for water-soluble gases that affect acidity or alkalinity; the gas dissolves in the water, and the resulting solution colors the litmus paper. For example, alkaline ammonia gas causes the litmus paper to change from red to blue.

Blue litmus paper turns red under acidic conditions, and red litmus paper turns blue under basic or alkaline conditions, with the color change occurring over the pH range 4.5-8.3 at 25 °C (77 °F).

Neutral litmus paper is purple.

Litmus can be prepared as an aqueous solution that functions similarly. Under acidic conditions, the solution is red, and under basic conditions, the solution is blue.

The properties of turning litmus paper blue, bitter taste, slippery feel, and neutralizing acids are true of bases.

An acidic solution has opposite qualities.

It has a pH lower than 7 and therefore turns litmus paper red.

It neutralizes bases, tastes sour, and does not feel slippery.

67. A is correct.

An electrolyte is a substance that dissociates into cations (i.e., positive ions) and anions (i.e., negative ions) when placed in solution. An electrolyte produces an electrically conducting solution when dissolved in a polar solvent (e.g., water). The dissolved ions disperse uniformly through the solvent.

If an electrical potential (i.e., voltage) is applied to such a solution, the cations of the solution migrate towards the electrode (i.e., an abundance of electrons). In contrast, the anions migrate towards the electrode (i.e., a deficit of electrons).

An acid is a substance that ionizes when dissolved in suitable ionizing solvents such as water.

If a high proportion of the solute dissociates to form free ions, it is a strong electrolyte.

If most of the solute does not dissociate, it is a weak electrolyte.

The more free ions present, the better the solution conducts electricity.

68. A is correct.

The 7 strong acids are HCl (hydrochloric acid), HNO_3 (nitric acid), H_2SO_4 (sulfuric acid), HBr (hydrobromic acid), HI (hydroiodic acid), $HClO_3$ (chloric acid), and $HClO_4$ (perchloric acid).

69. B is correct.

The activity series determines if a metal displaces another metal in the solution.

The reaction occurs if the added metal is above (i.e., activity series) the metal currently bonded with the anion.

An activity series ranks substances in their order of relative reactivity.

For example, magnesium metal can displace hydrogen ions from solution, so it is more reactive than elemental hydrogen:

$$Mg\,(s) + 2\,H^+\,(aq) \rightarrow H_2\,(g) + Mg^{2+}\,(aq)$$

Zinc can displace hydrogen ions from solution, so zinc is more reactive than elemental hydrogen:

$$Zn\,(s) + 2\,H^+\,(aq) \rightarrow H_2\,(g) + Zn^{2+}\,(aq)$$

Magnesium metal can displace zinc ions from solution:

$Mg\ (s) + Zn^{2+}\ (aq) \rightarrow Zn\ (s) + Mg^{2+}\ (aq)$

The metal activity series with the most active (i.e., most strongly reducing) metals appear at the top and the least active metals near the bottom.

Li: $2\ Li\ (s) + 2\ H_2O\ (l) \rightarrow LiOH\ (aq) + H_2\ (g)$

K: $2\ K\ (s) + 2\ H_2O\ (l) \rightarrow 2\ KOH\ (aq) + H_2\ (g)$

Ca: $Ca\ (s) + 2\ H_2O\ (l) \rightarrow Ca(OH)_2\ (s) + H_2\ (g)$

Na: $2\ Na\ (s) + 2\ H_2O\ (l) \rightarrow 2\ NaOH\ (aq) + H_2\ (g)$

The above can displace H_2 from water, steam, or acids

Mg: $Mg\ (s) + 2\ H_2O\ (g) \rightarrow Mg(OH)_2\ (s) + H_2\ (g)$

Al: $2\ Al\ (s) + 6\ H_2O\ (g) \rightarrow 2\ Al(OH)_3\ (s) + 3\ H_2\ (g)$

Mn: $Mn\ (s) + 2\ H_2O\ (g) \rightarrow Mn(OH)_2\ (s) + H_2\ (g)$

Zn: $Zn\ (s) + 2\ H_2O\ (g) \rightarrow Zn(OH)_2\ (s) + H_2\ (g)$

Fe: $Fe\ (s) + 2\ H_2O\ (g) \rightarrow Fe(OH)_2\ (s) + H_2\ (g)$

The above can displace H_2 from steam or acids

Ni: $Ni\ (s) + 2\ H^+\ (aq) \rightarrow Ni^{2+}\ (aq) + H_2\ (g)$

Sn: $Sn\ (s) + 2\ H^+\ (aq) \rightarrow Sn^{2+}\ (aq) + H_2\ (g)$

Pb: $Pb\ (s) + 2\ H^+\ (aq) \rightarrow Pb^{2+}\ (aq) + H_2\ (g)$

The above can displace H_2 from acids only

$H_2 > Cu > Ag > Pt > Au$

The above cannot displace H_2

70. D is correct.

The Arrhenius acid-base theory states that acids produce H^+ ions (protons) in H_2O solution, and bases produce OH^- ions (hydroxide) in H_2O solution.

$Al(OH)_3\ (s)$ is insoluble in water; therefore, it cannot function as an Arrhenius base.

71. A is correct.

By the Brønsted-Lowry acid-base theory:

An acid (reactant) dissociates a proton to become the conjugate base (product).

A base (reactant) gains a proton to become the conjugate acid (product).

The definition is expressed in terms of an equilibrium expression:

acid + base $\leftrightarrow$ conjugate base + conjugate acid

72. D is correct.

Henderson-Hasselbach equation:

$$pH = pK_a + \log(A^- / HA)$$

A buffer is an aqueous solution that consists of a weak acid and its conjugate base, or vice versa.

Buffered solutions resist changes in pH and are often used to keep the pH at a nearly constant value in many chemical applications. It does this by readily absorbing or releasing protons (H^+) and ^-OH.

When an acid is added to the solution, the buffer releases ^-OH and accepts H^+ ions from the acid.

73. C is correct.

The weakest acid has the smallest K_a (or largest pK_a).

The weakest acid has the strongest (i.e., least stable) conjugate base.

74. D is correct.

The balanced reaction:

$$2\ H_3PO_4 + 3\ Ba(OH)_2 \rightarrow Ba_3(PO_4)_2 + 6\ H_2O$$

There are 2 moles of H_3PO_4 in a balanced reaction.

However, acids are categorized by the number of H^+ per mole of acid.

For example:

HCl is a monoprotic acid (has one H^+ to dissociate).

H_2SO_4 is a diprotic acid (has two H^+ to dissociate).

H_3PO_4 is a triprotic acid (has three H^+ to dissociate).

75. B is correct.

Na^+ forms a strong base (NaOH) and S forms a weak acid (H_2S), it undergoes a hydrolysis reaction in water:

$$Na_2S\ (aq) + 2\ H_2O\ (l) \rightarrow 2\ NaOH\ (aq) + H_2S$$

Ionic equation for individual ions:

$$2\ Na^+ + S^{2-} + 2\ H_2O \rightarrow 2\ Na^+ + 2\ OH^- + H_2S$$

Removing Na^+ ions from each side of the reaction:

$$S^{2-} + 2\ H_2O \rightarrow 2\ OH^- + H_2S$$

Removing H_2O from each side of the reaction:

$$S^{2-} + H_2O \rightarrow OH^- + HS^-$$

76. A is correct.

HNO_3 is a strong acid, which means that $[HNO_3] = [H_3O^+] = 0.0765$.

pH = $-\log[H_3O^+]$

pH = $-\log(0.0765)$

pH = 1.1

77. C is correct.

Strong acids dissociate a proton to produce the weakest conjugate base (i.e., most stable anion).

Weak acids dissociate a proton to produce the strongest conjugate base (i.e., least stable anion).

Acetic acid (CH_3COOH) has a pK_a of about 4.8 and is considered a weak acid because it does not dissociate completely. Therefore, it does not readily dissociate into H^+ and CH_3COO^-.

Each of the listed strong acids forms an anion stabilized by resonance.

HBr (hydrobromic acid) has a pK_a of –9.

HNO_3 (nitrous acid) has a pK_a of –1.4.

H_2SO_4 (sulfuric acid) is a diprotic acid and has a pK_a of –3 and 1.99.

HCl (hydrochloric acid) has a pK_a of –6.3.

78. A is correct.

It is important to recognize the chromate ions:

Dichromate: $Cr_2O_7^{2-}$

Chromium (II): Cr^{2+}

Chromic/Chromate: CrO_4^-

79. B is correct.

The Brønsted-Lowry acid-base theory focuses on the ability to accept and donate protons (H^+).

A Brønsted-Lowry acid is a substance that donates a proton in an acid-base reaction.

A Brønsted-Lowry base is a substance that accepts a proton.

80. C is correct.

In the Arrhenius theory, acids dissociate in an aqueous solution to produce H^+ (hydrogen ions).

In the Arrhenius theory, bases dissociate in an aqueous solution to produce ^-OH (hydroxide ions).

Notes for active learning

Notes for active learning

Electrochemistry – Detailed Explanations

Practice Set 1: Questions 1–20

1. D is correct.

In electrochemical (i.e., galvanic) cells, oxidation occurs at the anode, and reduction occurs at the cathode.

Br^- is oxidized at the anode, not the cathode.

2. C is correct.

As ionization energy increases, it is more difficult for electrons to be released by a substance. This is because the release of electrons would increase the charge of the substance, which is the definition of oxidation.

Substances with higher ionization energy are more likely to be reduced. If a substance is reduced in a redox reaction, it is an oxidizing agent because it facilitates the oxidation of the other reactant.

Therefore, with the increase of ionization energy, a substance is more likely to be reduced or be a stronger oxidizing agent.

3. A is correct.

Half-reaction:

$$H_2S \rightarrow S_8$$

Balancing half-reaction in acidic conditions:

Step 1: Balance atoms except for H and O

$$8\ H_2S \rightarrow S_8$$

Step 2: To balance oxygen, add H_2O to the side with fewer oxygen atoms

There is no oxygen at all, so skip this step.

Step 3: To balance hydrogen, add H^+:

$$8\ H_2S \rightarrow S_8 + 16\ H^+$$

Balance charges by adding electrons to the side with a greater/more positive total charge

Total charge on left side: 0

Total charge on right side: $16(+1) = +16$

Add 16 electrons to right side:

$$8\ H_2S \rightarrow S_8 + 16\ H^+ + 16\ e^-$$

4. B is correct.

An electrolytic cell is a nonspontaneous electrochemical cell requiring electrical energy (e.g., battery) to initiate the reaction.

The anode is positive, and the cathode is the negative electrode.

For electrolytic and galvanic cells, oxidation occurs at the anode, while reduction occurs at the cathode.

Therefore, Co metal is produced at the cathode because it is a reduction product (from an oxidation number of +3 on the left to 0 on the right). Co will not be produced at the anode.

5. C is correct.

The anode in galvanic cells attracts anions.

Anions in solution flow toward the anode, while cations flow toward the cathode.

Oxidation (i.e., loss of electrons) occurs at the anode.

Positive ions are formed while negative ions are consumed at the anode.

Therefore, negative ions flow toward the anode to equalize the charge.

6. A is correct.

By convention, the reference standard for potential is always hydrogen reduction.

7. C is correct.

Calculate the oxidation numbers of species involved and look for the oxidized species (increase in oxidation number).

Cd's oxidation number increases from 0 on the left side of the reaction to +2 on the right.

8. D is correct.

The oxidation number of Fe increases from 0 on the left to +3 on the right.

9. D is correct.

Salt bridge contains both cations (positive ions) and anions (negative ions).

Anions flow towards the oxidation half-cell because the oxidation product is positively charged, and the anions are required to balance the charges within the cell.

The opposite is true for cations; they flow towards the reduction half-cell.

10. B is correct.

A redox reaction, or oxidation-reduction reaction, involves the transfer of electrons between two reacting substances. An oxidation reaction specifically refers to a substance losing electrons, and a reduction reaction refers to a substance gaining electrons.

The oxidation and reduction reactions alone are half-reactions because they occur together as a whole reaction.

The key to this question is *transferring* electrons between *two* species, referring to the whole redox reaction in its entirety, not just one half-reaction.

An electrochemical reaction takes place during the passage of electric current and does involve redox reactions. However, it is not the correct answer to the question posed.

11. A is correct.

The cell reaction is:

$$Co\ (s) + Cu^{2+}\ (aq) \rightarrow Co^{2+}\ (aq) + Cu\ (s)$$

Separate it into half reactions:

$$Cu^{2+}\ (aq) \rightarrow Cu\ (s)$$

$$Co\ (s) \rightarrow Co^{2+}\ (aq)$$

The potentials provided are written in this format:

$$Cu^{2+}\ (aq)\ |\ Cu\ (s)$$

$$+0.34\ V$$

It means that for the reduction reaction:

$Cu^{2+}\ (aq) \rightarrow Cu\ (s)$, the potential is +0.34 V

The reverse reaction or oxidation reaction is:

$Cu\ (s) \rightarrow Cu^{2+}\ (aq)$ has opposing potential value: –0.34 V

Obtain the potential values for each half-reactions.

Reverse sign for potential values of oxidation reactions:

$Cu^{2+}\ (aq) \rightarrow Cu\ (s) = +0.34\ V$ (reduction)

$Co\ (s) \rightarrow Co^{2+}\ (aq) = +0.28\ V$ (oxidation)

Determine the standard cell potential:

Standard cell potential = sum of half-reaction potential

Standard cell potential = 0.34 V + 0.28 V

Standard cell potential = 0.62 V

12. C is correct.

The other methods listed involve two or more energy conversions between light and electricity.

It is an electrical device that converts light energy directly into electricity by the photovoltaic effect (i.e., chemical and physical processes).

The operation of a photovoltaic (PV) cell has the following requirements:

1) Light is absorbed, which excites electrons.

2) Separation of charge carries opposite types.

3) The separated charges are transferred to an external circuit.

In contrast, solar panels supply heat by absorbing sunlight.

A photoelectrolytic / photoelectrochemical cell is a photovoltaic cell or a device that splits water directly into hydrogen and oxygen using only solar illumination.

13. A is correct.

Each half-cell contains an electrode; two half-cells are required to complete a reaction.

14. B is correct.

A: electrolysis can be performed on any metal, not only iron.

C: electrolysis will not boil water nor raise the ship.

D: it is probably unlikely that the gases would stay in the compartments.

15. C is correct.

Balanced reaction:

$$Zn\ (s) + CuSO_4\ (aq) \rightarrow Cu\ (s) + ZnSO_4\ (aq)$$

Zn is a stronger reducing agent (more likely to be oxidized) than Cu. This can be determined by each element's standard electrode potential (E°).

16. D is correct.

The positive cell potential indicates that the reaction is spontaneous; it favors the formation of products.

Cells that generate electricity spontaneously are considered galvanic cells.

17. B is correct.

A joule is a unit of energy.

18. C is correct.

Electrolysis of aqueous sodium chloride yields hydrogen and chlorine, with aqueous sodium hydroxide remaining in the solution.

Sodium hydroxide is a strong base, which means the solution will be basic.

19. C is correct.

Balanced equation:

$Ag^+ + e^- \rightarrow Ag\ (s)$

Formula to calculate deposit mass:

mass of deposit = (atomic mass × current × time) / 96,500 C

mass of deposit = [107.86 g × 3.50 A × (12 min × 60 s/min)] / 96,500 C

mass of deposit = 2.82 g

20. D is correct.

The anode is the electrode where oxidation occurs.

A salt bridge provides electrical contact between the half-cells.

The cathode is the electrode where reduction occurs.

A spontaneous electrochemical cell is called a galvanic cell.

==

Practice Set 2: Questions 21–40

==

21. A is correct.

Ionization energy (IE) is the energy required to release one electron from an element.

Higher IE means the element is more stable.

Elements with high IE usually only need one or two more electrons to achieve stable configuration (e.g., complete valence shell – 8 electrons, such as the noble gases, or 2 electrons as in hydrogen). It is more likely for these elements to gain an electron and reach stability than lose an electron. When an atom gains electrons, its oxidation number goes down, and it is reduced. Therefore, elements with high IE are easily reduced.

In oxidation-reduction reactions, species that undergo reduction are oxidizing agents because their presence allows the other reactant to be oxidized. Because elements with high IE are easily reduced, they are strong oxidizing agents.

Reducing agents are species that undergo oxidation (i.e., lose electrons).

Elements with high IE do not undergo oxidation readily and therefore are weak reducing agents.

22. D is correct.

Balancing a half-reaction in basic conditions:

The first few steps are identical to balancing reactions in acidic conditions.

Step 1: Balance atoms except for H and O

$C_8H_{10} \rightarrow C_8H_4O_4^{2-}$ C is balanced

Step 2: To balance oxygen, add H_2O to the side with fewer oxygen atoms

$C_8H_{10} + 4\ H_2O \rightarrow C_8H_4O_4^{2-}$

Step 3: To balance hydrogen, add H^+ to the opposing side of H_2O added in the previous step

$C_8H_{10} + 4\ H_2O \rightarrow C_8H_4O_4^{2-} + 14\ H^+$

This next step is the unique additional step for basic conditions.

Step 4: Add equal amounts of OH^- on both sides. The number of OH^- should match the number of H^+ ions. Combine H^+ and OH^- on the same side to form H_2O. If there are H_2O molecules on each side, subtract accordingly to end up with H_2O on one side only.

There are 14 H^+ ions on the right, so add 14 OH^- ions on both sides:

$C_8H_{10} + 4\ H_2O + 14\ OH^- \rightarrow C_8H_4O_4^{2-} + 14\ H^+ + 14\ OH^-$

Combine H^+ and OH^- ions to form H_2O:

$C_8H_{10} + 4\ H_2O + 14\ OH^- \rightarrow C_8H_4O_4^{2-} + 14\ H_2O$

H_2O molecules are on both sides, which cancel, and some H_2O remain on one side:

$$C_8H_{10} + 14\ OH^- \rightarrow C_8H_4O_4^{2-} + 10\ H_2O$$

Step 5: Balance charges by adding electrons to the side with a greater/more positive total charge

Total charge on left side: 14(–1) = –14

Total charge on the right side: –2

Add 12 electrons to the right side:

$$C_8H_{10} + 14\ OH^- \rightarrow C_8H_4O_4^{2-} + 10\ H_2O + 12\ e^-$$

23. A is correct.

In electrochemical (i.e., galvanic) cells, oxidation occurs at the anode, and reduction occurs at the cathode.

Therefore, CO_2 is produced at the anode because it is an oxidation product (carbon's oxidation number increases from 0 on the left to +2 on the right).

24. A is correct.

Cu^{2+} is being reduced (i.e., gains electrons), while Sn^{2+} is being oxidized (i.e., loses electrons).

Salt bridge contains both cations (positive ions) and anions (negative ions).

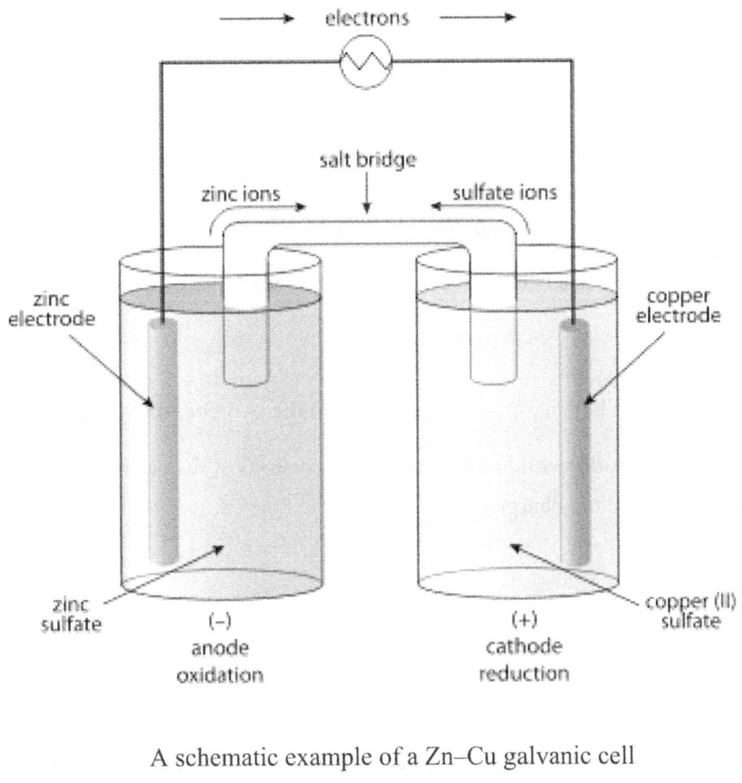

A schematic example of a Zn–Cu galvanic cell

Anions flow towards the oxidation half-cell because the oxidation product is positively charged, and the anions are required to balance the charges within the cell. The opposite is true for cations; they flow towards the reduction half-cell.

25. B is correct.

In all cells, reduction occurs at the cathode, while oxidation occurs at the anode.

26. C is correct.

ZnO is being reduced; the oxidation number of Zn decreases from +2 on the left to 0 on the right.

27. D is correct.

Consider the reactions:

$Ni\ (s) + Ag^+\ (aq) \rightarrow Ag\ (s) + Ni^{2+}\ (aq)$

Because it is spontaneous, Ni is more likely to be oxidized than Ag. Ni is oxidized, while Ag is reduced in this reaction.

Arrange the reactions so the metal that was oxidized in a reaction is reduced in the following reaction:

$Ni\ (s) + Ag^+\ (aq) \rightarrow Ag\ (s) + Ni^{2+}\ (aq)$

$Cd\ (s) + Ni^{2+}\ (aq) \rightarrow Ni\ (s) + Cd^{2+}\ (aq)$

$Al\ (s) + Cd^{2+}\ (aq) \rightarrow Cd\ (s) + Al^{3+}\ (aq)$

Lastly, in this reaction $Ag\ (s) + H^+\ (aq) \rightarrow$ no reaction should be first because Ag was not oxidized in this reaction, so it should be before a reaction where Ag is reduced.

$Ag\ (s) + H^+\ (aq) \rightarrow$ no reaction

$Ni\ (s) + Ag^+\ (aq) \rightarrow Ag\ (s) + Ni^{2+}\ (aq)$

$Cd\ (s) + Ni^{2+}\ (aq) \rightarrow Ni\ (s) + Cd^{2+}\ (aq)$

$Al\ (s) + Cd^{2+}\ (aq) \rightarrow Cd\ (s) + Al^{3+}\ (aq)$

The metal being oxidized in the last reaction has the highest tendency to be oxidized.

28. A is correct.

The purpose of the salt bridge is to balance the charges between the two chambers/half-cells.

Oxidation creates cations at the anode, while reduction reduces cations at the cathode. The ions in the bridge travel to those chambers to balance the charges.

29. B is correct.

Since $G° = -nFE$, when E° is positive, G is negative.

30. C is correct.

The electrons travel through the wires that connect the cells instead of the salt bridge. The function of the salt bridge is to provide ions to balance charges at the cathode and anode.

31. A is correct.

Electrochemistry is the branch of physical chemistry that studies chemical reactions at the interface of an ionic conductor (i.e., the electrolyte) and an electrode.

Electric charges move between the electrolyte and the electrode through a series of redox reactions, and chemical energy is converted to electrical energy.

32. B is correct.

Light energy from the sun causes the electron to move towards the silicon wafer, starting the electric generation process.

33. B is correct.

Oxidation number (or *oxidation state*) indicates the degree of oxidation (i.e., loss of electrons) in an atom.

If an atom is electron-poor, it has lost electrons and would have a positive oxidation number.

If an atom is electron-rich, it has gained electrons and would have a negative oxidation number.

34. B is correct.

Reduction potential measures a substance's ability to acquire electrons (i.e., undergo reduction).

The more positive the reduction potential, the more likely it is that the substance will be reduced. Reduction potential is generally measured in volts.

35. D is correct.

Batteries run down and need to be recharged, while fuel cells do not run down because they can be refueled.

36. A is correct.

A disproportionation reaction is when a species undergoes oxidation and reduction in the same reaction.

The first step in balancing this reaction is to write the substance undergoing disproportionation twice on the reactant side.

37. A is correct.

The terms spontaneous *electrochemical*, *galvanic*, and *voltaic* are synonymous.

An example of a voltaic cell is an alkaline battery – it generates electricity spontaneously.

Electrons flow from the anode (oxidation half-cell) to the cathode (reduction half-cell).

38. B is correct.

Reaction at the anode:

$$2\ H_2O \rightarrow O_2 + 4\ H^+ + 4\ e^-$$

Oxygen gas is released, and H^+ ions are added to the solution, which causes the solution to become acidic (i.e., lowers the pH).

39. D is correct.

Nonspontaneous electrochemical cells are electrolysis cells.

In electrolysis, the metal being reduced is produced at the cathode.

40. B is correct.

Electrolysis is nonspontaneous because it needs an electric current from an external source to occur.

Practice Set 3: Questions 41–60

---

41. C is correct.

Electronegativity indicates the tendency of an atom to attract electrons.

An atom that attracts electrons strongly would be more likely to pull electrons from another atom. As a result, this atom would be a strong oxidizer because it would cause other atoms to be oxidized when the electrons are pulled towards the atom with high electronegativity.

A strong oxidizing agent is a weak reducing agent – they are opposing attributes.

42. A is correct.

MnO_2 is being reduced into Mn_2O_3.

The oxidation number increases from +4 on the left to +3 on the right.

43. C is correct.

Electrolytic cell: needs electrical energy input

Battery: spontaneously produces electrical energy. A dry cell is a type of battery.

Half-cell: does not generate energy by itself.

44. B is correct.

The anode in galvanic cells attracts anions.

Anions in solution flow toward the anode, while cations flow toward the cathode.

Oxidation (i.e., loss of electrons) occurs at the anode.

Positive ions are formed, while negative ions are consumed at the anode.

Therefore, negative ions flow toward the anode to equalize the charge.

45. B is correct.

In electrochemical (i.e., galvanic) cells, oxidation occurs at the anode, and reduction occurs at the cathode.

Al metal is produced at the cathode because it is a reduction product (from an oxidation of +3 on the left to 0 on the right).

46. B is correct.

Anions in solution flow toward the anode, while cations flow toward the cathode.

Oxidation is the loss of electrons. Sodium is a group I element and therefore has a single valence electron. During oxidation, the Na becomes Na^+ with a complete octet.

47. A is correct.

A salt bridge contains both cations (positive ions) and anions (negative ions).

Anions flow towards the oxidation half-cell because the oxidation product is positively charged, and the anions are required to balance the charges within the cell.

The opposite is true for cations; they flow towards the reduction half-cell.

Anions in the salt bridge should flow from Cd to Zn half-cell.

48. B is correct.

This can be determined using the electrochemical series:

Equilibrium	**E°**
Li^+ (*aq*) + e^- ↔ Li (*s*)	–3.03 volts
K^+ (*aq*) + e^- ↔ K (*s*)	–2.92
*Ca^{2+} (*aq*) + 2 e^- ↔ Ca (*s*)	–2.87
*Na^+ (*aq*) + e^- ↔ Na (s)	–2.71
Mg^{2+} (*aq*) + 2 e^- ↔ Mg (s)	–2.37
Al^{3+} (*aq*) + 3 e^- ↔ Al (*s*)	–1.66
Zn^{2+} (*aq*) + 2 e^- ↔ Zn (*s*)	–0.76
Fe^{2+} (*aq*) + 2 e^- ↔ Fe (*s*)	–0.44
Pb^{2+} (*aq*) + 2 e^- ↔ Pb (*s*)	–0.13
2 H^+ (*aq*) + 2 e^- ↔ H_2 (*g*)	0.0
Cu^{2+} (*aq*) + 2 e^- ↔ Cu (*s*)	+0.34
Ag^+ (*aq*) + e^- ↔ Ag (*s*)	+0.80
Au^{3+} (*aq*) + 3 e^- ↔ Au (*s*)	+1.50

For a substance to act as an oxidizing agent for another substance, the oxidizing agent must be located below the substance being oxidized in the series.

Cu is being oxidized, which means only Ag^+ or Au^{3+} are capable of oxidizing Cu.

49. D is correct.

Anions in solution flow toward the anode, while cations flow toward the cathode.

Oxidation (i.e., loss of electrons) occurs at the anode.

Positive ions are formed while negative ions are consumed at the anode.

Therefore, negative ions flow toward the anode to equalize the charge.

50. B is correct.

Electrochemistry is the branch of physical chemistry that studies chemical reactions at the interface of an ionic conductor (i.e., the electrolyte) and an electrode.

Electric charges move between the electrolyte and the electrode through a series of redox reactions, and chemical energy is converted to electrical energy.

51. A is correct.

E° tends to be negative and G positive because electrolytic cells are nonspontaneous.

Electrons must be forced into the system for the reaction to proceed.

52. C is correct.

The ion needs to be oxidized to create chlorine gas (Cl_2) from chloride ion (Cl^-).

In electrolysis, oxidation occurs at the anode, which is the positive electrode.

The positively-charged electrode would absorb electrons from the ion and transfer them to the cathode for reduction.

53. B is correct.

In electrolytic cells:

The anode is a positively charged electrode where oxidation occurs.

The cathode is a negatively charged electrode where reduction occurs.

54. C is correct.

Nonspontaneous electrochemical cells are electrolysis cells.

In electrolysis, the metal that is being oxidized continuously dissolves at the anode.

55. D is correct.

Dry-cell batteries are the common disposable household batteries. They are based on zinc and manganese dioxide cells.

56. C is correct.

Disproportionation is a type of redox reaction in which a species is simultaneously reduced and oxidized to form two different products.

Unbalanced reaction: $HNO_2 \rightarrow NO + HNO_3$

Balanced reaction: $3\ HNO_2 \rightarrow 2\ NO + HNO_3 + H_2O$

$2\ H_2O \rightarrow 2\ H_2 + O_2$: decomposition

$H_2SO_3 \rightarrow H_2O + SO_2$: decomposition

$Mg + H_2SO_4 \rightarrow MgSO_4 + H_2$: single replacement

57. D is correct.

Electrolysis is the same process in reverse for the chemical process inside a battery.

Electrolysis is often used to separate elements.

58. A is correct.

Fuel cell automobiles are fueled by hydrogen, and the only emission is water (the statement above is reverse of the true statement).

59. A is correct.

Impure copper is oxidized so that it would happen at the anode. Then, pure copper would plate out (i.e., be reduced) on the cathode.

60. B is correct.

Electrolysis is a chemical reaction that results when electrical energy is passed through a liquid electrolyte.

Electrolysis utilizes direct electric current (DC) to drive an otherwise non-spontaneous chemical reaction. The voltage needed for electrolysis is called the decomposition potential

Practice Set 4: Questions 61–80

61. D is correct.

Calculate the oxidation numbers of species involved and identify the oxidized species (increased oxidation number). Zn's oxidation number increases from 0 on the left side of reaction I to +2 on the right.

62. C is correct.

The lack of an aqueous solution makes it a dry cell. An example of a dry cell is an alkaline battery.

63. B is correct.

Half-reaction: $C_2H_6O \rightarrow HC_2H_3O_2$

Balancing half-reaction in acidic conditions:

Step 1: Balance atoms except for H and O

$C_2H_6O \rightarrow HC_2H_3O_2$ (C is balanced)

Step 2: To balance oxygen, add H_2O to the side with fewer oxygen atoms

$C_2H_6O + H_2O \rightarrow HC_2H_3O_2$

Step 3: To balance hydrogen, add H^+ to the opposing side of H_2O added in the previous step

$C_2H_6O + H_2O \rightarrow HC_2H_3O_2 + 4\ H^+$

Step 4: Balance charges by adding electrons to the side with a greater/more positive total charge

Total charge on the left side: 0

Total charge on right side: 4(+1) = +4

Add 4 electrons to the right side:

$C_2H_6O + H_2O \rightarrow HC_2H_3O_2 + 4\ H^+ + 4\ e^-$

64. B is correct.

In electrochemical (i.e., galvanic) cells, oxidation occurs at the anode and reduction at the cathode.

CCl_4 is produced at the anode because it is an oxidation product (carbon's oxidation number increases from 0 on the left to +4 on the right).

65. C is correct.

Battery: spontaneously produces electrical energy. A dry cell is a type of battery.

Half-cell: does not generate energy by itself

Electrolytic cell: needs electrical energy input

66. D is correct.

Anions in solution flow toward the anode, while cations flow toward the cathode.

Oxidation (i.e., loss of electrons) occurs at the anode.

Positive ions are formed, while negative ions are consumed at the anode.

Therefore, negative ions flow toward the anode to equalize the charge.

67. A is correct.

A reducing agent is a reactant that is oxidized (i.e., loses electrons).

Therefore, the best reducing agent is most easily oxidized.

Reversing each of the half-reactions shows that the oxidation of Cr (*s*) has a potential of +0.75 V, which is greater than the potential (+0.13 V) for the oxidation of Sn^{2+} (*aq*).

Thus, Cr (*s*) is a stronger reducing agent.

68. B is correct.

A salt bridge contains both cations (positive ions) and anions (negative ions).

Anions flow towards the oxidation half-cell because the oxidation product is positively charged, and the anions are required to balance the charges within the cell.

The opposite is true for cations; they flow toward the reduction half-cell.

Anions in the salt bridge should flow from Cd to Zn half-cell.

69. B is correct.

When zinc is added to HCl, the reaction is:

$$Zn + HCl \rightarrow ZnCl_2 + H_2O,$$

which means that Zn is oxidized into Zn^{2+}.

The number provided in the problem is reduction potential, so to obtain the oxidation potential, flip the reaction:

$$Zn\ (s) \rightarrow Zn^{2+} + 2\ e^-$$

The $E°$ is inverted and becomes +0.76 V.

Because the $E°$ is higher than hydrogen's value (which is set to 0), the reaction occurs.

70. C is correct.

Calculate the mass of metal deposited in cathode:

Step 1: Calculate total charge using current and time

$$Q = \text{current} \times \text{time}$$

$$Q = 1\ A \times (10\ \text{minutes} \times 60\ \text{s/minute})$$

$Q = 600\ A \cdot s = 600\ C$

Step 2: Calculate moles of electron that has the same amount of charge

moles $e^- = Q / 96{,}500\ C/mol$

moles $e^- = 600\ C / 96{,}500\ C/mol$

moles $e^- = 6.22 \times 10^{-3}$ mol

Step 3: Calculate moles of metal deposit

Half-reaction of zinc ion reduction:

$Zn^{2+}\ (aq) + 2\ e^- \rightarrow Zn\ (s)$

moles of Zn = (coefficient Zn / coefficient e^-) × moles e^-

moles of Zn = (½) × 6.22 × 10^{-3} mol

moles of Zn = 3.11 × 10^{-3} mol

Step 4: Calculate mass of metal deposit

mass Zn = moles Zn × molecular mass of Zn

mass Zn = 3.11 × 10^{-3} mol × (65 g/mol)

mass Zn = 0.20 g

71. A is correct.

The oxidation number of Mg increases from 0 to +2, which means that Mg loses electrons (i.e., is oxidized).

Conversely, the oxidation number of Cu decreases, which means that it gains electrons (i.e., is reduced).

72. D is correct.

Electrolysis uses a direct electric current to provide electricity to a nonspontaneous redox process to drive the reaction. The direct electric current must be passed through an ionic substance or solution that contains electrolytes.

Electrolysis is often used to separate elements.

73. D is correct.

Oxidation is the loss of electrons, while reduction is the gain of electrons.

74. D is correct.

E° for the cell is always positive.

75. B is correct.

Galvanic cells are spontaneous and generate electrical energy.

76. D is correct.

In electrolytic cells:

The anode is a positively charged electrode where oxidation occurs.

The cathode is a negatively charged electrode where reduction occurs.

77. B is correct.

Balanced equation:

$Cu^{2+} + 2\ e^- \rightarrow Cu\ (s)$

Formula to calculate deposit mass:

mass of deposit = atomic mass × moles of electron

Since Cu has 2 electrons per atom, multiply the moles by 2:

mass of deposit = atomic mass × (2 × moles of electron)

mass of deposit = atomic mass × (2 × current × time) / 96,500 C

4.00 g = 63.55 g × (2 × 2.50 A × time) / 96,500 C

time = 4,880 s

time = (4,880 s × 1 min/60s × 1 hr/60 min) = 1.36 hr

78. C is correct.

Alkaline and dry cell are common disposable batteries and are non-rechargeable.

Fuel cells require fuel that is going to be consumed and are non-rechargeable.

79. A is correct.

Electrolytic cells do not occur spontaneously; the reaction only occurs with the addition of external electrical energy.

80. D is correct.

A redox reaction, or oxidation-reduction reaction, involves the transfer of electrons between two reacting substances. An oxidation reaction specifically refers to the substance that is losing electrons, and a reduction reaction refers to the substance gaining electrons.

The oxidation and reduction reactions alone are half-reactions because they occur together to form a whole reaction.

Therefore, half-reaction can represent a separate oxidation process or a separate reduction process.

Appendix

Periodic Table of the Elements

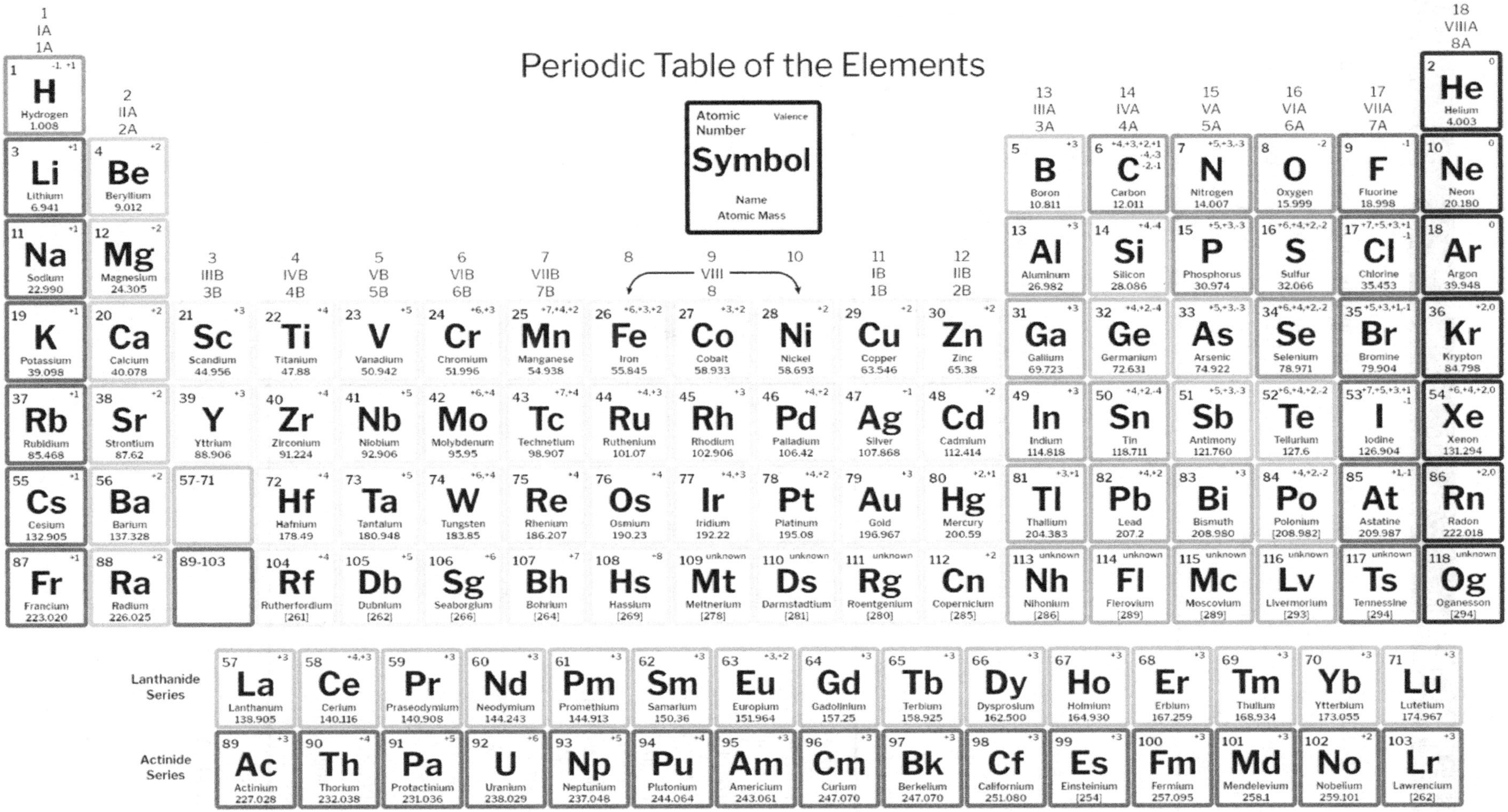

Common Chemistry Equations

Throughout the test the following symbols have the definitions specified unless otherwise noted.

L, mL	= liter(s), milliliter(s)	mm Hg	= millimeters of mercury
g	= gram(s)	J, kJ	= joule(s), kilojoule(s)
nm	= nanometer(s)	V	= volt(s)
atm	= atmosphere(s)	mol	= mole(s)

ATOMIC STRUCTURE

$E = h\nu$

$c = \lambda\nu$

E = energy

ν = frequency

λ = wavelength

Planck's constant, $h = 6.626 \times 10^{-34}$ J s

Speed of light, $c = 2.998 \times 10^{8}$ m s^{-1}

Avogadro's number $= 6.022 \times 10^{23}$ mol^{-1}

Electron charge, $e = -1.602 \times 10^{-19}$ coulomb

EQUILIBRIUM

$$K_c = \frac{[C]^c[D]^d}{[A]^a[B]^b}, \text{ where } a\,A + b\,B \rightleftarrows c\,C + d\,D$$

$$K_p = \frac{(P_C)^c(P_D)^d}{(P_A)^a(P_B)^b}$$

$$K_a = \frac{[H^+][A^-]}{[HA]}$$

$$K_b = \frac{[OH^-][HB^+]}{[B]}$$

$$K_w = [H^+][OH^-] = 1.0 \times 10^{-14} \text{ at } 25°C$$

$$= K_a \times K_b$$

$$pH = -\log[H^+],\ pOH = -\log[OH^-]$$

$$14 = pH + pOH$$

$$pH = pK_a + \log\frac{[A^-]}{[HA]}$$

$$pK_a = -\log K_a,\ pK_b = -\log K_b$$

<u>Equilibrium Constants</u>

K_c (molar concentrations)

K_p (gas pressures)

K_a (weak acid)

K_b (weak base)

K_w (water)

KINETICS

$$\ln[A]_t - \ln[A]_0 = -kt$$

$$\frac{1}{[A]_t} - \frac{1}{[A]_0} = kt$$

$$t_{1/2} = \frac{0.693}{k}$$

k = rate constant

t = time

$t_{1/2}$ = half-life

GASES, LIQUIDS, AND SOLUTIONS

$$PV = nRT$$

$$P_A = P_{total} \times X_A, \text{ where } X_A = \frac{\text{moles A}}{\text{total moles}}$$

$$P_{total} = P_A + P_B + P_C + \ldots$$

$$n = \frac{m}{M}$$

$$K = °C + 273$$

$$D = \frac{m}{V}$$

$$KE \text{ per molecule} = \frac{1}{2}mv^2$$

Molarity, M = moles of solute per liter of solution

$$A = abc$$

P = pressure
V = volume
T = temperature
n = number of moles
m = mass
M = molar mass
D = density
KE = kinetic energy
v = velocity
A = absorbance
a = molar absorptivity
b = path length
c = concentration

Gas constant, $R = 8.314 \text{ J mol}^{-1} \text{ K}^{-1}$
$= 0.08206 \text{ L atm mol}^{-1} \text{ K}^{-1}$
$= 62.36 \text{ L torr mol}^{-1} \text{ K}^{-1}$
1 atm = 760 mm Hg
= 760 torr
STP = 0.00°C and 10^5 Pa

THERMOCHEMISTRY/ ELECTROCHEMISTRY

$$q = mc\Delta T$$

$$\Delta S° = \sum S° \text{ products} - \sum S° \text{ reactants}$$

$$\Delta H° = \sum \Delta H_f° \text{ products} - \sum \Delta H_f° \text{ reactants}$$

$$\Delta G° = \sum \Delta G_f° \text{ products} - \sum \Delta G_f° \text{ reactants}$$

$$\Delta G° = \Delta H° - T\Delta S°$$

$$= -RT \ln K$$

$$= -nFE°$$

$$I = \frac{q}{t}$$

q = heat
m = mass
c = specific heat capacity
T = temperature
$S°$ = standard entropy
$H°$ = standard enthalpy
$G°$ = standard free energy
n = number of moles
$E°$ = standard reduction potential
I = current (amperes)
q = charge (coulombs)
t = time (seconds)

Faraday's constant, F = 96,485 coulombs per mole of electrons

$$1 \text{ volt} = \frac{1 \text{ joule}}{1 \text{ coulomb}}$$

Notes for active learning

Solubility Rules for Ionic Compounds

Use the following rules to *predict the solubility of ionic compounds* in water. These rules generally relate to the compound's lattice and hydration energies of the individual ions.

One of the factors determining the *lattice energy* is the magnitudes of charges on the ions (i.e., the greater the magnitude of the charges, the greater the lattice energy). Simplified, the greater the lattice energy, the lower the solubility.

The rules focus on the magnitudes of the charges and ignore other influences on the lattice energies, such as hydration energy. In addition, the rules disregard the effect of temperature.

The solubility rules are listed in order of *decreasing importance*. For example, if the first rule applies, disregard the subsequent rules. Similarly, if the second rule is applicable, disregard the third rule. Thus, only use the third rule when rules one and two are not applicable.

The following ions are nearly always soluble: $C_2H_3O_2^-$, ClO_3^-, ClO_4^-. K^+, Na^+, NH_4^+, and NO_3^-.

General principle: Hydrides (H^-) decompose in water to yield H_2 and ^-OH. Many metal oxides react with water to produce hydroxides (^-OH).

***Rule 1*: compounds containing a +1 or –1 ion are typically soluble**

Examples include: $AlCl_3$, $FeCl_2$, K_2SO_4, NaBr, Rb_3PO_4

Exceptions (Insoluble): OH^- (other than with cations that produce strong bases)

Hg_2^{2+} (other than with NO_3^-, $C_2H_3O_2^-$, ClO_3^-, and ClO_4^-)

Pb^{2+}, Hg^{2+} with most –1 ions

IB metals (column 11)

***Rule 2*: compounds containing a +3 or –3 or higher ion are typically NOT soluble**

Examples include: $AlPO_4$, $Ca_3(PO_4)_2$, TiO2

Exceptions (Soluble): cations combined with sulfate or dichromate

***Rule 3*: compounds containing –2 ion are typically NOT soluble**

Examples include:

$CaCO_3$, FeS, ZnC_2O_4

Exceptions (Soluble):

Dichromates and sulfates (insoluble if combined with Ba, Ca, Hg, Pb, and Sr)

There are exceptions to the solubility rules besides those listed. However, the ions normally encountered follow these rules.

Solubility examples for ionic compounds

AgBr	Insoluble	Rule 1	Ag^+
AgCl	Insoluble	Rule 1	$Ag^+ + Cl^-$
$AgIO_3$	Insoluble	Rule 1	$Ag^+ + IO_3^-$
AgOH	Insoluble	Rule 1	weak base
Ag_2S	Insoluble	Rule 1	Ag^+
Hg_2Cl_2	Insoluble	Rule 1	Hg_2^{2+}
$PbCl_2$	Insoluble	Rule 1	Pb^{2+}
$AgNO_3$	Soluble	Rule 1	$NO_3^- + Ag^+$
$Ba(NO_3)_2$	Soluble	Rule 1	NO_3^-
$Ba(OH)_2$	Soluble	Rule 1	strong base
NaCl	Soluble	Rule 1	$Na^+ + Cl^-$
$NaClO_3$	Soluble	Rule 1	$Na^+ + ClO_3^-$
NaOH	Soluble	Rule 1	strong base
Na_3PO_4	Soluble	Rule 1	Na^+
$(NH_4)_2CO_3$	Soluble	Rule 1	NH_4^+
NH_4IO_3	Soluble	Rule 1	$NH_4^+ + IO_3^-$
$FeAsO_4$	Insoluble	Rule 2	$Fe^{3+} + AsO_4^{3-}$
$Fe_2(SO_4)_3$	Soluble	Rule 2	SO_4^{2-}
$BaCO_3$	Insoluble	Rule 3	CO_3^{2-}
$BaSO_4$	Insoluble	Rule 3	$Ba^{2-} + SO_4^{2-}$
$MgSO_4$	Insoluble	Rule 3	SO_4^{2-}
$PbCrO_4$	Insoluble	Rule 3	CrO_4^{2-}
ZnS	Insoluble	Rule 3	S^{2-}
$Ba(IO_3)_2$	Insoluble	exception	

Glossary of Chemistry Terms

A

Absolute entropy (of a substance) – the increase in the entropy of a substance as it goes from a perfectly ordered crystalline form at 0 K (where its entropy is zero) to the temperature in question.

Absolute zero – the zero point on the absolute temperature scale; –273.15 °C or 0 K; theoretically, the temperature at which molecular motion ceases (i.e., the system does not emit or absorb energy, atoms at rest).

Absorption spectrum – spectrum associated with absorption of electromagnetic radiation by atoms (or other species), resulting from transitions from lower to higher energy states.

Accuracy – how close a value is to the actual value; see *precision*.

Acid – a substance that produces H^+ (*aq*) ions in an aqueous solution and gives a pH of less than 7.0; strong acids ionize entirely or almost entirely in dilute aqueous solution; weak acids ionize only slightly. It turns litmus red.

Acid dissociation constant – an equilibrium constant for dissociating a weak acid.

Acid rain – rainwater with a pH of less than 5.7; caused by the gases NO_2 (vehicle exhaust fumes) and SO_2 (from burning fossil fuels) dissolving in the rain. It kills fish, wildlife, and trees and destroys buildings and lakes.

Acidic salt – contains an ionizable hydrogen atom; does not necessarily produce acidic solutions.

Actinides – the fifteen chemical elements that are between actinium (89) and lawrencium (103).

Activated complex – a structure forming because of a collision between molecules while new bonds form.

Activation energy – the amount of energy that reactants must absorb in their ground states to reach the transition state needed for a reaction can occur.

Active metal – a metal with low ionization energy that loses electrons readily to form cations.

Activity (of a component of ideal mixture) – a dimensionless quantity whose magnitude is equal to the molar concentration in an ideal solution; equal to partial pressure in an ideal gas mixture; 1 for pure solids or liquids.

Activity series – a listing of metals (and hydrogen) in order of decreasing activity.

Actual yield – the amount of a specified pure product obtained from a given reaction; see *theoretical yield.*

Addition reaction – a reaction in which two atoms or groups of atoms are added to a molecule, one on each side of a double or triple bond.

Adhesive forces – forces of attraction between a liquid and another surface.

Adsorption – the adhesion of a species onto the surfaces of particles.

Aeration – the mixing of air into a liquid or a solid.

Alcohol – hydrocarbon derivative containing a ~OH group attached to a carbon atom, not in an aromatic ring.

Alkali metals – the elements of Group IA on the periodic table (e.g., Na, K, Rb).

Alkaline battery – a dry cell in which the electrolyte contains KOH.

Alkaline earth metals – group IIA metals on the periodic table; see *earth metals.*

Allomer – a substance that has a different composition than another but the same crystalline structure.

Allotropes – elements with different structures (therefore different forms), such as carbon (e.g., diamonds, graphite, and fullerene).

Allotropic modifications (allotropes) – different forms of the same element in the same physical state.

Alloy – a mixture of metals. For example, bronze is an alloy formed from copper and tin.

Alloying – mixing metal with other substances (usually other metals) to modify its properties.

Alpha (α) particle – a helium nucleus; helium ion with 2+ charge; an assembly of two protons and two neutrons.

Amorphous solid – a non-crystalline solid with no well-defined ordered structure.

Ampere – unit of electrical current; one ampere equals one coulomb per second.

Amphiprotism – the ability of a substance to exhibit amphiprotic by accepting donated protons.

Amphoterism – the ability to react with both acids and bases; to act as either an acid or a base.

Amplitude – the maximum distance that medium particles carrying the wave move from their rest position.

Anion – a negative ion; an atom or group of atoms that has gained one or more electrons.

Anode – in a cathode ray tube, the positive electrode (electrode at which oxidation occurs); the positive side of a dry cell battery or a cell.

Antibonding orbital – a molecular orbital higher in energy than any of the atomic orbitals from which it is derived; lends instability to a molecule or ion when populated with electrons; denoted with star (*) superscript.

Artificial transmutation – an artificially induced nuclear reaction caused by the bombardment of a nucleus with subatomic particles or small nuclei.

Associated ions – short-lived species formed by the collision of dissolved ions of opposite charges.

Atmosphere – a unit of pressure; the pressure supports a column of mercury 760 mm high at 0 °C.

Atom – a chemical element in its smallest form; it comprises neutrons and protons within the nucleus and electrons circling the nucleus; it is the smallest part of an element that can exist.

Atomic mass unit (amu) – one-twelfth of the mass of an atom of the carbon-12 isotope; used for stating atomic and formula weights; known as a dalton.

Atomic number – represents an element corresponding with the number of protons within the nucleus; the number of protons in the nucleus of the atom.

Atomic orbital (*AO*) – a region or volume in space where the probability of finding electrons is highest.

Atomic radius – radius of an atom.

Atomic weight – weighted average of the masses of the constituent isotopes of an element; the relative masses of atoms of different elements.

Aufbau (or *building up*) principle – describes the order in which electrons fill orbitals in atoms.

Autoionization – an ionization reaction between identical molecules.

Avogadro's Law – at the same temperature and pressure, equal volumes of gases contain the same number of molecules.

Avogadro's number (N_A) – the number (6.022×10^{23}) of atoms, molecules, or particles found in precisely 1 mole of a substance.

B

Background radiation – extraneous to an experiment; usually the low-level natural radiation from cosmic rays and trace radioactive substances present in the environment.

Band – a series of very closely spaced nearly continuous molecular orbitals that belong to the crystal as a whole.

Band of stability – band containing nonradioactive nuclides in a plot of neutrons *vs.* their atomic number.

Band theory of metals – the theory that accounts for the bonding and properties of metallic solids.

Barometer – a device used to measure the pressure in the atmosphere.

Base – a substance that produces ^{-}OH (*aq*) ions in an aqueous solution; accepts a proton and has a high pH; strongly soluble bases are soluble in water and are entirely dissociated; weak bases ionize only slightly; a typical example of a base is sodium hydroxide (NaOH). It turns litmus blue.

Basic anhydride – the oxide of a metal that reacts with water to form a base.

Basic salt – a salt containing an ionizable OH group.

Beta (β) particle – an electron emitted from the nucleus when a neutron decays to a proton and an electron.

Binary acid – a binary compound in which H is bonded to one or more electronegative nonmetals.

Binary compound – a compound consisting of two elements; it may be ionic or covalent.

Binding energy (nuclear binding energy) – the energy equivalent ($E = mc^2$) of the mass deficiency of an atom (where E is the energy in joules, m is the mass in kilograms, and c is the speed of light in m/s^2).

Boiling – the phase transition of liquid vaporizing.

Boiling point – the temperature at which the vapor pressure of a liquid is equal to the applied pressure; the *condensation point.*

Boiling point elevation – the increase in the boiling point of a solvent caused by the dissolution of a nonvolatile solute.

Bomb calorimeter – a device used to measure the heat transfer between a system and its surroundings at constant volume.

Bond – the attraction and repulsion between atoms and molecules is a cornerstone of chemistry.

Bond energy – the amount of energy necessary to break one mole of bonds in a substance, dissociating the substance in its gaseous state into atoms of its elements in the gaseous state.

Bond order – half the number of electrons in bonding orbitals minus half the electrons in antibonding orbitals.

Bonding orbital – a molecular orbit lower in energy than any of the atomic orbitals from which it is derived; lends stability to a molecule or ion when populated with electrons.

Bonding pair – pair of electrons involved in a covalent bond.

Boron hydrides – binary compounds of boron and hydrogen.

Born-Haber cycle – a series of reactions (and the accompanying enthalpy changes) which, when summed, represents the hypothetical one-step reaction by which elements in their standard states are converted into crystals of ionic compounds (and the accompanying enthalpy changes).

Boyle's Law – at a constant temperature, the volume occupied by a definite mass of a gas is inversely proportional to the applied pressure.

Breeder reactor – a nuclear reactor that produces more fissionable nuclear fuel than it consumes.

Brønsted-Lowrey acid – a chemical species that donates a proton.

Brønsted-Lowrey base – a chemical species that accepts a proton.

Buffer solution – resists change in pH; contains either a weak acid and a soluble ionic salt of the acid or a weak base and a soluble ionic salt of the base.

Buret – a piece of volumetric glassware, usually graduated in 0.1 mL intervals, used to deliver solutions for titrations in a quantitative (drop-like) manner; also spelled *burette*.

C

Calorie – the amount of heat required to raise the temperature of one gram of water from 14.5 °C to 15.5 °C; 1 calorie = 4.184 joules.

Calorimeter – a device used to measure the heat transfer between a system and its surroundings.

Canal ray – a stream of positively charged particles (cations) that moves toward the negative electrode in cathode ray tubes; observed to pass through canals in the negative electrode.

Capillary – a tube having a very small inside diameter.

Capillary action – the drawing of a liquid up the inside of a small-bore tube when adhesive forces exceed cohesive forces; the depression of the surface of the liquid when cohesive forces exceed the adhesive forces.

Catalyst – a chemical compound used to change the rate (to speed or slow it) of a regenerated reaction (i.e., not consumed) at the end of the reaction.

Catenation – the bonding of atoms of the same element into chains or rings (i.e., the ability of an element to bond with itself).

Cathode – the electrode at which reduction occurs; in a cathode ray tube, the negative electrode.

Cathodic protection – protection of a metal (making a cathode) against corrosion by attaching it to a sacrificial anode of more easily oxidized metal.

Cathode ray tube – a closed glass tube containing gas under low pressure, with electrodes near the ends and a luminescent screen near the positive electrode; produces cathode rays when a high voltage is applied.

Cation – a positive ion; an atom or group of atoms that lost one or more electrons.

Cell potential – the potential difference, E_{cell}, between oxidation and reduction half-cells under nonstandard conditions; the force in a galvanic cell pulls electrons through a reducing agent to an oxidizing agent.

Central atom – an atom in a molecule or polyatomic ion bonded to more than one other atom.

Chain reaction – a reaction that, once initiated, sustains itself and expands; a reaction in which reactive species, such as radicals, are produced in more than one step; these reactive species propagate the chain reaction.

Charles' Law – at constant pressure, the volume occupied by a definite mass of gas is directly proportional to its absolute temperature.

Chemical bonds – the attractive forces holding atoms together in elements or compounds.

Chemical change – when one or more new substances are formed.

Chemical equation – description of a chemical reaction by placing the formulas of the reactants on the left of an arrow and the formulas of the products on the right.

Chemical equilibrium – a state of dynamic balance in which the rates of forward and reverse reactions are equal; there is no net change in concentrations of reactants or products while a system is at equilibrium.

Chemical kinetics – the study of rates and mechanisms of chemical reactions and factors they depend on.

Chemical periodicity – the variations in properties of elements with their position in the periodic table.

Chemical reaction – the change of one or more substances into another or multiple substances.

Cloud chamber – a device for observing the paths of speeding particles as vapor molecules condense on them to form fog-like tracks.

Cobalt chloride paper – water test; water changes the color from blue to pink.

Coefficient of expansion – the ratio of the change in the length or the volume of a body to the original length or volume for a unit change in temperature.

Cohesive forces – the forces of attraction among particles of a liquid.

Colligative properties – physical properties of solutions that depend upon the number but not the kind of solute particles present.

Collision theory – theory of reaction rates that states that effective collisions between reactant molecules must occur for the reaction to occur.

Colloid – a heterogeneous mixture in which solute-like particles do not settle out (e.g., many kinds of milk).

Combination reaction – two substances (elements or compounds) combine to form one compound.

Combustible – classification of liquid substances that burn based on flashpoints; any liquid having a flashpoint at or above 37.8 °C (100 °F) but below 93.3 °C (200 °F), except any mixture having components with flashpoints of 93.3 °C (200 °F) or higher, the total of which makes up 99% or more of the volume of the mixture.

Combustion (or *burning*) – an exothermic reaction between an oxidant and fuel with heat and often light.

Common ion effect – suppression of ionization of a weak electrolyte by the presence in the same solution of a strong electrolyte containing one of the same ions as the weak electrolyte.

Complex ions – ions resulting from coordinating covalent bonds between simple ions and other ions or molecules.

Composition stoichiometry – describes the quantitative (mass) relationships among elements in compounds.

Compound – a substance of two or more chemically bonded elements in fixed proportions; can be decomposed into constituent elements.

Compressed gas – a single or mixture of gases having (in a container) an absolute pressure exceeding 40 psi at 21.1 °C (70 °F).

Compression – an area in a longitudinal wave where the particles are closer and pushed in.

Concentration – the amount of solute per unit volume, the mass of solvent or solution.

Condensation – the phase change from gas to liquid.

Condensed phases – the liquid and solid phases; phases in which particles interact strongly.

Condensed states – the solid and liquid states.

Conduction – heat transfer between substances in direct contact with each other (i.e., must be touching); when particles of a hotter substance vibrate, these molecules bump into nearby particles and transfer some energy.

Conduction band – a vacant or partially filled band of energy levels just higher in energy than a filled band; a band within which, or into which, electrons must be promoted to allow electrical conduction to occur in a solid.

Conductor – material that allows electric flow more freely.

Conjugate acid-base pair – in Brønsted-Lowry terms, a reactant and a product that differ by a proton, H^+.

Conformations – structures of a compound that differ by the extent of their rotation about a single bond.

Continuous spectrum – contains wavelengths in a specified region of the electromagnetic spectrum.

Control rods – rods of materials such as cadmium or boron steel that act as neutron absorbers (not merely moderators), used in nuclear reactors to control neutron fluxes and therefore fission rates.

Conjugated double bonds – double bonds separated from each other by one single bond –C=C–C=C–

Contact process – the industrial process for sulfur trioxide and sulfuric acid production from sulfur dioxide.

Convection – the physical flow of matter when heat flows by energized molecules from one place to another through the movement of fluids. The transfer of heat through a liquid or a gas when molecules of the liquid or gas move and carry the heat.

Coordinate covalent bond – a covalent bond with shared electrons furnished by the same species; a bond between a Lewis acid and a Lewis base.

Coordination compound or complex – a compound containing coordinate covalent bonds.

Coordination number – the number of donor atoms coordinated to metal; the number of nearest neighbors of an atom or ion in describing crystals.

Coordination sphere – the metal ion and its coordinating ligands but no uncoordinated counter-ions.

Corrosion – oxidation of metals (e.g., rusting) in the presence of air and moisture.

Coulomb – the SI unit of electrical charge; unit symbol – C.

Covalent bond – a force of attraction (chemical bond) formed by the sharing of electron pairs between two atoms.

Covalent compounds – compounds made of two or more nonmetal atoms bonded by sharing valence electrons.

Critical mass – the minimum mass of a particular fissionable nuclide in a given volume required to sustain a nuclear chain reaction.

Critical point – the combination of critical temperature and critical pressure of a substance.

Critical pressure – the pressure required to liquefy a gas (vapor) at its *Critical temperature*.

Critical temperature – the temperature above which a gas cannot be liquefied; the temperature above which a substance cannot exhibit distinct gas and liquid phases.

Crystal – a solid that is packed with ions, molecules or atoms in an orderly fashion.

Crystal field stabilization energy – a measure of the net energy of stabilization gained by a metal ion's nonbonding d electrons due to complex formation.

Crystal field theory – theory of bonding in transition metal complexes in which ligands and metal ions are treated as point charges; a purely ionic model; ligand point charges represent the crystal (electrical) field perturbing the metal's d orbitals containing nonbonding electrons.

Crystal lattice – a pattern of arrangement of particles in a crystal.

Crystal lattice energy – the amount of energy that holds a crystal together; the energy change when a mole of solid forms fom its constituent molecules or ions (for ionic compounds) in their gaseous state (always negative).

Crystalline solid – a solid characterized by a regular, ordered arrangement of particles.

Curie (Ci) – the basic unit to describe the intensity of radioactivity in a sample of material; one curie equals 37 billion disintegrations per second or approximately the amount of radioactivity given off by 1 gram of radium.

Current – a flow of charged particles, such as electrons or ions, moving through an electrical conductor or space. It is measured as the net rate of flow of electric charge; the unit is Ampere (A).

Cuvette – glassware used in spectroscopic experiments; usually made of plastic, glass, or quartz and should be as clean and transparent as possible.

Cyclotron – a device for accelerating charged particles along a spiral path.

D

Dalton's Law (or the *law of partial pressures*) – the pressure exerted by a mixture of gases is the sum of the partial pressures of the individual gases.

Daughter nuclide – nuclide produced in nuclear decay.

Debye – the unit used to express dipole moments.

Degenerate – in orbitals, describes orbitals of the same energy.

Deionization – the removal of ions; in the case of water, mineral ions such as sodium, iron and calcium.

Deliquescence – substances that absorb water from the atmosphere to form liquid solutions.

Delocalization – in reference to electrons, bonding electrons distributed among more than two atoms bonded; occurs in species that exhibit resonance.

Density – mass per unit volume; $D = M \times V$.

Deposition – settling particles within a solution; the direct solidification of vapor by cooling; see *sublimation*.

Derivative – a compound that can be imagined arising from a parent compound by replacing one atom with another atom or group of atoms; used extensively in organic chemistry to identify compounds.

Detergent – a soap-like emulsifier with a sulfate, SO_3, or a phosphate group instead of a carboxylate group.

Deuterium – an isotope of hydrogen whose atoms are twice as massive as ordinary hydrogen; deuterium atoms contain a proton and a neutron in the nucleus.

Dextrorotatory – refers to an optically active substance that rotates plane-polarized light clockwise, also known as *dextro*.

Diagonal similarities – chemical similarities in the Periodic Table of Elements of Period 2 to elements of Period 3 one group to the right, especially evident toward the left of the periodic table.

Diamagnetism – weak repulsion by a magnetic field.

Differential Scanning Calorimetry (DSC) – a technique for measuring temperature, direction, and magnitude of thermal transitions in a sample material by heating/cooling and comparing the amount of energy required to maintain its rate of temperature increase or decrease with an inert reference material under similar conditions.

Differential Thermal Analysis (DTA) – a technique for observing the temperature, direction and magnitude of thermally induced transitions in a material by heating/cooling a sample and comparing its temperature with an inert reference material under similar conditions.

Differential thermometer – a thermometer used to measure very small temperature changes accurately.

Dilution – the process of reducing the concentration of a solute in a solution, usually by mixing with more solvent.

Dimer – molecule formed by combining two smaller (identical) molecules.

Dipole – electric or magnetic separation of charge; charge separation between two covalently bonded atoms.

Dipole-dipole interactions – attractive electrostatic forces between polar molecules (i.e., between molecules with permanent dipoles).

Dipole moment – the product of the distance separating opposite charges of an equal magnitude of charge; a measure of the polarity of a bond or molecule; a measured dipole refers to the dipole moment of an entire molecule.

Dispersing medium – the solvent-like phase in a colloid.

Dispersed phase – the solute-like species in a colloid.

Displacement reactions – reactions in which one element displaces another from a compound.

Disproportionation reactions – redox reactions in which the oxidizing agent and the reducing agent are the same species.

Dissociation – in an aqueous solution, the process by which a solid ionic compound separates into its ions.

Dissociation constant – equilibrium constant for dissociating a complex ion into a simple ion and coordinating species (ligands).

Dissolution or solvation – the spread of ions in a monosaccharide.

Distilland – the material in a distillation apparatus that is to be distilled.

Distillate – the material in a distillation apparatus collected in the receiver.

Distillation – separating a liquid mixture into its components based on differences in boiling points; the process in which components of a mixture are separated by boiling away the more volatile liquid; the vaporization of a liquid by heating and then the condensation of the vapor by cooling.

Domain – a cluster of atoms in a ferromagnetic substance, which align in the same direction in the presence of an external magnetic field.

Donor atom – a ligand atom whose electrons are shared with a Lewis acid.

***d*-orbitals** – beginning in the third energy level, a set of five degenerate orbitals per energy level, higher in energy than s and p orbitals of the same energy level.

Dosimeter – a small, calibrated electroscope worn by laboratory personnel to measure incident ionizing radiation or chemical exposure.

Double bond – covalent bond resulting from the sharing of four electrons (two pairs) between two atoms.

Double salt – solid consisting of two co-crystallized salts.

Doublet – two peaks or bands of about equal intensity appearing close on a spectrogram.

Downs cell – electrolytic cell for the commercial electrolysis of molten sodium chloride.

DP number – the degree of polymerization; the average number of monomer units per polymer unit.

Dry cells (voltaic cells) – ordinary batteries for appliances (e.g., flashlights, radios).

Dumas method – a method used to determine the molecular weights of volatile liquids.

Dynamic equilibrium – an equilibrium in which the processes occur continuously with no net change.

E

Earth metal – highly reactive elements in group IIA of the periodic table (includes beryllium, magnesium, calcium, strontium, barium, and radium); see *alkaline earth metal.*

Effective collisions – a collision between molecules resulting in a reaction; one in which the molecules collide with proper relative orientations and sufficient energy to react.

Effective molality – the sum of the molalities of solute particles in a solution.

Effective nuclear charge – the nuclear charge experienced by the outermost electrons of an atom; the actual nuclear charge minus the effects of shielding due to inner-shell electrons (e.g., a set of dx_2-y_2 and dz_2 orbitals); those d orbitals within a set with lobes directed along the x, y and z-axes.

Electrical conductivity – the measure of how easily an electric current can flow through a substance.

Electric charge – a measured property (coulombs) that determines electromagnetic interaction.

Electrochemical cell – using a chemical reaction's current; electromotive force is made.

Electrochemistry – the study of chemical changes produced by electrical current and electricity production by chemical reactions.

Electrodes – surfaces upon which oxidation and reduction half-reactions occur in electrochemical cells; a conductor dips into an electrolyte and allows the electrons to flow to and from the electrolyte.

Electrode potentials – potentials, E, of half-reactions as reductions versus the standard hydrogen electrode.

Electrolysis – 1) occurs in electrolytic cells; chemical decomposition occurs by passing an electric current through a solution containing ions. 2) producing a chemical change using electricity; used to split up water into H and O_2.

Electrolyte – a substance (i.e., anions, cations) which, when dissolved in water, conducts electricity. An ionic solution that conducts a certain amount of current and split categorically as weak and strong.

Electrolytic cells – electrochemical cells in which electrical energy causes nonspontaneous redox reactions to occur (i.e., forced to occur by applying an outside source of electrical energy).

Electrolytic conduction – electrical current passes by ions through a solution or pure liquid.

Electromagnetic radiation – energy propagated using electric and magnetic fields that oscillate in directions perpendicular to the direction of travel of the energy; a type of wave that can go through vacuums as well as material; classified as a "self-propagating wave."

Electromagnetism – fields of an electric charge and electric properties that change how particles move and interact.

Electromotive force – a device that gains energy as electric charges pass through it.

Electromotive series – the relative order of tendencies for elements and their simple ions to act as oxidizing or reducing agents; also known as the "activity series."

Electron – a subatomic particle having a mass of 0.00054858 amu and a charge of 1–.

Electron affinity – the amount of energy absorbed in the process in which an electron is added to a neutral isolated gaseous atom to form a gaseous ion with a 1– charge; it has a negative value if energy is released.

Electron configuration – the specific distribution of electrons in atomic orbitals of atoms or ions.

Electron-deficient compounds –contain at least one atom (other than H) that shares fewer than eight electrons.

Electron shells – an orbital around the atom's nucleus with a fixed number of electrons (usually two or eight).

Electronic transition – the transfer of an electron from one energy level to another.

Electronegativity – a measure of the relative tendency of an atom to attract electrons to itself when chemically combined with another atom.

Electronic geometry – the geometric arrangement of orbitals containing the shared and unshared electron pairs surrounding the central atom or polyatomic ion.

Electrophile – positively charged or electron-deficient.

Electrophoresis – a technique for separating ions by migration rate and direction of migration in an electric field.

Electroplating – a metal is covered with another metal layer using electricity; plating a metal onto a (cathodic) surface by electrolysis.

Element – a substance that cannot be decomposed into simpler substances by chemical means; defined by its *atomic number*. A substance that cannot be split into simpler substances by chemical means.

Eluant (or eluent) – the solvent used in the process of elution, as in liquid chromatography.

Eluate – a solvent (or mobile phase) which passes through a chromatographic column and removes the sample components from the stationary phase.

Emission spectrum – the emission of electromagnetic radiation by atoms (or other species) resulting from electronic transitions from higher to lower energy states.

Empirical formula – gives the simplest whole-number ratio of atoms of each element present in a compound; also known as the simplest formula.

Emulsifying agent – a substance that coats the particles of the dispersed phase and prevents coagulation of colloidal particles; an emulsifier.

Emulsion – colloidal suspension of a liquid in a liquid.

Endothermic – describes processes that absorb heat energy (H).

Endothermicity – the absorption of heat by a system as the process occurs.

Endpoint – the point at which an indicator changes color and a titration is stopped.

Energy – a system's ability to do work.

Enthalpy (*H*) – the heat content of a specific amount of substance; E= PV.

Entropy (*S*) – a thermodynamic state or property that measures the degree of disorder (i.e., randomness) of a system; the amount of energy not available for work in a closed thermodynamic system (usually denoted by S).

Enzyme – a protein that acts as a catalyst in biological systems.

Equation of state – an expression describes the behavior of matter in a given state; the van der Waals equation describes the behavior of the gaseous state.

Equilibrium or chemical equilibrium – a state of dynamic balance with the rates of forward and reverse reactions equal; the state of a system when neither forward nor reverse reaction is thermodynamically favored.

Equilibrium constant – a quantity that characterizes the equilibrium position for a reversible reaction; its magnitude is equal to the mass action expression at equilibrium; equilibrium, "K," varies with temperature.

Equivalence point – the point when chemically equivalent amounts of reactants have reacted.

Equivalent weight – an oxidizing or reducing agent whose mass gains (oxidizing agents) or loses (reducing agents) 6.022×10^{23} electrons in a redox reaction.

Evaporation – vaporization of a liquid below its boiling point.

Evaporation rate – the rate at which a particular substance will vaporize (evaporate) compared to the rate of a known substance such as ethyl ether, especially useful for health and fire-hazard considerations.

Excited state – any state other than the ground state of an atom or molecule; see *ground state*.

Exothermic – describes processes that release heat energy (H).

Exothermicity – the release of heat by a system as a process occurs.

Explosive – a chemical or compound that causes a sudden, almost instantaneous release of pressure, gas, heat, and light when subjected to sudden shock, pressure, high temperature, or applied potential.

Explosive limits – the range of concentrations over which a flammable vapor mixed with the proper ratios of air will ignite or explode if a source of ignition is provided.

Extensive property – a property that depends upon the amount of material in a sample.

Extrapolate – to estimate the value of a result outside the range of a series of known values; a technique used in standard additions calibration procedure.

F

Faraday constant – a unit of electrical charge widely used in electrochemistry and equal to ~ 96,500 coulombs; represents 1 mole of electrons, or the Avogadro number of electrons: 6.022×10^{23} electrons.

Faraday's law of electrolysis – a two-part law that Michael Faraday published about electrolysis. 1. the mass of a substance altered at an electrode during electrolysis is directly proportional to the quantity of electricity transferred at that electrode. 2. the mass of an elemental material altered at an electrode is directly proportional to the element's equivalent weight; one equivalent weight of a substance is produced at each electrode during the passage of 96,487 coulombs of charge through an electrolytic cell.

Fast neutron – a neutron ejected at high kinetic energy in a nuclear reaction.

Ferromagnetism – the ability of a substance to become permanently magnetized by exposure to an external magnetic field.

Flashpoint – the temperature at which a liquid will yield enough flammable vapor to ignite; there are various recognized industrial testing methods; therefore, the method used must be stated.

Fluorescence – absorption of high energy radiation by a substance and subsequent emission of visible light.

First Law of Thermodynamics – the amount of energy in the universe is constant (i.e., energy is neither created nor destroyed in ordinary chemical reactions and physical changes); known as the Law of Conservation of Energy.

Fluids – substances that flow freely; gases and liquids.

Flux – a substance added to react with the charge, or a product of its reduction; in metallurgy, it is usually added to lower a melting point.

Foam – colloidal suspension of a gas in a liquid.

Formal charge – a method of counting electrons in a covalently bonded molecule or ion; it counts bonding electrons as though they were equally shared between the two atoms.

Formula – a combination of symbols that indicates the chemical composition of a substance.

Formula unit – the smallest repeating unit of a substance; the molecule for nonionic substances.

Formula weight – the mass of one formula unit of a substance in atomic mass units.

Fossil fuels – formed from the remains of plants and animals that lived millions of years ago.

Fractional distillation – when a fractioning column is used in a distillation apparatus to separate the components of a liquid mixture with different boiling points.

Fractional precipitation – removal of some ions from a solution by precipitation while leaving other ions with similar properties in the solution.

Free energy change – the indicator of the spontaneity of a process at constant T and P (e.g., if ΔG is negative, the process is spontaneous).

Free radical – a highly reactive chemical species carrying no charge and having a single unpaired electron in an orbital.

Freezing – phase transition from liquid to solid.

Freezing point depression – the decrease in the freezing point of a solvent caused by the presence of a solute.

Frequency – the number of repeating points on a wave that passes a given observation point per unit time; the unit is 1 hertz = 1 cycle per 1 second.

Fuel – any substance that burns in oxygen to produce heat.

Fuel cells – a voltaic cell that converts the chemical energy of a fuel and an oxidizing agent directly into electrical energy continuously.

G

Gamma (γ) ray – a highly penetrating type of nuclear radiation similar to x-ray radiation, except that it comes from within the nucleus of an atom and has higher energy; energy-wise, very similar to cosmic rays except that cosmic rays originate from outer space.

Galvanic cell – battery made up of electrochemical with two different metals connected by a salt bridge.

Galvanizing – placing a thin layer of zinc on a ferrous material to protect the underlying surface from corrosion.

Gangue – sand, rock and other impurities surrounding the mineral of interest in an ore.

Gas – a state of matter in which the particles have no definite shape or volume, though they fill their container.

Gay-Lussac's Law – the expression used for each of the two relationships named after the French chemist Joseph Louis Gay-Lussac concerning the properties of gases; more usually applied to his law of combining volumes.

Geiger counter – a gas-filled tube that discharges electrically when ionizing radiation passes through it.

Gel – colloidal suspension of a solid dispersed in a liquid; a semi-rigid solid.

Gibbs (free) energy – the thermodynamic state function of a system that indicates the amount of energy available for the system to do useful work at constant T and P; a value that indicates the spontaneity of a reaction (usually denoted by G).

Graham's Law – the rates of effusion of gases are inversely proportional to the square roots of their molecular weights or densities.

Ground state – the lowest energy state or most stable state of an atom, molecule, or ion; see *excited state*.

Group – a vertical column in the periodic table; known as a family.

H

Haber process – a process for the catalyzed industrial production of ammonia from N_2 and H_2 at high temperature and pressure.

Half-cell – the compartment in which the oxidation or reduction half-reaction occurs in a voltaic cell.

Half-life – the time required for half of a reactant to be converted into product(s); the time required for half of a given sample to undergo radioactive decay.

Half-reaction – the oxidation or the reduction part of a redox reaction.

Halogens – group VIIA elements: F, Cl, Br, I; halogens are nonmetals.

Hard water – water high in dissolved minerals that is it difficult to form lather with soap.

Heat – a form of energy that flows between two samples of matter because of their temperature differences.

Heat capacity – the amount of heat required to raise the temperature of a mass one degree Celsius.

Heat of condensation – the amount of heat that must be removed from one gram of vapor at its condensation point to condense the vapor with no change in temperature.

Heat of crystallization – the amount of heat that must be removed from one gram of a liquid at its freezing point to freeze it with no change in temperature.

Heat of fusion – the amount of heat required to melt one gram of a solid at its melting point with no change in temperature; usually expressed in J/g; the molar heat of fusion is the amount of heat required to melt one mole of a solid at its melting point with no change in temperature and is usually expressed in kJ/mol.

Heat of solution – the amount of heat absorbed in forming a solution that contains one mole of the solute; the value is positive if heat is absorbed (endothermic) and negative if heat is released (exothermic).

Heat of vaporization – the amount of heat required to vaporize one gram of a liquid at its boiling point with no change in temperature; usually expressed in J/g; the molar heat of vaporization is the amount of heat required to vaporize one mole of liquid at its boiling point with no change in temperature and is usually expressed as ion kJ/mol.

Heisenberg uncertainty principle – states that it is impossible to accurately determine the *momentum* and *position* of an electron simultaneously.

Henry's Law – the gas pressure above a solution is proportional to the concentration of the gas in the solution.

Hess' Law of heat summation – the enthalpy change for a reaction is the same whether it occurs in one step or a series of steps.

Heterogeneous catalyst – exist in a different phase (solid, liquid, or gas) from the reactants; a contact catalyst.

Heterogeneous equilibria – equilibria involving species in more than one phase.

Heterogeneous mixture – a mixture that does not have uniform composition and properties throughout.

Heteronuclear – consisting of different elements.

High spin complex – crystal field designation for an outer orbital complex; t_{2g} and e_g orbitals are singly occupied before pairing occurs.

Homogeneous catalyst – in the same phase (solid, liquid, or gas) as the reactants.

Homogeneous equilibria – when *reagents* and *products* are of the same phase (i.e., gases, liquids, or solids).

Homogeneous mixture – a mixture which has uniform composition and properties throughout.

Homologous series – compounds with each member differing from the next by a specific number and kind of atoms.

Homonuclear – consisting of only one element.

Hund's rule – orbitals of a given sublevel must be occupied by single electrons before pairing begins; see *Aufbau (or building up) principle.*

Hybridization – mixing atomic orbitals to form a new set of atomic orbitals with the same electron capacity and properties and energies intermediate between the original unhybridized orbitals.

Hydrate – a solid compound that contains a definite percentage of bound water.

Hydrate isomers – crystalline complexes that differ in whether water exists inside or outside the coordination sphere.

Hydration – the reaction of a substance with water.

Hydration energy – the energy change accompanying the hydration of a mole of gas and ions.

Hydride – a binary compound of hydrogen.

Hydrocarbons – compounds that contain only carbon and hydrogen.

Hydrogen bond – a relatively strong dipole-dipole interaction (but still considerably weaker than the covalent or ionic bonds) between molecules containing hydrogen directly bonded to a small, highly electronegative atom, such as N, O or F.

Hydrogenation – the reaction in which hydrogen adds across a double or triple bond.

Hydrogen-oxygen fuel cell – hydrogen is the fuel (reducing agent) and oxygen is the oxidizing agent.

Hydrolysis – the reaction of a substance with water or its ions.

Hydrolysis constant – an equilibrium constant for a hydrolysis reaction.

Hydrometer – a device used to measure the densities of liquids and solutions.

Hydrophilic colloids – colloidal particles that repel water molecules.

I

Ideal gas – a hypothetical gas that obeys the postulates of the kinetic-molecular theory.

Ideal gas law – the product of pressure and the volume of an ideal gas is directly proportional to the number of moles of the gas and the absolute temperature. PV = nRT

Ideal solution – obeys Raoult's Law strictly.

Immiscible liquids – do not mix to form a solution (e.g., oil and water).

Indicators – for acid-base titrations, organic compounds that exhibit different colors in solutions of different acidities, determine the point at which the reaction between two solutes is complete.

Inert pair effect – characteristic of the post-transition minerals; the tendency of the electrons in the outermost atomic *s* orbital to remain un-ionized or unshared in compounds of post-transition metals.

Inhibitory catalyst – an inhibitor; a catalyst that decreases the rate of reaction.

Inner orbital complex – valence bond designation for a complex in which the metal ion utilizes d orbitals for one shell inside the outermost occupied shell in its hybridization.

Inorganic chemistry – a part of chemistry concerned with inorganic (non carbon-based) compounds.

Insulator – a material that resists the flow of electric current or heat transfer; does not allow heat to flow easily.

Insoluble compound – a substance that will not dissolve in a solvent, even after mixing.

Integrated rate equation – an expression giving the concentration of a reactant remaining after a specified time; has a different mathematical form for different orders of reactants.

Intermolecular forces – forces between individual particles (atoms, molecules, ions) of a substance.

Ion – a molecule that has gained or lost electrons; an atom or a group of atoms carries an electric charge (Na^+).

Ion product for water – equilibrium constant for water ionization; $K\text{w} = [H_3O^+]\cdot[^-OH] = 1.00 \times 10^{-14}$ at 25 °C.

Ionic bond – electrostatic attraction between oppositely charged ions, resulting from a transfer of electrons.

Ionic bonding – chemical bonding resulting from transferring electrons from one atom or group.

Ionic compounds – compounds containing predominantly ionic bonding.

Ionic geometry – arrangement of atoms (not lone pairs of electrons) about the central atom of a polyatomic ion.

Ionization – the breaking up of a compound into separate ions; in an aqueous solution, the process by which a molecular compound reacts with water and forms ions.

Ionization constant – equilibrium constant for the ionization of a weak electrolyte.

Ionization energy – the minimum amount of energy required to remove the most loosely held electron of an isolated gaseous atom or ion.

Ion exchange – a method of removing hardness from water, and it replaces the positive ions that cause the hardness with H^+ ions.

Ionization isomers – result from the interchange of ions inside and outside the coordination sphere.

Isoelectric – having the same electronic configurations.

Isomers – different substances with the same formula.

Isomorphous – refers to crystals having the same atomic arrangement.

Isotopes – two or more forms of atoms of the same element with different masses; atoms containing the same number of protons but different numbers of neutrons.

IUPAC – acronym for "International Union of Pure and Applied Chemistry."

J

Joule (J) – a unit of energy in the SI system; one joule is 1 $kg \cdot m^2/s^2$, which is 0.2390 calories.

K

K capture – absorption of a *K* shell (n = 1) electron by a proton as it is converted to a neutron.

Kelvin – a unit of measure for temperature based upon an absolute scale.

Kinetics – a sub-field of chemistry specializing in reaction rates.

Kinetic energy (*KE*) – energy that matter processes by its motion.

Kinetic-molecular theory – a theory that attempts to explain macroscopic observations on gases in microscopic or molecular terms.

L

Lanthanides – elements 57 (lanthanum) through 71 (lutetium); grouped because of their similar behavior in chemical reactions.

Lanthanide contraction – a decrease in the radii of the elements following the lanthanides compared to what would be expected if there were no *f*-transition metals.

Latent heat – the energy absorbed or released when a substance changes state without changing temperature.

Lattice – unique arrangement of atoms or molecules in a crystalline liquid or solid.

Law of combining volumes (Gay-Lussac's Law) – at constant temperature and pressure, the volumes of reacting gases (and any gaseous products) can be expressed as ratios of small whole numbers.

Law of conservation of energy – energy cannot be created or destroyed; it can only change form.

Law of conservation of matter – there is no detectable change in the quantity of matter during an ordinary chemical reaction.

Law of conservation of matter and energy – the amount of matter and energy in the universe is fixed.

Law of definite proportions (law of constant composition) – different samples of a pure compound contain the same elements in the same proportions by mass.

Law of partial pressures (or *Dalton's Law*) – the pressure exerted by a mixture of gases is the sum of the partial pressures of the individual gases.

Laws of thermodynamics – physical laws which define quantities of thermodynamic systems describe how they behave and (by extension) set certain limitations such as perpetual motion.

Lead storage battery – secondary voltaic cell used in most automobiles.

Leclanche cell – a common type of *dry cell.*

Le Chatelier's principle – states that a system at equilibrium, or striving to attain equilibrium, responds in such a way as to counteract any stress placed upon it; if stress (change of conditions) is applied to a system at equilibrium, the system will shift in the direction that reduces stress.

Leveling effect – acids stronger than the acid characteristic of the solvent reacts with the solvent produce that acid; a similar statement applies to bases. The strongest acid (base) that can exist in a given solvent is the acid (base) characteristic of the solvent.

Levorotatory – an optically active substance rotates plane-polarized light counterclockwise, known as a *levo.*

Lewis acid – any species that can accept a share in an electron pair.

Lewis base – any species that can make available a share in an electron pair.

Lewis dot formula (electron dot formula) – representation of a molecule, ion or formula unit by showing atomic symbols and only outer shell electrons.

Ligand – a Lewis base in a coordination compound.

Light – that portion of the electromagnetic spectrum visible to the naked eye; known as "visible light."

Limiting reactant – a substance that stoichiometrically limits the number of product(s) that can be formed.

Linear accelerator – a device used for accelerating charged particles along a straight line path.

Line spectrum – an atomic emission or absorption spectrum.

Linkage isomers – a particular ligand bonds to a metal ion through different donor atoms.

Liquid – a state of matter which takes the shape of its container.

Liquid aerosol – colloidal suspension of a liquid in gas.

London dispersion forces – very weak and very short-range attractive forces between short-lived temporary (induced) dipoles; known as "dispersion forces."

Lone pair – pair of electrons residing on one atom and not shared by other atoms; unshared pair.

Low spin complex – crystal field designation for an inner orbital complex; contains electrons paired t_{2g} orbitals before e_g orbitals are occupied in octahedral complexes.

Lubricant – a substance capable of reducing friction (i.e., force that opposes the direction of motion).

M

Magnetic field – a space around a magnet where magnetism can be detected.

Magnetic quantum number – quantum mechanical solution to a wave equation designating the orbital within a given set (*s*, *p*, *d*, *f*) in which an electron resides.

Manometer – a two-armed barometer.

Mass (*m*) – a measure of the amount of matter in an object; mass is usually measured in grams or kilograms.

Mass action expression – for a reversible reaction, aA + bB cC + dD; the product of the concentrations of the products (species on right), each raised to the power corresponding to its coefficient in the balanced chemical equation, divided by the product of the concentrations of reactants (species on left), each raised to the power corresponding to its coefficient in the balanced equation; at equilibrium the mass action expression is equal to K.

Mass deficiency – the amount of matter converted into energy when an atom forms from constituent particles.

Mass number – the sum of the numbers of protons and neutrons in an atom; an integer.

Mass spectrometer – an instrument that measures the charge-to-mass ratio of charged particles.

Matter – anything that has mass and occupies space.

Mechanism – the sequence of steps by which reactants are converted into products.

Melting point – the temperature at which liquid and solid coexist in equilibrium.

Meniscus – the shape assumed by the surface of a liquid in a cylindrical container.

Melting – the phase change from a solid to a liquid.

Metal – a chemical element that is a good conductor of electricity and heat and forms cations and ionic bonds with nonmetals; elements below and to the left of the stepwise division (metalloids) in the upper right corner of the periodic table; about 80% of known elements are metals.

Metallic bonding – bonding within metals due to the electrical attraction of positively charged metal ions for mobile electrons that belong to the crystal.

Metallic conduction – conduction of electrical current through a metal or along a metallic surface.

Metalloid – a substance with the properties of metals and nonmetals (B, Al, Si, Ge, As, Sb, Te, Po and At).

Metathesis reactions – reactions in which two compounds react to form two new compounds, with no changes in oxidation number; reactions in which the ions of two compounds exchange partners.

Method of initial rates – method of determining the rate-law expression by carrying out a reaction with different initial concentrations and analyzing the resultant changes in initial rates.

Methylene blue – a heterocyclic aromatic chemical compound with the molecular formula $C_{16}H_{18}N_3SCl$.

Miscible liquids – mix to form a solution (e.g., alcohol and water).

Miscibility – the ability of one liquid to mix with (dissolve in) another liquid.

Mixture – two or more different substances mingled together but not chemically combined. A sample of matter composed of two or more substances, each of which retains its identity and properties.

Moderator – a substance (e.g., deuterium, oxygen, paraffin) capable of slowing fast neutrons upon collision.

Molality – a concentration expressed as a number of moles of solute per kilogram of solvent.

Molarity – the number of moles of solute per liter of solution.

Molar solubility – the number of moles of a solute that dissolves to produce a liter of a saturated solution.

Mole – a measurement of an amount of substance; a single mole contains approximately 6.022×10^{23} units or entities; abbreviated mol.

Molecule – a chemically bonded number of electrically neutral atoms.

Molecular equation – a chemical reaction in which formulas are written as if substances existed as molecules; only complete formulas are used.

Molecular formula – indicates the actual number of atoms present in a molecule of a molecular substance.

Molecular geometry – the arrangement of atoms (not lone pairs of electrons) around a central atom of a molecule or polyatomic ion.

Molecular orbital (*MO*) – resulting from the overlap and mixing of atomic orbitals on different atoms (i.e., a region where an electron can be found in a molecule, as opposed to an atom); an MO belongs to the molecule.

Molecular orbital theory – a theory of chemical bonding based upon postulated molecular orbitals.

Molecular weight – the mass of one molecule of a nonionic substance in atomic mass units.

Molecule – the smallest particle of a compound capable of stable, independent existence.

Mole fraction – the number of moles of a component in a mixture divided by the number of moles in the mixture.

Monoprotic acid – can form only one hydronium ion per molecule; may be strong or weak.

Mother nuclide – nuclide that undergoes nuclear decay.

N

Native state – refers to the occurrence of an element in an uncombined or free state in nature.

Natural radioactivity – spontaneous decomposition of an atom.

Neat – conditions with a liquid reagent or gas performed with no added solvent or co-solvent.

Nernst equation – corrects standard electrode potentials for nonstandard conditions.

Net ionic equation – results from canceling spectator ions and eliminating brackets from a total ionic equation.

Neutralization – the reaction of an acid with a base to form a salt and water; usually, the reaction of hydrogen ions with hydrogen ions to form water molecules.

Neutrino – particle that can travel at speeds close to the speed of light; created due to radioactive decay.

Neutron – a neutral unit or subatomic particle that has no net charge and a mass of 1.0087 amu.

Nickel-cadmium cell (NiCd battery) – a dry cell where the anode is Cd, the cathode is NiO_2, and the electrolyte is basic.

Nitrogen cycle – the complex series of reactions by which nitrogen is slowly but continually recycled in the atmosphere, lithosphere, and hydrosphere.

Noble gases – elements of the periodic Group 0; He, Ne, Ar, Kr, Xe, Rn; known as "rare gases;" formerly called "inert gases."

Nodal plane – a region in which the probability of finding an electron is zero.

Nonbonding orbital – a molecular orbital derived only from an atomic orbital of one atom; lends neither stability nor instability to a molecule or ion when populated with electrons.

Nonelectrolyte – a substance whose aqueous solutions do not conduct electricity.

Nonmetal – an element that is not metallic.

Nonpolar bond – a covalent bond in which electron density is symmetrically distributed.

Nuclear – of or about the atomic nucleus.

Nuclear binding energy – the energy equivalent of the mass deficiency; the energy released in forming an atom from the subatomic particles.

Nuclear fission – when a heavy nucleus splits into nuclei of intermediate masses and protons are emitted.

Nuclear magnetic resonance spectroscopy – a technique that exploits the magnetic properties of specific nuclei; helpful in identifying unknown compounds.

Nuclear reaction – involves a change in the composition of a nucleus and can emit or absorb a tremendous amount of energy.

Nuclear reactor – a system in which controlled nuclear fission reactions generate heat energy on a large scale which is subsequently converted into electrical energy.

Nucleons – particles comprising the nucleus; protons, and neutrons.

Nucleus – the very small and dense, positively charged center of an atom containing protons and neutrons, as well as other subatomic particles; the net charge is positive.

Nuclides – refers to different atomic forms of elements; in contrast to isotopes, which refer only to different atomic forms of a single element.

Nuclide symbol – designation for an atom A/Z E, in which E is the symbol of an element, Z is its atomic number, and A is its mass number.

Number density – a measure of the concentration of countable objects (e.g., atoms, molecules) in space; the number per volume.

O

Octahedral – molecules and polyatomic ions with one atom in the center and six atoms at the corners of an octahedron.

Octane number – a number that indicates how smoothly a gasoline burns.

Octet rule – during bonding, atoms tend to reach an electron arrangement with eight electrons in the outermost shell. Many representative elements attain at least a share of eight electrons in their valence shells when they form molecular or ionic compounds; there are some limitations.

Open sextet – species with only six electrons in the highest energy level of the central element (many Lewis acids).

Orbital – may refer to an atomic orbital or a molecular orbital.

Organic chemistry – the chemistry of substances that contain carbon-hydrogen bonds.

Organic compound – substances that contain carbon.

Osmosis – when solvent molecules pass through a semi-permeable membrane from a dilute solution into a more concentrated solution.

Osmotic pressure – the hydrostatic pressure produced on the surface of a semi-permeable membrane.

Outer orbital complex – valence bond designation for a complex in which the metal ion utilizes d orbitals in the outermost (occupied) shell in hybridization.

Overlap – the interaction of orbitals on different atoms in the same region of space.

Oxidation – the addition of oxygen or the loss of electrons. An algebraic increase in the oxidation number; may correspond to a loss of electrons.

Oxidation numbers – quantitative values used as mechanical aids in writing formulas and balancing equations; for single-atom ions, they correspond to the charge on the ion; more electronegative atoms are assigned negative oxidation numbers; known as *oxidation states*.

Oxidation-reduction reactions – reactions in which oxidation and reduction occur; known as *redox reactions*.

Oxide – a binary compound of oxygen.

Oxidizing agent – the substance that oxidizes another substance and is reduced.

P

Pairing – a favorable interaction of two electrons with opposite m values in the same orbital.

Pairing energy – the energy required to pair two electrons in the same orbital.

Paramagnetism – attraction toward a magnetic field, stronger than diamagnetism but still weak compared to ferromagnetism.

Partial pressure – the force exerted by one gas in a mixture of gases.

Particulate matter – fine, divided solid particles suspended in polluted air.

Pauli exclusion principle – no electrons in the same atom may have identical sets of four quantum numbers.

Percentage ionization – the percentage of the weak electrolyte that will ionize in a solution of given concentration.

Percent by mass – 100% times the actual yield divided by the theoretical yield.

Percent composition – the mass percent of each element in a compound.

Percent purity – the percent of a specified compound or element in an impure sample.

Period – the elements in a horizontal row of the periodic table.

Periodicity – regular periodic variations of properties of elements with their atomic number (and position in the periodic table).

Periodic Law – the properties of the elements are periodic functions of their atomic numbers.

Periodic table – an arrangement of elements by increasing atomic numbers, emphasizes periodicity.

Peroxide – a compound with oxygen in –1 oxidation state; metal peroxides contain the peroxide ion, O_2^{2-}.

pH – the measure of acidity (or basicity) of a solution; negative logarithm of the concentration (mol/L) of the H_3O^+ $[H^+]$ ion; scale is commonly used over a range 0 to 14.

Phase diagram – shows the equilibrium temperature-pressure relationships for different phases of a substance.

pH scale – a scale from 0 to 14. If the pH of a solution is 7 it is neutral; if the pH of a solution is less than 7 it is acidic; if the pH of a solution is greater than 7 it is basic.

Permanent hardness – hardness (relative to lathering soap) in water that cannot be removed by boiling; caused by calcium sulfate.

Photoelectric effect – emission of an electron from the surface of a metal caused by impinging electromagnetic radiation of specific minimum energy; the current increases with increasing radiation intensity.

Photon – a carrier of electromagnetic radiation of all wavelengths, such as gamma rays and radio waves; known as *quantum of light*.

Physical change – when a substance changes from one physical state to another, but no substances with different composition are formed; physical change may involve a phase change (e.g., melting, freezing) or another physical change such as crushing a crystal or separating one volume of liquid into different containers; does not produce a new substance.

Plasma – a physical state of matter that exists at extremely high temperatures in which molecules are dissociated, and most atoms are ionized.

Polar bond – a covalent bond with an unsymmetrical distribution of electron density.

Polarimeter – a device used to measure optical activity.

Polarization – the buildup of a product of oxidation or reduction of an electrode, preventing further reaction.

Polydentate – refers to ligands with more than one donor atom.

Polyene – a compound that contains more than one double bond per molecule.

Polymerization – the combination of many small molecules to form large molecules.

Polymer – a large molecule consisting of chains or rings of linked monomer units, usually characterized by high melting and boiling points.

Polymorphous – refers to substances that can crystallize in more than one crystalline arrangement.

Polyprotic acid – forms two or more hydronium ions per molecule; often, at least one ionization step is weak.

Positron – a nuclear particle with the mass of an electron but opposite charge (positive).

Potential difference (or *voltage*) – the force which moves the electrons around the circuit; the unit is Volt (V).

Potential energy (*PE*) – energy stored in a body or a system due to its position in a force field or configuration.

Power – the rate at which energy is converted from one form to another; the unit is Watts (W). Power = voltage × current (P = VI).

Precipitate – an insoluble solid formed by mixing in solution the constituent ions of a slightly soluble solution.

Precision – how close the results of multiple experimental trials are; see *accuracy*.

Pressure – force per unit area; unit is Pascal (Pa).

Primary standard – a known high degree of purity substance that undergoes one invariable reaction with the other reactant of interest.

Primary voltaic cells – voltaic cells that cannot be recharged; no further chemical reaction is possible once the reactants are consumed.

Products – chemicals produced (from reactants) in a chemical reaction.

Proton – a subatomic particle having a mass of 1.0073 amu and a charge of +1, found in the nuclei of atoms.

Protonation – the addition of a proton (H^+) to an atom, molecule or ion.

Pseudobinaryionic compounds – contain more than two elements but are named like binary compounds.

Q

Quanta – the minimum amount of energy emitted by radiation.

Quantum mechanics – the study of how atoms, molecules, subatomic particles, etc. behave and are structured; a mathematical method of treating particles based on quantum theory, which assumes that energy (of small particles) is not infinitely divisible.

Quantum numbers – numbers that describe the energies of electrons in atoms; derived from quantum mechanical treatment.

Quarks – elementary particles and a fundamental constituent of matter, combining to form hadrons (i.e., protons and neutrons).

R

Radiation – 1) heat transfer through invisible rays, which travel outwards from the hot object without a medium. 2) high-energy particles or rays emitted during the nuclear decay processes.

Radical – an atom or group of atoms that contains one or more unpaired electrons; usually a very reactive species.

Radioactive dating – method of dating ancient objects by determining the ratio of amounts of mother and daughter nuclides present in an object and relating the ratio to the object's age via half-life calculations.

Radioactive tracer – a small amount of radioisotope replacing a nonradioactive isotope of the element in a compound whose path (e.g., in the body) or whose decomposition products are monitored by detection of radioactivity; known as a "radioactive label."

Radioactivity – the spontaneous disintegration of atomic nuclei.

Raoult's Law – the vapor pressure of a solvent in an ideal solution decreases as its mole fraction decreases.

Rate-determining step – the slowest step in a mechanism; the step that determines the overall reaction rate.

Rate-law expression – equation relating the reaction rate to the concentrations of the reactants and the specific rate of the constant.

Rate of reaction – the change in the concentration of a reactant or product per unit time.

Reactants – substances consumed in a chemical reaction; react together in a chemical reaction.

Reaction quotient – the mass action expression under any set of conditions (not necessarily equilibrium); its magnitude relative to K determines the direction in which the reaction must occur to establish equilibrium.

Reaction ratio – the relative amounts of reactants and products involved in a reaction; may be the ratio of moles, millimoles, or masses.

Reaction stoichiometry – describes the quantitative relationships among substances participating in chemical reactions.

Reactivity series (or activity series) – an empirical, calculated, and structurally analytical progression of a series of metals, arranged by "reactivity" from highest to lowest; used to summarize information about the reactions of metals with acids and water, double displacement reactions and the extraction of metals from ores.

Reagent – a substance (or compound) added to a system to cause a chemical reaction or to visualize if a reaction occurs; the terms reactant and reagent are often used interchangeably; however, a *reactant* is more specifically a substance consumed during a chemical reaction.

Reducing agent – a substance that reduces another substance and is itself oxidized.

Reduction – the removal of oxygen or the gaining of electrons.

Resonance – the concept in which two or more equivalent dot formulas for the same arrangement of atoms (resonance structures) are necessary to describe the bonding in a molecule or ion.

Reverse osmosis – forcing solvent molecules to flow through a semi-permeable membrane from a concentrated solution into a dilute solution by applying greater hydrostatic pressure on the concentrated side than the osmotic pressure opposing it.

Reversible reaction – proceses that do not go to completion and occur in the forward and reverse direction.

S

Saline solution – a general term for NaCl (i.e., sodium chloride) in water.

Salt – when a metal replaces the hydrogen of an acid.

Salts – ionic compounds composed of anions and cations.

Salt bridge – a U-shaped tube containing an electrolyte, connects the two half-cells of a voltaic cell.

Saturated solution –no more solute will dissolve at that temperature.

***s*-block elements** – group 1 and 2 elements (alkali and alkaline metals), including hydrogen and helium.

Schrödinger equation – quantum state equation representing the behavior of an electron around an atom; describes the wave function of a physical system evolving.

Second Law of Thermodynamics – the universe tends toward a state of greater disorder in spontaneous processes.

Secondary standard – a solution that has been titrated against a primary standard; a standard solution.

Secondary voltaic cells – voltaic cells that can be recharged; original reactants can be regenerated by reversing the direction of the current flow.

Semiconductor – a substance that does not conduct electricity at low temperatures but will do so at higher temperatures.

Semi-permeable membrane – a thin partition between two solutions through which specific molecules can pass but others cannot.

Shielding effect – electrons in filled sets of *s*, *p* orbitals between the nucleus and outer shell electrons shield the outer shell electrons somewhat from the effect of protons in the nucleus; known as the "screening effect."

Sigma (σ) bonds – bonds resulting from the head-on overlap of atomic orbitals. The region of electron sharing is along and (cylindrically) symmetrical to the imaginary line connecting the bonded atoms.

Sigma orbital – molecular orbital resulting from the head-on overlap of two atomic orbitals.

Single bond – covalent bond resulting from the sharing of two electrons (one pair) between two atoms.

Sol – a suspension of solid particles in a liquid; artificial examples include sol-gels.

Solid – one of the states of matter, where the molecules are packed close, resistance to movement/deformation and volume change.

Solubility product constant – equilibrium constant that applies to the dissolution of a slightly soluble compound.

Solubility product principle – the solubility product constant expression for a slightly soluble compound is the product of the concentrations of the constituent ions, each raised to the power that corresponds to the number of ions in one formula unit.

Solute – the dispersed (i.e., dissolved) phase of a solution; the solution is mixed into the solvent (e.g., NaCl in saline water).

Solution – a homogeneous mixture of multiple substances; comprised of solutes and solvents; a mixture of a solute (usually a solid) and a solvent (usually a liquid).

Solvation – the process by which solvent molecules surround and interact with solute ions or molecules.

Solvent – the dispersing medium of a solution (e.g., H_2O in saline water).

Solvolysis – the reaction of a substance with the solvent in which it is dissolved.

***s*-orbital** – a spherically symmetrical atomic orbital; one per energy level.

Specific gravity – the ratio of the density of a substance to the density of water.

Specific heat – the amount of heat required to raise the temperature of one gram of substance one degree Celsius.

Specific rate constant – an experimentally determined (proportionality) constant, which is different for different reactions, and which changes only with temperature; k in the rate-law expression: Rate = k [A] × [B].

Spectator ions – ions in a solution that do not participate in a chemical reaction.

Spectral line – any of several lines corresponding to definite wavelengths of an atomic emission or absorption spectrum; marks the energy difference between two energy levels.

Spectrochemical series – arrangement of ligands in order of increasing ligand field strength.

Spectroscopy – the study of radiation and matter, such as X-ray absorption and emission spectroscopy.

Spectrum – display of component wavelengths (colors) of electromagnetic radiation.

Speed of light – the speed at which radiation travels through a vacuum (299,792,458 m/sec).

Square planar – describes molecules and polyatomic ions with one atom in the center and four atoms at the corners of a square.

Square planar complex – relationship with metal in the center of a square plane, with ligand donor atoms at each of the four corners.

Standard conditions for temperature and pressure (STP) – a standardization used to compare experimental results (25 °C and 100.000 kPa).

Standard electrodes – half-cells in which the oxidized and reduced forms of a species are present at the unit activity (1.0 M solutions of dissolved ions, 1.0 atm partial pressure of gases, pure solids, and liquids).

Standard electrode potential – by convention, the potential (Eo) of a half-reaction as a reduction relative to the standard hydrogen electrode when species are present at unit activity.

Standard entropy – the absolute entropy of a substance in its standard state at 298 K.

Standard molar enthalpy of formation – the amount of heat absorbed in forming one mole of a substance in a specified state from its elements in their standard states.

Standard molar volume – the space occupied by one mole of an ideal gas under standard conditions; 22.4 liters.

Standard reaction – a process where the numbers of moles of reactants in the balanced equation, in their standard states, are entirely converted to the numbers of moles of products in the balanced equation, at their standard state.

State of matter – a homogeneous, macroscopic phase (e.g., gas, plasma, liquid, solid) in increasing concentration.

Stoichiometry – the quantitative relationships among elements and compounds undergoing chemical changes.

Strong electrolyte – a substance that conducts electricity well in a dilute aqueous solution.

Strong field ligand – ligand that exerts a strong crystal or ligand electrical field and generally forms low spin complexes with metal ions when possible.

Structural isomers – compounds that contain the same number and kinds of atoms with different geometry.

Subatomic particles – comprise an atom (e.g., protons, neutrons, electrons).

Sublimation – the direct vaporization of a solid by heating without passing through the liquid state; a phase transition from solid to limewater fuel or gas.

Substance – any matter, specimens with the same chemical composition and physical properties.

Substitution reaction – a reaction in which another atom or group of atoms replaces an atom or a group of atoms.

Supercooled liquids – liquids that, when cooled, apparently solidify but continue to flow very slowly under the influence of gravity.

Supercritical fluid – a substance at a temperature above its critical temperature.

Supersaturated solution – contains a higher than saturation concentration of solute; slight disturbance or seeding causes crystallization of excess solute.

Suspension – a heterogeneous mixture in which solute-like particles settle out of the solvent-like phase sometime after their introduction. A mixture of a liquid and a finely divided insoluble solid.

T

Talc – a mineral representing the Mohs Scale composed of hydrated magnesium silicate with the chemical formula $H_2Mg_3(SiO_3)_4$ or $Mg_3Si_4O_{10}(OH)_2$.

Temperature – a measure of heat intensity (i.e., the hotness or coldness of a sample); a measure of the kinetic energy of an object. Unit is degrees, and scales are Celsius, Fahrenheit, and Kelvin.

Temporary hardness – hardness in water that can be removed by boiling; caused by calcium hydrogen carbonate.

Ternary acid – a ternary compound containing H, O and another element, often a nonmetal.

Ternary compound – a compound consisting of three elements; may be ionic or covalent.

Tetrahedral – a term used to describe molecules and polyatomic ions with one atom in the center and four atoms at the corners of a tetrahedron.

Theoretical yield – the maximum amount of a specified product that could be obtained from specified amounts of reactants, assuming complete consumption of the limiting reactant according to only one reaction and complete recovery of the product; see *actual yield.*

Theory – a model describing the nature of a phenomenon.

Thermal conductivity – a property of a material to conduct heat (often noted as k).

Thermal cracking – decomposition by heating a substance in the presence of a catalyst and the absence of air.

Thermochemistry – the study of absorption/release of heat within a chemical reaction; studies heat energy associated with chemical reactions and physical transformations.

Thermodynamics – studying the effects of changing temperature, volume, or pressure (or work, heat, and energy) on a macroscopic scale.

Thermodynamic stability – when a system is in its lowest energy state with its environment (equilibrium).

Thermometer – a device that measures the average energy of a system.

Thermonuclear energy – energy from nuclear fusion reactions.

Third Law of Thermodynamics – entropy of a pure crystalline substance at absolute zero temperature is zero.

Titration – a procedure in which one solution is added to another solution until the chemical reaction between the two solutes is complete; the concentration of one solution is known, and that of the other is unknown. The process of adding one solution to a measured amount of another to find out exactly how much of each is required to react.

Torr – a unit to measure pressure; 1 Torr is equivalent to 133.322 Pa or 1.3158×10^{-3} atm.

Total ionic equation – the expression for a chemical reaction written to show the predominant form of species in aqueous solution or contact with water.

Transition elements (metals) – B Group elements except IIB in the periodic table; sometimes called transition elements, elements with incomplete *d* sub-shells; the *d*-block elements.

Transition state theory –reactants pass through high-energy transition states before forming products.

Transuranic element – an atomic number greater than 92; none of the transuranic elements are stable.

Triple bond – the sharing of three pairs of electrons within a covalent bond (e.g., N_2).

Triple point – where the temperature and pressure of three phases are the same; water has a unique phase diagram.

Tyndall effect – results from light scattering by colloidal particles (a mixture where one substance is dispersed evenly throughout another) or by suspended particles.

U

Uncertainty –any measurement that involves estimating any amount that cannot be precisely reproducible.

Uncertainty principle – knowing the location of a particle makes the momentum uncertain, while knowing the momentum of a particle makes the location uncertain.

Unit cell – the smallest repeating unit of a lattice.

Unit factor – statements used in converting between units.

Universal (or ideal) gas constant – proportionality constant in the ideal gas law (0.08206 L·atm/(K·mol)).

UN number – a four-digit code used to note hazardous and flammable substances.

Unsaturated hydrocarbons – hydrocarbons that contain double or triple carbon-carbon bonds.

V

Valence bond theory – proposes that covalent bonds are formed when atomic orbitals on different atoms overlap and the electrons are shared.

Valence electrons – outermost electrons of atoms; usually those involved in bonding.

Valence shell electron pair repulsion theory (VSEPR) – assumes electron pairs are arranged around the central element of a molecule or polyatomic ion with maximum separation (and minimum repulsion) among regions of high electron density.

Valency – the number of electrons an atom wants to gain, lose, or share to have a full outer shell.

Van der Waals' equation – a quantitative relationship of a state extending the ideal gas law to real gases by including two empirically determined parameters, which are specific for different gases.

Van der Waals force – one of the forces (attraction/repulsion) between molecules.

Van't Hoff factor – the ratio of moles of particles in solution to moles of solute dissolved.

Vapor – when a substance is below the critical temperature in the gas phase.

Vaporization – the phase change from liquid to gas.

Vapor pressure – the particle pressure of vapor at the surface of its parent liquid.

Viscosity – the resistance of a liquid to flow (e.g., oil has a higher viscosity than water).

Volt – one joule of work per coulomb; the unit of electrical potential transferred.

Voltage – the potential difference between two electrodes; a measure of the chemical potential for a redox reaction.

Voltaic cells – electrochemical cells in which spontaneous chemical reactions produce electricity; known as *galvanic cells.*

Voltmeter – an instrument that measures the cell potential.

Volumetric analysis – measuring the volume of a solution (of known concentration) to determine the substance's concentration within the solution; see *titration*.

W

Water equivalent – the amount of water absorbing the same heat as the calorimeter per degree of temperature increases.

Weak electrolyte – a substance that conducts electricity poorly in a dilute aqueous solution.

Weak field ligand – a ligand exerting a weak crystal or ligand field and generally forms high spin complexes with metals.

X

X-ray – electromagnetic radiation between gamma and UV rays.

X-ray diffraction – a method for establishing structures of crystalline solids using single wavelength X-rays and studying the diffraction pattern.

X-ray photoelectron spectroscopy – a spectroscopic technique used to measure the composition of a material.

Y

Yield – the amount of product produced during a chemical reaction.

Z

Zone melting – remove impurities from an element by melting and slowly traveling it down an ingot (cast).

Zone refining – a method of purifying a metal bar by passing it through an induction heater; this causes impurities to move along a melted portion.

Zwitterion (formerly called a dipolar ion) – a neutral molecule with a positive and negative electrical charge; multiple positive and negative charges can be present, distinct from dipoles at different locations within that molecule; known as *inner salts.*

Advanced Placement (AP) prep books by Sterling Test Prep

Biology Practice Questions

Biology Review

Physics 1 Practice Questions

Physics 1 Review

Physics 2 Practice Questions

Physics 2 Review

Environmental Science

Psychology

U.S. History

World History

European History

U.S. Government and Politics

Comparative Government and Politics

Human Geography

Visit our Amazon store

College Level Examination Program (CLEP)

Biology Review

Biology Practice Questions

Chemistry Review

Chemistry Practice Questions

Introductory Business Law Review

College Algebra Practice Questions

College Mathematics Practice Questions

History of the United States I Review

History of the United States II Review

Western Civilization I Review

Western Civilization II Review

Social Sciences and History Review

American Government Review

Introductory Psychology Review

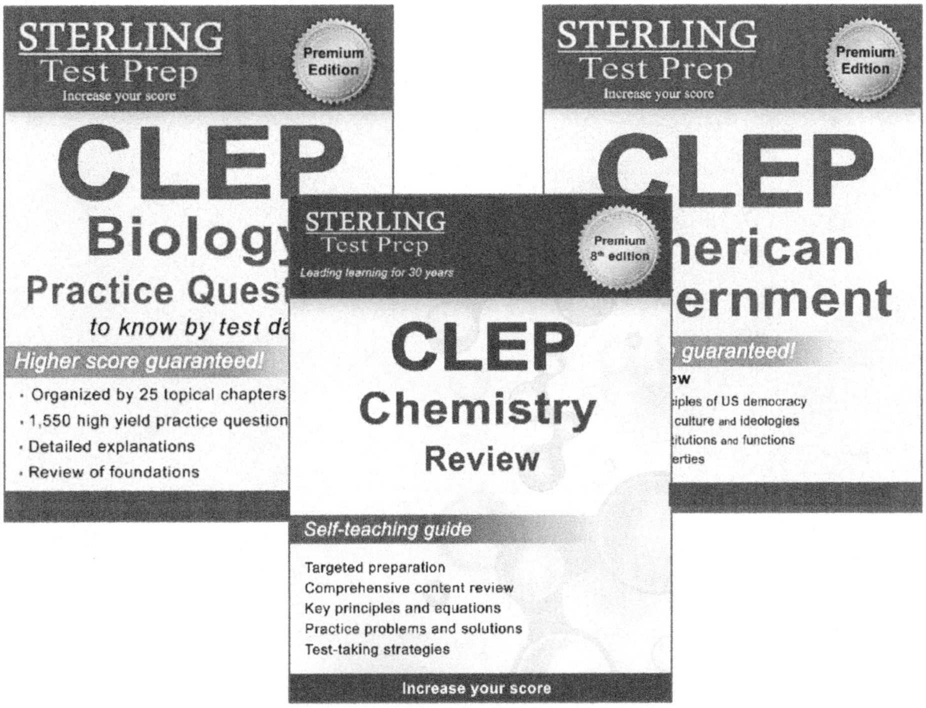

Visit our Amazon store